Chironomidae Larvae

Biology and Ecology of the aquatic Orthocladiinae

Chironomidae Larvae of the Netherlands and Adjacent Lowlands

Biology and Ecology of the aquatic Orthocladiinae

Prodiamesinae – Diamesinae – Buchonomyiinae – Podonominae – Telmatogetoninae

H.K.M. Moller Pillot

KNNV Publishing

CONTENTS

I G OR PR DI BU PO TE T R A I

1 INTRODUCTION

This third part of our work on Chironomidae larvae covers the Orthocladiinae and all the small subfamilies. Systematically, it would have been better to include the Podonominae in Part I of this series, together with the related Tanypodinae. New keys to the larvae of Chironomini will be published by Henk Vallenduuk elsewhere.We are not sure whether we will be able to produce a book on the Tanytarsini, because their ecology is insufficiently known; neither do we know whether keys to Orthocladiinae larvae will be published in this series.
The terrestrial Orthocladiinae will be treated in articles. An article on one genus (*Smittia*) has already been published (Moller Pillot, 2008). As almost all terrestrial species can sometimes be found in aquatic environments, their occurrence in water bodies are treated briefly in this book.

Details about the scope of the information in this book and the origin of the data used can be found in Part II of the series (Moller Pillot, 2009: 6). Because the Orthocladiinae inhabit flowing water more than the Chironomini, we have consulted more references from regions beyond the scope of this book to give a more comprehensive view of the ecology of the species. Nevertheless, species living only in montane or boreal regions are omitted or only briefly mentioned. A growing problem is that in the modern literature, references to the original data are frequently omitted, which meant that we had to contact the authors personally to obtain them.
As in Parts I and II, the primary aim of this book is to give a picture of where the different species can live. However, the Orthocladiinae inhabit dynamic environments more than other chironomids. Stream ecosystems in particular are highly dynamic, by which all benthic populations are continuously redistributed (Townsend & Hildrew, 1976; Townsend et al., 1987; Armitage et al., 2001). This means that characteristic species are often scarce or absent (Fretwell, 1972) and may settle and survive in localities that are not their optimum habitat (Moller Pillot, 2003). Describing the living conditions on the basis of known records is therefore very risky. Readers should bear this in mind when using this book and interpreting their own data.

STRUCTURE OF THE BOOK

The subfamilies are treated separately; within each subfamily the genera and species are ordered alphabetically. An important part of the biology and ecology is described under the genus name and not repeated under the species names. In the tables in Chapter 9 the species of all subfamilies are arranged together in alphabetical order. In these tables we omitted the species about which we had insufficient data.

2 GENERAL ASPECTS OF THE SYSTEMATICS, BIOLOGY AND ECOLOGY OF THE AQUATIC ORTHOCLADIINAE

Important aspects of the ecology of chironomid larvae have been discussed in Part I, Chapter 2. Some aspects of the ecology of the Chironomini, treated in Chapter 2 of Part II apply more or less to the Orthocladiinae and will not always be summarised in this book.

2.1 SYSTEMATICS

For a long time the subfamily Orthocladiinae was divided into two tribes: Orthocladiini and Metriocnemini (e.g. Goetghebuer, 1940–50; Pinder, 1978). The former author included *Corynoneura* and *Thienemanniella* in a separate subfamily Corynoneurinae (Goetghebuer, 1939). Saether (1977a, 1979a) stated that some genera take a more plesiomorphic position, which led to a provisional division into eight tribes. The most plesiomorphic genera are *Stilocladius* and some genera often combined as the *Brillia* group, followed by the Corynoneurini and then the Orthocladiini and Metriocnemini (Cranston et al., 2012). In general this division has been universally followed, but the most recent keys and ecological publications do not use the division into tribes.

To make identification of larvae easier, Pankratova (1970) and Moller Pillot (1984) included the subfamilies Diamesinae and Prodiamesinae in their books on Orthocladiinae. There is no doubt about the separate status of these three subfamilies. In this book they are treated separately (Chapters 3, 4 and 5).

2.2 IDENTIFICATION

Langton & Pinder (2007: key to adult males) and Langton & Visser (2003: key to exuviae) are the most up to date keys to these stages. The latter publication also gives descriptions. In Chapters 3 to 8 we only mention whether it was possible or not to identify the adult male of a species using Pinder (1978) when this is necessary for interpreting the literature. In most cases, the key to exuviae by Langton (1991) is still very useful. In the text we mention when a species is not known from the British Isles, and therefore absent from Langton & Pinder (2007).

Identification of fourth instar larvae

A key to genera with descriptions of each genus is given in Cranston et al. (1983). Very good descriptions and figures of all genera can be found on Peter Cranston's website: http://chirokey.skullisland.info/#taxa. The keys to species by Cranston (1982) and Moller Pillot (1984) have to be revised. A very good key with short descriptions and very good figures is given by Schmid (1993), but this book is out of print. A significant part of the information given in these books has been incorporated into the key by Janecek (2007). A useful key in the Slovak language is made by Bitušik (2000). The keys by Chernovskij (1949) and Pankratova (1970) are sometimes mentioned in this book because they have been used for many years by most workers in Eastern Europe.
Some valuable characters are rarely used in the current keys to larvae. Fig. 1 shows that the shapes of the ventromental plates, and especially the bases of them in combination with the location of the setae submenti, differ widely between genera or species.

It seems probable that head length and head width are not influenced by water temperature, or only very slightly, although the total length of fourth instar larvae of the winter generation can be much larger (e.g. Hannesdóttir et al., 2012).

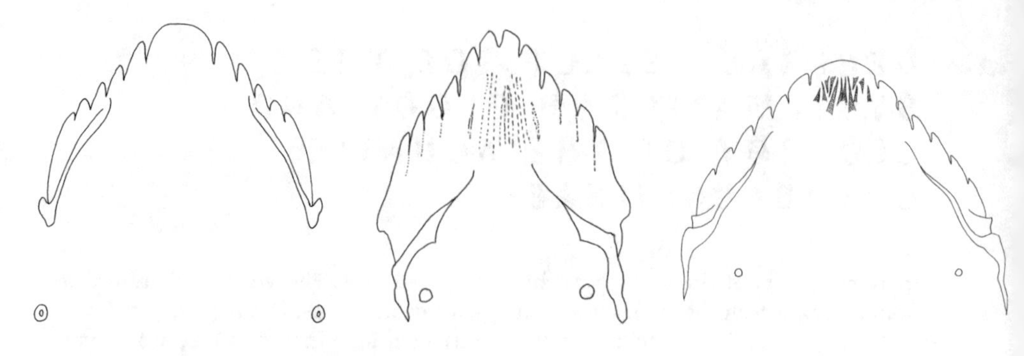

1 *Mentum of Paracricotopus niger, Eukiefferiella claripennis and Cricotopus bicinctus*

Younger larvae

Hitherto all existing keys apply only to fourth instar larvae. However, as in other subfamilies of the Chironomidae, most characters apply also to the third and second instar. The stated morphometric characters are specific to fourth instar larvae, unless otherwise mentioned. As a rule, head length and head width in the third instar are about 60% of those of the fourth instar, and in the second instar are about 60% of those of the third instar. However, the first antennal segment is relatively shorter in younger instars and so the antennal ratio is also lower in younger instars. Younger larvae usually have relatively longer setae, but this may vary even within a genus. For information about factors causing intraspecific differences in body size, see Vallenduuk & Moller Pillot (2007: 18).

The descriptions of the species in Moller Pillot (1984) sometimes include the head length of the third instar larva. More exact measurements of head length and head width of third and often also second instar larvae can be found in Schmid (1993). Only very few authors describe all instars of a species separately; such descriptions are mentioned in the species descriptions in this book.

First instar larvae usually have a quite deviating mentum. Other characters of the fourth instar are also not useful for identifying first instar larvae (e.g. first instar *Cricotopus* larvae have a simple l_4). Schmid (1993: 493–501) contains figures of the mentum and antenna of 13 species of the Orthocladiinae. Some other authors give figures and descriptions of the first instar of one species (e.g. Rodova, 1966: *Cricotopus sylvestris*; Zelentsov, 1980: *Psectrocladius* s.s.).

2.3 DISTRIBUTION

Most of the data on the distribution of chironomid species in Europe are taken from Fauna Europaea (Saether & Spies, 2010). For the Netherlands different sources have been used; Limnodata.nl gives all data supplied by the water authorities. Because Fauna Europaea and Limnodata are often incomplete and include some misidentifications, we have added many of our own data and omitted many doubtful records, usually without further comment. Nevertheless, some mistakes in the text are inevitable. The author is not responsible for the checklist published under his name in Beuk (2002: 109–128).

A complete picture of the distribution of species can rarely be given. This is because in the Netherlands identification is mainly restricted to larvae and in many other countries too few scientists work on chironomids. Inhabitants of special water types (e.g. temporary water) are poorly known because most investigations have been restricted to lakes and streams.

2.4 LIFE CYCLE

The general life history of the Chironomidae has been described in Vallenduuk & Moller Pillot (2007: 9–14). In contrast to the Chironomini, most Orthocladiinae are not thermophilous and emerge early in spring, and in regions with a maritime climate even in winter. There is no general rule about the season of maximum abundance, but many species can be scarce in summer.
Many studies have been carried out on the life histories or emergence patterns of Orthocladiinae (e.g. Ringe, 1974; Mackey, 1977; Tokeshi, 1986; Lindegaard & Mortensen, 1988; Reiff, 1994). We therefore have much information about the periods in which older larvae, pupae and adults of most species are present. However, there are very few detailed studies of the life cycle of a single species. As a consequence we are usually uncertain whether the larvae go into diapause or not and which factors are responsible for the observed life histories. The explanations given in literature are often based only on indications and will be mentioned as such in this book. As Goddeeris (1989) pointed out, life cycles can be elucidated only if the whole population can be studied, including the early instars. Because most Orthocladiinae live in flowing water this is usually very difficult: in many cases larvae migrate from one microhabitat to another or are transported by drift. If small larvae were absent from the samples because of net mesh size or because only one microhabitat was investigated, it is not possible to decide whether there is a diapause of young larvae or delayed hatching of eggs.

As a rule, there is no fixed **number of generations**. When a species has a **diapause**, development stops in one larval stage at a certain day length, for instance in late summer or in autumn. Most probably such species have no other moments of synchronisation. The number of generations depends on the growth rate of the larvae, which is influenced by environmental factors, such as temperature and food. For example, Hannesdóttir et al. (2012) stated for *Eukiefferiella claripennis* one generation a year in a cold stream (7 °C) and two in a warm stream (23 °C) in Iceland. See also under the genus *Orthocladius*. When there are three or more generations a year, the latest larvae of the second or third generation can overlap with first larvae of the next generation.

Differences between the **numbers of larvae** in different generations seem to be the result of environmental factors, such as temperature (see under *Orthocladius*), food, predation and competition (Thienemann, 1954; Matěna, 1989). Verberk et al. (2005: 271) suggested that early in the year, especially in eutrophic environments, a predictable pulse of food availability occurs and later in the year temporal fluctuations in food availability can be expected. The same applies to predation and competition. In some cases, drift plays a dominant role (Gendron & Laville, 2000). See further Vallenduuk & Moller Pillot (2007: 12–13) and Moller Pillot (2009: 9).

The **duration of development** and the time of emergence of a generation can be different, even between two streams or lakes in the same region. For example, Ringe (1974: diagram 20, p. 297) stated that the second generation of *Parametriocnemus stylatus* in the Breitenbach in Germany emerged in high summer and in the Rohrwiesenbach in early autumn, possibly under the influence of food conditions. Reiff (1994) found great differences between emergences in lakes in Bavaria in many species.

Orthocladiinae inhabit a wide range of biotopes and show very **different life cycle strategies**. For example, Tokeshi (1986) observed almost continuous emergence of *Tvetenia calvescens*, but in *Orthocladius rhyacobius* a long egg-dormancy period from late spring until mid winter to exploit abundant food resources, particularly diatoms, in spring. It is possible that in such cases the eggs do not hatch at high temperatures or under the influence of other factors. If some of the eggs

do hatch during summer, a small second generation is possible. (For a variation on this strategy, see *Pseudodiamesa branickii*, subfamily Diamesinae).
Many species seem to have a diapause in second and third instar in autumn, possibly to avoid emergence in winter, or to realise a synchronisation in spring to avoid concurrence, or to improve the chance of mating. Moreover, the need for food during diapause is reduced and the larvae are less active, which reduces predation. In such cases, most probably the proper diapause (complete interruption of development) is in the third instar, but development in the second instar may slow down as well, also because of low temperatures (Goddeeris, pers. comm.).
Species of temporary water (e.g. in the genera *Hydrobaenus* and *Trissocladius*) go into dormancy in summer to survive desiccation. Emergence in winter is less rare in species living in flowing water and in terrestrial species than in species of stagnant water, probably because they have fewer problems with ice cover.

One of the **consequences** of the life cycles for water authorities is that larger (e.g. fourth instar) larvae can be scarce or absent during part of the year and that exuviae and adults can be collected only in certain periods, depending on the species. The fluctuations in numbers of larvae during summer are relatively low in many Orthocladiinae, because they have multivoltine life cycles with overlapping generations (Armitage et al., 1995). When the numbers of Orthocladiinae emergences are much lower in summer (Lehmann, 1971; Lesage & Harrison, 1980; Hayes & Murray, 1989) factors other than life cycle will be the cause (see above); such decreasing numbers are often not universal (see e.g. Ringe, 1974).

In contrast to most Chironomini, the further development of many Orthocladiinae and Diamesinae species of running water starts as early as mid winter. Therefore, the **instar of hibernation** given in table 4 in Chapter 9 applies to the end of the autumn (early December).

2.5 OVIPOSITION

As far as is known, Orthocladiinae usually deposit their eggs on firm substrata, such as macrophytes and stones near the water's edge, as do most other Chironomidae (Pankratova, 1970: 24; Nolte, 1993: 60). Darby (1962: 194) supposed that dense emergent vegetation can be responsible for a scarcity of chironomids because of the absence of open water, which attracts the females. This is supported by several data (e.g. Frigge & Olde Lohuis, unpublished), but there is very little information about this.

In many species each female seems to produce only one egg mass, but Darby (1962) observed that a female of *Cricotopus bicinctus* probably deposited about five strings of eggs in a short time (see under this species). Repeated oviposition has been confirmed only in taxa with a terrestrial development (Nolte, 1993: 44), but exact information about most species is lacking. The females of *Cricotopus* die within a few hours of oviposition (Kettisch, 1936–1938; LeSage & Harrison, 1980).

Munsterhjelm (1920) and Nolte (1993) described and figured the egg masses of many species. The number of eggs per egg mass depends on nutrition, the development time of the larvae (Nolte, 1993: 56) and temperature: at lower temperatures egg masses contain more eggs (Dettinger-Klemm, 2000: 305). For this reason the numbers given in table 4 in Chapter 9 have to be used with caution.

2.6 GENDER RATIO

Parthenogenesis is a widespread phenomenon in terrestrial Orthocladiinae, but as far as we know it is rare in the aquatic species of the subfamily. In the genus *Limnophyes*, Schleuter (1985) found many more females in temporary water than Ringe (1974) in permanent brooks. Ringer stated as a rule slightly more females than males in most species. In some species a more striking dominance of females has been stated, for example in *Brillia bifida* (as *B. modesta*) (Ringe, 1974; Caspers, 1980). There is a number of possible causes of aberrant sex ratios; see Moller Pillot (2008).

2.7 FEEDING

Like other subfamilies, the Orthocladiinae show large differences in the **mouthparts** of the larvae, indicating important differences in feeding. In contrast to Chironomini, the ventromental plates are small and never striated, suggesting that filter feeding does not play an important role (see Moller Pillot, 2009: 12). The mandible has no dorsal tooth, possibly because the larvae have another manner of scraping than Chironomini. Also, the structure of the salivary glands differs obviously from those of the Chironomini (Kurazhskovskaya, 1969), suggesting that even detritivorous species do not have the same possibilities for digesting food as Chironomini larvae. It is remarkable that the larvae of *C. sylvestris* and *C. trifasciatus* display no differences in mouthparts, although the former consumes mainly algae and detritus, and the latter only fresh plant tissue. In this case one has to conclude that the differences lie mainly in the possibilities for ingesting food.

Darby (1962: 131) observed larvae of *Cricotopus bicinctus* feeding on *Spirogyra*, beginning at one free end of the filament. The action of the mandibles forces the filament over the mental teeth, breaking or piercing the cell walls to release the cell contents as food. When feeding on coarser fragments of plant material, the mental plate functions by abrasive action. The mental teeth of those individuals showed much greater wear than the mandibular teeth.

In contrast to Chironomini, the larvae of Orthocladiinae are rarely (or never?) filter-feeders. Many species are **grazers/scrapers** or **deposit-collectors** (collectors/gatherers). Scrapers prefer runs and riffles in streams, deposit-collectors prefer stretches with slow-flowing water (pools) (Fesl, 2002; Syrovátka et al., 2009). Scrapers (e.g. most species of *Orthocladius*) have stout mandibles with strong teeth, which gradually wear down with age (Soponis, 1977; Lamberti & Moore, 1984). They feed mainly on algae and other periphyton on plants and stones, but this layer also contains much detritus, bacteria, protozoans, fungi and free-living algae so that the difference with deposit-collectors is gradual (Karlström, 1978).

Lamberti & Moore (1984) emphasise that fine detritus alone is an extremely poor food source. Bacteria, fungi and other microorganisms associated with fine detritus may be much more important (see also Moore, 1979: 321). Bacteria have a very high protein content. Goedkoop & Johnson (1992) stated that in oligotrophic lakes bacteria are the most important food source for chironomids, but that in eutrophic lakes degradation of phytoplankton is much more important. Comparable differences can be expected between oligotrophic woodland brooks and eutrophic lower courses, which often flow through more open landscapes.
Diatoms have a relatively low protein content, but the presence of puncta in the cell wall permits many species to rapidly digest the cytoplasm. For some species of Orthocladiinae diatoms are such an important food source that their life cycle can be attuned to the diatom peaks of abundance (e.g. in *Orthocladius*). Tokeshi (1986a: 503) stated that no species investigated by him exploited the larger food particles like Closterium and the diatoms Synedra and Nitzschia. Kawecka & Kownacki (1974) stated that other characters of diatoms also influence their selection by Chironomidae larvae. The importance of protozoans as food has been stressed by Tarkowska-Kukuryk & Mieczan (2008).
Storey (1987: 344) argued that for first instar larvae **nitrogenous food** could determine the carrying capacity of the environ-

ment; diatoms can play an important role in this because of their size. Walentowicz & McLachlan (1980) stated that in a stretch of a small upper course with input of peat particles with low nitrogen content, the percentage of Orthocladiinae decreased and of Tanytarsini increased. The latter group used the peat particles for tube-building and feeding. At a downstream site, where the nitrogen content of the particles had increased, the percentage of Orthocladiinae had increased again.

Very few Orthocladiinae are true **shredders**, able to consume and digest living plant tissue. The most important examples are *Cricotopus brevipalpis* and *C. trifasciatus*. *Brillia* species often live on submerged wood, but most probably these larvae feed only on decaying material. As Anderson & Sedell (1979) pointed out, fungal and bacterial activity in dead leaves are important to shredder feeding, not the reverse. The range of food quality available to collectors is possibly greater than that for shredders. Orthocladiinae larvae can also consume living or dead **animal food**; in contrast to some Diamesinae they are never true predators, but use these food sources only incidentally (Konstantinov, 1958; Zinchenko et al., 1986; own data).

For further aspects of feeding, see Moller Pillot (2009: 12–14).

2.8 MICROHABITAT

OPPORTUNISM

As a rule, Orthocladiinae larvae can be found in microhabitats where other factors, such as food, oxygen availability and protection against current, are optimal. This means that the same species can live in different microhabitats in different water types. In contrast to many Chironomini species (see Moller Pillot, 2009: 11), only few species seem to change microhabitats between summer and winter, possibly because freezing does not play an important role in running water.

TUBES

Many Orthocladiinae species build a larval tube, usually fastened to the substrate. The larvae use salivary secretion, but the majority of species incorporate particles of detritus, diatoms, filamentous algae, etc. into the tube. Tubes built entirely of salivary secretion are the rule in species like *Synorthocladius semivirens* and *Eukiefferiella clypeata* and can be found in many other species, especially when suitable particles are not available (Brennan & McLachlan, 1979; Zinchenko et al., 1986). Armitage et al. (1995) give a summary of the literature.
Sometimes the tube is more or less transportable, because it is not fastened to the substrate; this is not exceptional in the genus *Psectrocladius* (own data), but has been observed also in other species (Zinchenko et al., 1986). Better transportable tubes as in Trichoptera are constructed by very few Orthocladiinae, for example *Heterotanytarsus apicalis*.
Larvae of many semiterrestrial and terrestrial species are free-living. *Corynoneura* and *Thienemanniella* walk freely on firm substrate. As a rule such species make a tube for pupation only.
Hershey (1987) suggested that the tubes evolved as an anti-predator adaptation and have been adapted later (e.g. in many Chironomini) for filter feeding. She found that larvae that spent more time out of their tube were predated more often.

BURROWING

Orthocladiinae are collected deeper in the substrate less often than Chironomini. However, they can be found up to 40 cm deep (Schmid, 1992, 1993a). Older larvae are found more superficially (except in winter); many young larvae live at a depth of 20 to 30 cm, especially in diapause. Schmid's results were determined by the grain size and oxygen supply in the bottom of alpine streams. In lowland brooks with loamy-sandy bottoms, Orthocladiinae are often absent from deeper layers.

EXPOSED WATER BOTTOMS

Many Orthocladiinae, terrestrial species in particular, live in the wet bottom above the water level. Aquatic larvae can also be found above the water level and in some species it is not known whether the eggs are deposited there and if young larvae (or even older larvae) live mostly above or below the water level (e.g. *Chaetocladius piger*, *Limnophyes pentaplastus*). In some cases the larvae live in a pond or ditch, but the pupa creeps out of the water (e.g. *Metriocnemus* species).

VASCULAR PLANTS AND MOSSES

Very few Orthocladiinae consume living macrophyte tissue or mosses (see section 2.7). Dying leaves are consumed by more species, probably because many nutritious substances become available as the leaves die. Most Orthocladiinae living on plants collect or scrape algae, other microorganisms and detritus from the leaves (Mason & Bryant, 1975; Tokeshi, 1986a). Therefore many species are as common on artificial (plastic) plants as on living plants (Higler, unpublished). Their life cycles are often correlated with the abundance of their food; related species have different life cycles to avoid concurrence (e.g. in the genus *Orthocladius*).

As a rule, the larvae of different species prefer different areas within vegetation and different places on the plants (e.g. Tokeshi & Pinder, 1985). Species also show differences in microdistribution within moss cushions (Lindegaard et al., 1975; Nolte, 1989).

Many investigations suggest that invertebrates living among vegetation are less vulnerable to predation. However we found no literature proving this for chironomid larvae.

WOOD

Many Orthocladiinae living on wood can also be found on stones and plants. However, some species are true xylophages (e.g. *Orthocladius lignicola*) and other species are more common on wood because they feed on microorganisms living on decaying wood (e.g. *Brillia longifurca*). More detailed information can be found in Cranston & Oliver (1988), Armitage et al. (1995) and Hoffmann & Hering (2000). The tree species seems to be of minor importance (Spänhoff et al., 2000). Relatively few Orthocladiinae are found on wood in shady sites, probably because of a lack of periphyton (Klink, 2011). Some species are found on wood near the water surface because the oxygen supply is better there.

2.9 DENSITIES

Part II of this series (Moller Pillot, 2009), on the tribe Chironomini, includes impressions, for many species, of the larval densities found by one or more investigators. The Orthocladiinae inhabit plants, wood and stones more than the Chironomini. The densities per m^2 of bottom surface are consequently much more variable. Some authors therefore use densities per dm^2 of plant surface or per gram dry weight of the plant. However, larval densities vary considerably between the central parts and the margins of plant and moss vegetation (e.g. Lindegaard et al., 1975). They can also be quite different on different plant species (e.g. Tokeshi & Pinder, 1985); mosses in particular can be densely colonised (see Nolte, 1991). Moreover, as in bottom dwellers, densities fluctuate throughout the life cycle (see Part I: Vallenduuk and Moller Pillot, 2007: 16–17). For this reason, in this book densities are mentioned for only a few species.

An example can illustrate this difficulty. We found about 10 larvae of *Corynoneura scutellata*/dm^2 grass leaf surface in a stream in January. Balushkina (1987) collected about 18 larvae/dm^2 bottom surface of (?) the same species in a lake in Russia. Without knowing the density of plants, such data are completely incomparable, quite apart from margin effects and seasonal effects. Of course, within a single study it is possible to compare the densities of different species or the densities of a species in different microhabitats, years or seasons.

Fluctuations in densities of larvae during the year can be due to the life cycle (see section 2.4), but may also be caused by food availability, predators, parasites, etc. (Vallenduuk & Moller Pillot, 2007: 16 ff.; see also under *Eukiefferiella ilkleyensis* and *Parametriocnemus stylatus*: Life cycle). In many cases the reason for lowered densities remains obscure (e.g. LeSage & Harrison, 1980). Fluctuations in the numbers of emergences are influenced to an even greater extent by synchronisation. Hayes & Murray (1989) studied the variation in exuviae numbers in a lowland river in Ireland and concluded that a single sample taken in summer may be inadequate for the purpose of river monitoring.

2.10 WATER TYPE

Tolerances and preferences in general

As stressed in Part I of this series (Vallenduuk & Moller Pillot, 2007: 19), stated tolerances and preferences have to be used with caution. Correlations between the occurrence of a species and an environmental factor can play a role in one stream system and hardly or not at all in another. For example, a species requiring a good oxygen supply may prefer lower temperatures, but occurs at higher temperature at a high current velocity or at lower saprobities, or in large rivers only where wood or stones are available near the water surface. Optimal food or oxygen conditions are different in different microhabitats and at different combinations of temperature, current velocity or depth. Moreover, animals are more tolerant of a less optimal factor if all other factors are optimal.

Where optimal localities are available, a species will also be found in places where conditions are less suitable (even in other water types!) due to short-distance dispersal of larvae and adults. Usually a species is absent from suitable localities when such refugia are not available within a short distance, from which remigration is possible after local disappearance. This aspect is more important in a dynamic (irregular changing) than in a stable environment (Moller Pillot, 2003). For instance, some species are most numerous in temporary water, but survive drought in low numbers in permanent water bodies. The presence or absence of a species at a given time depends mainly on variable factors such as weather, food, predation, etc.

The relations given in the text and tables have no overall significance, but can help with interpreting data, especially in more or less comparable situations.

Stream typology

In this book we have not used a stream typology, even though it can be easier to form an idea of the ecology of a species when it is characterised by a stream type. There are different reasons for not using a typology here:

- Combinations of stream properties are different in different regions. Current velocity is not always correlated with stones on the bottom or upper courses with shade or low temperature. Also, individual streams within one region produce highly characteristic communities (Singh & Harrison, 1984: 246).
- This book describes the ecology of individual species. For some species one factor is relatively important, but does not change the whole community, such as the presence of wood for *Brillia longifurca* or shade for the genus *Cricotopus*.
- Within one stretch of stream the most important factors may vary, such as current velocity, microhabitats, shade, etc.
- There is an increasing tendency to use existing typologies as a basis for standardisation. In nature, biodiversity is based on the continued existence of differences, even between streams of the same type. At this moment we do not know the causes of most of these differences.

In this book the term 'lowland stream' is used frequently, but with a different

meaning than in the central European literature. Braukmann (1984), when describing streams mainly in the southwestern and western part of Germany, gives for Flachlandbäche a slope of 1 to 10 m/km. Smissaert (1959), investigating the brooks and streams in southern Limburg (the Netherlands), considered streams with a slope of 3 to 10 m/km to belong to the 'Geul-type' and stated that these streams are comparable with streams of the Untere Salmonidenregion in Illies (1952).
Because the classification of Smissaert is usable in the Netherlands, we will use the term 'lowland stream' only for streams with a slope of less than 3 m/km and a mean current velocity less than 75 cm/sec. Other terms will be used only when citing literature.

CURRENT

We use the current velocity as measured 10 cm below the water surface. In fact, the current at the spot where the larva lives is more important. The depth of the stream, the density of the vegetation and other factors also plays a role in determining the actual current velocity. Syrovatka et al. (2009) use a more complex hydraulic parameter (Froude number). These authors also stated that the amount of sedimented POM (particulate organic matter) was identified as an important variable. Sedimentation is dependent on current velocity, depth and the presence of sheltered places.

It is necessary to bear in mind that for the animals mainly three aspects of current are important: food, oxygen and not to be dragged down. In every stream, current velocity varies during the year. If the velocity is too high, the larvae will be swept away and dragged down, whereas at very low velocities the oxygen content or the microhabitat can become unfavourable and many larvae will leave the site and drift away. Sometimes a species is not absent where the current is too fast, because there are sheltered places where enough food is available (Syrovatka & Brabec, 2010). Species inhabiting plants or silt are particularly susceptible to being swept away during a catastrophic flood (Gendron & Laville, 2000).

A rheophilous species can occur frequently at low current velocities in a stream or a region with many riffles, whereas the same species will be absent in a lowland region where the optimum stream velocities are absent (see above under Tolerances and preferences).

In many cases the influence of current velocity cannot in practice be separated from other factors, which vary with altitude in both mountain and lowland streams: temperature, food quality and availability of microhabitats such as stones, gravel and plants. We have tried to find information in the literature and in our own data (as far as possible) to describe the tolerance of different current velocities of every species, independent of other factors.

Many rheophilous species can live in streams as well as in the wash zone of lakes. This aspect has been left out of consideration in table 7 in Chapter 9. See further under Dimensions below.

Dimensions

In large **rivers** the current velocity in the middle of the stream is relatively much faster than in small brooks. Owing to this the presence of water plants, silt and animals is not comparable. Many Orthocladiinae live in large rivers mainly on or beneath wood or stones. In the descriptions of the species we have often written that the larvae are collected or not collected in large rivers. One should bear in mind that the absence of wood or the presence of stones is usually not the natural situation.

In large **lakes** the oxygen content in the littoral zone is often better than in ponds and pools. Many rheophilous species can quite often be found also in large lakes (in some cases only near the shore where the influence of waves is greatest); they do not need the current, but only enough oxygen. Water movement also reduces silt deposition and settlement of plants, and such pioneer situations can be inhabited by rheophilous species. In northern lakes even more rheophilous species can be found, because oxygen demand is lower at lower temperatures. The main species found in the profundal zone of eutrophic lakes are Chironominae with haemoglobin; the profundal zone of oligotrophic and boreal lakes can also support a number of Orthocladiinae (see e.g. Saether, 1979).

TEMPERATURE

Without doubt temperature has much influence on the main processes of life, such as decomposition, development time, etc., and is the most important factor influencing the feeding of chironomids (Moore, 1979: 321). Brabec (2000) wrote that temperature seems to be the main regulator of community composition. Many authors call species cold-stenothermic or eurythermic (e.g. Brundin, 1949; Lehmann, 1971). Lindegaard & Brodersen (1995: 263) suggested that temperature may be the key factor in the distribution of Diamesinae and cold-stenothermic Orthocladiinae, while other factors may be more important for Chironomini. Brundin (1956a) wrote: 'Es ist auch klar, dass das Temperaturoptimum bei verschiedenen Arten sehr verschieden ist, so dass z. B. viele arktische Arten mit sehr niedrigen Temperatursummen auskommen können.'
Although it is clear, that many species are adapted to fast or slow decomposition of organic material or limited by a certain availability of oxygen, we found that indications such as cold-stenothermic are often unfounded. In most cases the presence of a species is dependent on a mixture of factors and temperature is rarely dominant within these factors. Many species that live in cold springs in the mountains also live in much warmer springs of lowland brooks. This is also illustrated by the investigations by Hamerlik & Brodersen (2010) of city fountains, where often 'cold as well as warm water species were present and temperature was not an important factor determining the species distributions'. Again, the lower temperature optima of many Orthocladiinae given by Rossaro (1991) are usually not valid in other regions and seem to be based mainly on the fact that these species in the running waters of Italy are confined to upper courses because of other conditions found in these upper courses.

The relation between temperature and oxygen supply is complex. Verberk et al. (2011) emphasise that while the solubility of oxygen is lower at high temperatures, the diffusion rate is higher, so that actually more oxygen is available to an organism in warmer habitats. The problem for the animal is that its oxygen demand rises with temperature. As later investigations have shown, the animal can tolerate higher temperatures when oxygen supply is high (Verberk, pers. comm.). This is important in hygropetric environments and for larvae living near the water surface.

Regarding the influence of temperature on food availability, high temperature and hardening both quicken the rate of decomposition, which may lead to species having a different temperature optimum in more acid water.

SHADE

The influence of shade has been treated in Moller Pillot (2009: 15). Most probably Orthocladiinae are more influenced by shade than most other subfamilies. For instance, *Cricotopus* larvae are relatively scarce in shaded brooks and streams and *Parametriocnemus* is especially common in wooded landscapes. In most cases it is not known whether flying and swarming of the adults or food and microhabitat of the larvae are the main factor. Apart from these functional relations, it is possible that larvae prefer shaded or unshaded parts of a stream, as has been stated by Higler (1975) for Trichoptera. According to Luferov (1971), Orthocladiinae prefer more light than Chironomini.

PERMANENCE

As has been explained in Moller Pillot (2009: 16), the information in our books about the occurrence of species in temporary water applies only to isolated water bodies in which the bottom is desiccated to at least 10 cm deep in summer. When larvae survive in deeper layers of gravel and sand in brooks and streams we consider this also to be surviving drought in that water body. In other circumstances (e.g. impermeable bottoms) the presence of many species is usually dependent on egg-laying females during the wet periods. As Wotton et al. (1992) point out, small size and a short life cycle, as in *Cricotopus sylvestris* and *Psectrocladius limbatellus*, are characteristics of species inhabiting temporary habitats. These species survive drought periods in permanent water bodies. Some Orthocladiinae species have a diapause in summer and are able to survive completely dry periods in the same place. An example is *Trissocladius brevipalpis*, which has a summer dormancy in second instar. Such species are often obligate inhabitants of temporary water. However, in many cases it is sufficient when a riparian zone is dry in summer.

2.11 ALTITUDE

Many investigations show that the occurrence of a species in a stream system can be correlated with altitude or at least with longitudinal zonation (e.g. Laville & Vinçon, 1991; Punti et al., 2009; Milošević et al., 2012). However, what has been described in the section on Temperature applies here too, and to an even greater degree because altitude itself has no influence at all. The important factors here are temperature, current velocity, dimensions, shading, trophic conditions, silting, permanence, etc. Already Gendron & Laville (1997) stated that there data suggest a clinal rather than a zonal pattern. Species such as *Heleniella ornaticollis* or *Eukiefferiella brevicalcar*, stated as being species of high altitude zones in the references above, can be found at an altitude of 100 m in the Netherlands. All running waters exhibit a longitudinal zonation, but this generalised concept takes on different expressions according to regional or local conditions (Higler & Statzner, 1988). For instance, in lowland brooks the current velocity often increases downstream and permanence often plays a major role. See also under Stream typology (in section 2.10).

2.12 pH

The influence of pH has also been treated in Vallenduuk & Moller Pillot (2007) and, more extensively, in Moller Pillot (2009). A small number of Orthocladiinae species are strongly acidophilic (although usually not so strongly as suggested in Orendt, 1999). On the other hand many species do not tolerate a pH lower than 5 and some species are basophilic (e.g. *Psectrocladius barbimanus*). Such species are usually also characteristic of high conductivity (Ruse, 2002). Very few studies have been made of fast-flowing acid streams, and so we know hardly anything about acidity tolerance of species living in fast-flowing streams.

OTHER FACTORS CORRELATED WITH pH

Some species (e.g. *Heterotrissocladius marcidus*), which can live in acid as well as circumneutral woodland brooks, are absent from the majority of Dutch acid waters. An important reason for this may be that many acid upper courses on Pleistocene sandy soils became very poor in calcium and trace elements, probably because of transport downstream in winter. A shortage of available nutrients has been proven to be a reason for the absence of many oligochaetes in Dutch bog remnants (van Duinen et al., 2006). Mineral shortages as a result of acidification also lead to a decline in insect diversity in heathland (Vogels et al., 2011). In moorland pools transport of minerals is less important, but here also calcium, magnesium and manganese have declined (van Kleef 2010: 24). In addition, the large amounts of ammonium-N and the accumulation of organic material in Dutch moorland pools make the detritus less suitable as food, or worsens the microhabitat. Another problem is that most of the Dutch oligotrophic waters are acid, which makes it almost impossible to separate the influence of these two factors in investigations in the Netherlands (Verberk et al., 2012: 27). In lakes and moorland pools humic conditions also play a role. See section 2.17.

2.13 TROPHIC CONDITIONS

STREAMS

As mentioned in Vallenduuk & Moller Pillot (2007: 20) there is most probably no direct relation between nutrient content and the presence of chironomid larvae. Some Orthocladiinae are dependent on the availability of certain types of algae for food and/or specific organic matter decomposition processes, which in turn are related to trophic state and pH.

In upper courses heterotrophic processes (mainly decomposition of allochthonous material) are usually dominant, especially in woodland. Natural woodland brooks are not completely oligosaprobic, because there is much (usually slow) decomposition of this material (see under Saprobity below). Further downstream, or with increase of light, autotrophic processes (production by algae and water plants) increase, and not only as a result of anthropogenic influences. Algae and water plants produce more easily decomposing material, which in turn enables heterotrophic processes to become dominant again further downstream. The general effect of eutrophication in nutrient-poor heterotrophic streams is to shift the longitudinal succession of both autotrophic and heterotrophic processes upstream (Odum, 1956).

LAKES AND POOLS

For further aspects of trophic conditions, especially in lakes and pools, see Moller Pillot (2009). Because the influence of trophic conditions on animals is indirect (via metabolic processes) the role of temperature is very important. Therefore, it is possible that a species will prefer more eutrophic water in Scandinavia than in the Netherlands.

TROPHIC CONDITIONS IN THE NETHERLANDS

In the Netherlands oligotrophic water is scarce. In Holocene regions almost all lakes, ditches and pools are eutrophic; in Pleistocene regions oligotrophic pools are usually acid. Our data about oligotrophic water bodies are mainly borrowed from foreign literature.

2.14 SAPROBITY

In general, the reader is referred to Vallenduuk & Moller Pillot (2007: 20) and Moller Pillot (2009: 17–18). More so than the Chironomini, the Orthocladiinae are inhabitants of flowing water and the larvae live more on plants, stones and wood than on the bottom of streams. The consequences for the assessment of pollution of the stream as a whole are twofold. First, the Chironomini mainly live among and feed on the deposited organic matter on the bottom, which makes them more exposed to the toxic products in the silt. Second, in flowing water there is a permanent drift of larvae. A consequence of this is that species that are unable to **complete their whole life cycle** at a polluted site may be brought down from cleaner upper courses. This is especially important if part of the life cycle is spent in the bottom. We regularly found larvae and pupae of more or less intolerant Orthocladiinae below a sewage outfall. In cases where this seems to happen more regularly, we have marked the presence of the larvae with a (+) in table 5 on p. 282. Investigations into the influence of sewage effluent pay little attention to this problem (e.g. Bazzanti & Bambacigno, 1987; Pinder & Farr, 1987; Wilson, 1989).

Another question is that within a stretch of a stream, saprobity depends on the microhabitat. Gołowin (1968) stated that dead flow zones contain periphyton communities typical of more polluted water compared with the periphyton found in the zones of faster flow. For chironomid larvae, both abiotic conditions and food are different in different microhabitats. Chironomids reflect the conditions of the microhabitat more than the exact water quality and therefore give more information about the total system than that obtained from chemical and plankton analyses.

As argued by Verberk et al. (2012), hardly any organic silt is deposited in fast-flowing streams and therefore such streams can be almost oligosaprobic, even when the water is very eutrophic. We do not use the term oligosaprobity in this book when in such cases a thin layer of organic silt can be found on and under stones, between vegetation and near the banks.

2.15 OXYGEN

The relation between oxygen availability and temperature is very complex and has been treated in the section on Temperature above. Most important is that oxygen shortage for animals is caused mainly because, as temperatures rise, oxygen consumption increases more quickly than oxygen supply. This is the most important reason why inhabitants of organic bottoms, such as *Prodiamesa olivacea*, survive better in polluted streams in winter.

In much of the literature (and also Limnodata.nl) the data on oxygen contents are daytime measurements. The differences between day and night can be very large, especially in small stagnant water bodies and in unshaded brooks and streams. Walshe (1948) has shown that survival of larvae in anaerobic conditions is very different between individuals of the same species. Problems can be expected mainly, if at night temperatures hardly fall. Larvae may then leave the bottom and go into drift.

The only species of Orthocladiinae that is blood red is *Trissocladius brevipalpis*. However, Czeczuga (1962) stated that even larvae which are not perceptibly red can contain a small amount of haemoglobin. This is not known for Orthocladiinae species. However, she emphasised that apart from haemoglobin, the presence of a large amount of glycogen and its use in the vital processes can play an important role in an environment with a low oxygen concentration.

2.16 AMMONIUM

It is probable that many chironomids are intolerant of NH_3 (Ruse, unpublished); this is therefore important for bottom dwellers, especially at high pHs. However, the specific influence of NH_3 on Orthocladiinae (in comparison with other pollutants) has never been demonstrated. Because few Orthocladiinae live in silty bottoms, they will hardly be exposed to NH_3. When rearing chironomids in small vessels it may be necessary to refresh the water, as we experienced when rearing *Corynoneura* larvae.

Acid moorland pools in the Netherlands can be very rich in ammonium. The dissociation of NH_4^+ decreases at low pH values, making it less toxic. High ammonium contents do not have a direct influence on many (or all?) chironomids. At least some species of the genus *Psectrocladius* (*bisetus*, *oligosetus*, *platypus*) and *Paralimnophyes longiseta* seem to be equally abundant in moorland pools with and without much ammonium.

2.17 HUMIC ACID

In peaty environments often much organic matter washes into water bodies from the surrounding land. Brundin (1949: 540) pointed out that in such more humic water fewer chironomid species and fewer individuals are found. The key to chironomid associations in relation to trophic conditions by Saether (1979) is not usable for mesohumic and polyhumic lakes. According to Leentvaar (1959), such brown coloured (dystrophic) water bodies have a low phytoplankton development because of light absorption and inhibition of growth by humic acid. Some metal ions display a positive correlation with humic acids (Vogels et al., 2011). As Saether writes, the mesohumic and polyhumic lakes have not been investigated to a sufficient degree to draw any conclusions about this factor. However, we can say that humic acid can decrease (or sometimes increase) the toxicity of trace metals, low pH, etc. (Winner, 1984; Kraak et al., 2000; Moller Pillot, 2009: 16).

Polyhumic lakes and pools are relatively rich in *Psectrocladius* species. *Heterotanytarsus apicalis*, *Cricotopus sylvestris*, *Limnophyes asquamatus*, *Paralimnophyes longiseta* and *Corynoneura celeripes* are also well-known species of this water type (e.g. Brundin, 1949; Saether, 1979; Thienemann, 1954; own data).

2.18 TRACE METALS

Most investigations about the harmful influence of trace metals on chironomids apply to Chironominae (see Moller Pillot, 2009: 20). As far as Orthocladiinae have been investigated, the larvae appear to be more resistant than Chironominae to contamination by trace metals (Pinder, 1986: 14; Arslan & Emiroglu, unpublished). Moreover they are usually less exposed to high concentrations (see sections 2.14 and 2.16).

Trace elements are also nutrients. The absence or scarcity of many of them could play an important role in very nutrient-poor landscapes. In the Netherlands this can be expected especially in the acid bogs, moorland pools and upper courses in the sandy Pleistocene regions. See under pH (section 2.12).

2.19 SALINITY

For several reasons (fluctuations, seepage, oxygen content), tolerance to salt concentrations is not the same in different regions (see e.g. Vallenduuk & Moller Pillot, 2007: 20; Moller Pillot, 2009: 20). Where available, not only data from the Netherlands, but also from the Camargue or the Baltic region will be given.

2.20 PARASITISM AND PREDATION

Tokeshi & Townsend (1987: 843) suggested that in the epiphytic chironomid community, predators and parasitoids as a rule do not play an important role. However, LeSage & Harrison (1980a) stated that a large proportion of *Cricotopus* imagos were infested with mermithid nematode parasites. The level of parasitism differed greatly between species: in one year up to 67% in *C. triannulatus* and only 5% in *C. bicinctus*. Also, many larvae had been infested by mites, as has been stated by many other authors (e.g. Kouwets & Davids, 1984; Stur et al., 2005). Infestation by Protozoa has been hardly investigated in Orthocladiinae; Thienemann (1954: 305) mentions only infestation of two genera by *Thelohania chironomi*.

Armitage et al. (1995: 428) suggested that predation is not the most important factor affecting numbers of chironomid larvae, although sometimes a substantial influence has been stated. Most probably fishes are in many cases the most important predators and therefore the choice of microhabitat (often between plants) and the tube-dwelling behaviour in most Orthocladiinae may be a significant factor in avoiding predation. Hershey (1987) found that a larger proportion of the larvae that spent more time out of the tube were consumed. Swanson & Hammer (1983) suggested that the thick algal mat covering submerged shrubs may provide an important refuge for *Cricotopus ornatus* larvae. Klink (2010) has shown that invasive aliens like the shrimp *Dikero-*

gammarus villosus can lead to extinction of many species in a river. For instance, between 1992 and 2003, species of *Cricotopus*, *Rheocricotopus*, *Tvetenia* and *Synorthocladius* disappeared from the river IJssel in the Netherlands. Townsend & Hildrew (1979) found in a small stream in southern England that the numbers of small chironomid larvae were reduced by up to 84% through predation by the caddis fly larva *Plectrocnemia conspersa* and the alderfly larva *Sialis fuliginosa*. The absence of many parasites and predators in pioneer situations could play an important role for certain species of chironomids, but this aspect has not been thoroughly investigated.

2.21 DISPERSAL

Adult flying and dispersal has been treated at some length in Vallenduuk & Moller Pillot (2007: 8–9). It appeared that egg-depositing females can be scarce at distances of 3000 m (Moller Pillot, 2003) or out of their specific habitat (Delettre et al., 1992), but that dispersal over much larger distances is possible. Van Kleef (2010) suggested that chironomids in general are not dispersal limited. However, van Kleef & Esselink (2005) stated that the settlement of *Psectrocladius bisetus*, *oligosetus* and *psilopterus* is highly dependent on the presence of populations within one or two kilometres. Probably these species have a low dispersal power. On the other hand, between 2000 and 2010, *Cardiocladius fuscus* settled in some lowland streams in the Netherlands a few years after the construction of fish ladders, in areas many kilometres from the nearest populations of this species.
In a study of city fountains, Hamerlik & Brodersen (2010) concluded that the distance to natural sources determined the species composition of chironomids in the fountains. But they pointed out that the same species were able to colonise waters in the Azores islands located 1500 km from the European coast, which also underlines their good colonisation potential. Moreover, many species have been collected on the Faeroe Islands, among which very small Orthocladiinae such as *Corynoneura fittkaui* and *lobata* (Pedersen, 1971). This means that desiccation of such small adults is not always a problem, as has been suggested in Vallenduuk & Moller Pillot (2007).

3 SYSTEMATICS, BIOLOGY AND ECOLOGY OF GENERA AND SPECIES OF THE ORTHOCLADIINAE

This Chapter contains written descriptions. For a numerical evaluation see the tables in Chapter 9. Information given under the genus name is not repeated under the entries for the species. Only additional species-specific information is given under the species names.

Aagaardia Saether, 2000

The two European species are known only from Scandinavia and are not treated in this book.

Acamptocladius Brundin, 1956

SYSTEMATICS AND IDENTIFICATION

Two species have been described in Europe: *A. submontanus* (Edwards, 1932) and *A. reissi* Cranston & Saether, 1982. Males, females and larvae are keyed in Cranston & Saether (1982). Both pupae are described in Langton & Visser (2003). However, the differences between the two species are small in all stages and we found specimens of all stages that were more or less intermediate. These species may be synonymous (see also Bitušík, 2000: 53).

DISTRIBUTION IN EUROPE AND THE NETHERLANDS

Both species of *Acamptocladius* have been reported from just a few countries, but these findings are scattered over northern, southern, eastern and western parts of Europe (Saether & Spies, 2010). In the Netherlands there are records from at least seven localities in the provinces of Noord-Brabant, Drenthe and Overijssel (e.g. Duursema, 1996; own data). Exuviae from Overijssel (Bergvennen) have been identified by P. Langton as *A. reissi*.

LIFE CYCLE

Adults, pupae and prepupae have been collected in spring and late summer, suggesting at least two generations a year. At the end of November we found all larvae in second and third instar, probably in diapause.

MICROHABITAT

The larvae are found between plants or *Sphagnum* and on bottoms with organic matter (own data).

WATER TYPE AND pH

Cranston & Saether (1982) reported occurrence of the larvae in moorland lakes, a peat pool and some acid ball clay pools (pH 4.0–6.3). All Dutch records are from acid moorland pools with pH (so far as is known) between 4 and 5. In the Pripyat region in Belarus we collected a male and some larvae in several pools at pH values from 6.1 to 7.5.

Acricotopus lucens (Zetterstedt, 1850)

Acricotopus lucidus Brundin, 1949: 695; Pankratova, 1970: 207–208, fig. 127

SYSTEMATICS AND IDENTIFICATION

In the Palaearctic region only one species of the genus is known (Hirvenoja, 1973). All stages can be easily recognised. The adult male has a rather high antennal ratio (about 3); the AR of approx. 2 mentioned by Hirvenoja (1973) is a mistake. In the larva the sclerotisation of the posterior side of the procercus and the dark occipital sclerite are especially striking. The function of the beard near the ventromental plates is unknown.

DISTRIBUTION IN EUROPE AND THE NETHERLANDS

A. lucens has been collected across almost the whole of Europe, but seems to be absent from some parts of the Mediterranean area (Rossaro, 1982; Saether & Spies, 2010). In the Netherlands the species is common in most of the country, but scarce in the province of Zeeland (Nijboer & Verdonschot, unpublished; Limnodata.nl).

LIFE CYCLE

The adults fly from March to October (Hirvenoja, 1973; own data). The life cycle is not known in detail, but most probably there are several generations in spring and summer and a larval diapause in winter. There are some records of fourth instar larvae in January.

SWARMING

We observed once a small swarm of this species at the water's edge near the vegetation, at a height of 50 cm.

FEEDING

Based on a study of larval gut contents, the larvae seem to prefer diatoms and filamentous algae as food, but they also feed on detritus (A. Klink, unpublished; own data).

MICROHABITAT

The larvae live mainly on plants, sometimes near the water surface on floating leaves of *Glyceria*, *Nuphar*, *Lemna* and *Azolla* or between filamentous algae. Higler (1977) found the larvae scarcely on *Stratiotes*. They also live on organic bottoms and on stones, and are often collected in pools and ditches with much detritus, especially on peaty bottoms (Hirvenoja, 1973: 74; own data).

WATER TYPE

Current

The larvae live in stagnant and slowly flowing water; see table 7 on p. 286. Records from fast-flowing streams, as in Orendt (2002a), are undoubtedly larvae that have been carried from slow-flowing stretches or oxbow lakes. Moller Pillot (2003) stated that *A. lucens* was a typical habitat-shifting species in a small lowland ecosystem, in winter surviving in (often temporary) ditches and in summer shifting to the upper course of a lowland brook when the water flow there decreased, pH increased and vegetation became denser. Further downstream the larvae were only common in drift samples.

Dimensions
A. lucens lives mainly in small pools and ditches and in small lakes with much vegetation. The species is rare or absent in larger lakes (Kreuzer, 1940; Brundin, 1949; Mol et al., 1982; Orendt, 1993; Steenbergen, 1993; Reiff, 1994). In flowing water it can be found in small upper courses; downstream the larvae and exuviae often arrive by drift (see above under Current). The larvae are absent from large rivers.

Shade
There are no records from heavily shaded pools or ditches. Schleuter (1985) collected the species in two woodland pools, but only in clearings. It is not known if the adults fly in dense woods; the presence of water plants or food may also be influenced by shade.

Permanence
A. lucens is relatively common in temporary water (Verdonschot et al., 1992; Moller Pillot, 2003; Hamerlik & Brodersen, 2010). We found the larvae in a pool in Belarus which was dry and frozen in winter. Most probably recolonisation takes place mainly by flying females.

SOIL
Steenbergen (1993) collected significantly more larvae in water bodies on peaty soil, fewer on sand and relatively rarely on clay (cf. Bitušik, 2000: 54).

pH
A. lucens has been collected scarcely where pH values are lower than 6 and rarely in very acid water with a pH lower than 5, but there are several records with pH values between 5 and 5.5 (Friday, 1987; Leuven et al., 1987; Buskens, 1989a; van Kleef, unpublished). At very high pH values (> 8) the species seems to be less common (Steenbergen, 1993).

TROPHIC CONDITIONS, SAPROBITY AND OXYGEN
A. lucens has been collected mainly in water with low orthophosphate and chlorophyll-a contents and a low to moderately low BOD (Steenbergen, 1993; Limnodata.nl). However, the larvae are also often found in heavily polluted ditches with very unstable oxygen content and with a BOD of 10 or more, together with large numbers of *Chironomus* and *Psectrotanypus varius*. In such cases *A. lucens* probably always lives near the water surface (van Gijsen & Claassen, 1978: 80; H. Smit, unpublished; own data). Peters et al. (1988) collected them mainly in rather heavily polluted lowland brooks. Table 5 (p. 282) gives the mean probability of finding the larvae.
A. lucens seems to belong to the group of species which can be found in very clear water and in very oxygen poor conditions. This has been stated also for *Guttipelopia guttipennis* and *Zavreliella marmorata* (Moller Pillot & Buskens, 1990: 9; Vallenduuk & Moller Pillot, 2007: 88; Moller Pillot, 2009: 238). Such species are often adapted (but not confined) to polyhumic lakes. As Brundin (1949: 858) pointed out, such lakes can be found in oligotrophic environments, but often have a very low oxygen content.

SALINITY
In Western Europe the larvae are not rare in oligohaline water, but very scarce in more saline pools and ditches (Kreuzer, 1940; Steenbergen, 1993; Limnodata.nl). Krebs (1981, 1984, 1990) collected the species rarely in brackish water in the province of Zeeland. In the Baltic Sea it is not rare at more than 2000 mg Cl/l (Paasivirta, 2000).

Allocladius Kieffer, 1913

SYSTEMATICS AND IDENTIFICATION

Ferrington & Saether (2011) have divided the genus *Pseudosmittia* s. ampl. into three genera. Identification of these genera and species is only possible using this article. The four European species of *Allocladius* are known as adult and pupa, and three of them also as larva.

Allocladius arenarius (Strenzke, 1960)

Pseudosmittia arenaria Moller Pillot, 1984: 148–149, fig. VI.44 a, e (pro parte)
Pseudosmittia arenaria agg. Moller Pillot & Buskens, 1990: 30, 73 (pro parte)
Pseudosmittia gracilis Nolte, 1993: 33, 42 (nec aliis)

IDENTIFICATION

A. arenarius is apparently a parthenogenetic species. The two subspecies *arenarius* and *flavus* are possibly only forms without systematic value (Ferrington & Saether, 2011: 75). The females are not separable from those of *A. nanseni*. In most public ations these two species have not been separated. Pupae and larvae can be identified using Ferrington & Saether (ibid.). According to these authors the larvae are characterised by 7 to 8 claws on the posterior parapods. We stated sometimes 9 claws.

DISTRIBUTION IN EUROPE AND THE NETHERLANDS

The species has been identified only in a few countries in western and central Europe (Saether & Spies, 2010). In the Netherlands *A. arenarius* has been recorded throughout the country (Moller Pillot & Buskens, 1990: 73). We have verified six of the approx. 50 records and five of them belonged to this species (in the western and eastern part of the country). In one case a larva seemed to be *A. nanseni*.

LIFE CYCLE

We found fourth instar larvae during the whole year and stated emergence from February until autumn. The female produces more than one egg mass (Nolte, 1993: 42–43).

OR

MICROHABITAT AND ECOLOGY

The larvae live mainly in wet soils, often near water bodies, at the water's edge or a little higher (Strenzke, 1960; own data). For many days after inundation they can be found in the water column, creeping around on plants near the surface or on the bottom (they cannot swim). See also *A. nanseni*. They are caught in aquatic macroinvertebrate samples more frequently than most species of *Pseudosmittia* s.s., because of their active behaviour in water and their relative large size (about 4 mm).
In the Netherlands the species has been collected very often in wet grassland and brackish pools, but also along and in ponds, lakes, ditches and lowland brooks, in Pleistocene as well as in Holocene parts of the country. Most of the approx. 50 records of the aggregate from the Netherlands will apply to this species (see above under Distribution).

SALINITY

A. arenarius is most common in brackish environments, but is found everywhere also in fresh water (cf. table 6 on p. 284). According to Strenzke (1960: 420) the larvae develop normally at salinity up to 8‰ (about 4000 mg Cl/l). This author found no ecological difference between the subspecies *arenarius* and *flavus*.

Allocladius bothnicus (Tuiskunen, 1984)

Lindebergia bothnica Cranston et al., 1989: 208, fig. 9.42; Klink & Moller Pillot, 1996
Pseudosmittia bothnica Saether & Ferrington, 2003: 1, 3

IDENTIFICATION
Male, pupa and larva have been keyed and described by Ferrington & Saether (2011). The larva differs from the other species of *Allocladius* by having only three inner teeth on the mandible. The female has not been described, but is present in the collections of Klink and Moller Pillot.

DISTRIBUTION IN EUROPE AND THE NETHERLANDS
A. bothnicus has been recorded only in Finland, Russia and the Netherlands (Saether & Spies, 2010). In the Netherlands it has been found at five localities in different parts of the country (leg. A. Klink, H. Moller Pillot). Probably the species is widely distributed.

ECOLOGY
Klink (unpublished) found many larvae in a newly excavated channel of the river Rhine, where they lived mainly submerged in the sandy bottom. Elsewhere we reared the species from pioneer situations along Pleistocene pools and dead river branches just above the water level.

Allocladius longicrus (Kieffer, 1921)

A. longicrus is a terrestrial living species able to survive extreme drought in situ (Delettre, 1984). It is not treated here.

Allocladius nanseni (Kieffer, 1926)

Pseudosmittia nanseni Steinhart, 1998: 27–112

IDENTIFICATION
See under *A. arenarius*. Full-grown larvae are a little larger than that species, at about 5 mm (Steinhart, 1998; Ferrington & Saether, 2011). The posterior parapods have 11 claws. Males exist, but in many cases only parthenogenetic females are found (Steinhart, 1998: 52).

DISTRIBUTION IN EUROPE AND THE NETHERLANDS
A. nanseni has been collected in northern and western Europe, but there are only very few verified identifications (Saether & Spies, 2010). Occurrence of the species in the Netherlands is very probable, but this has so far not been reliably stated (see under *A. arenarius*).

ECOLOGY
Steinhart (1998) collected *A. nanseni* larvae from temporary flooded parts of the floodplain of the river Oder, but none from permanent water bodies. However, the larvae could be reared under aquatic as well as terrestrial conditions. Pupation takes place above the water surface. According to Ferrington & Saether (2011) the species appears to be truly aquatic at least in part of its range. It lives in fresh and brackish environments.

Brillia Kieffer, 1913

IDENTIFICATION AND MORPHOLOGY

All stages of the two European species of this genus can be identified very easily. After moulting the black patch on the postmentum of the larva of *B. bifida* is absent for a short period. The small central tooth of the mentum may be absent and in both species it may be divided. Such specimens do not belong to a separate species as suggested by Pankratova (1970: 133) (*B. pallida*). Within each species the weight of the adult and the length of the head of the fourth instar larva may vary considerably, possibly under the influence of external factors (Caspers, 1980a; Schmid, 1993).

Brillia bifida (Kieffer, 1909)

Brillia modesta Crisp & Lloyd, 1954: 273, 296, 305; Cranston, 1982: 46, fig. 16; Langton, 1991: 90

DISTRIBUTION IN EUROPE AND THE NETHERLANDS

B. bifida is distributed throughout Europe (Saether & Spies, 2010). In the Netherlands the species is a common inhabitant of the most eastern and southeastern parts of the country and the Veluwe. It has been collected locally in the dune region and the provinces of Drenthe and Noord-Brabant (Moller Pillot & Buskens, 1990; Limnodata.nl).

LIFE CYCLE

Emergence has been stated in Germany, the British Isles and Ireland during the whole year, although from November to March only in very small numbers (Thienemann, 1926; Crisp & Lloyd, 1954; Fahy, 1973; Ringe, 1974; Caspers, 1980a). Caspers suggested that slow or fast developing larvae can be found in different places within one brook. For further details see table 4 on p. 280.

GENDER RATIO

Ringe (1974) and Caspers (1980) collected many more females than males of this species in brooks in Germany.

FEEDING

Cranston et al. (1989) mentioned that in general the *Brillia modesta* group feeds on decaying leaves (see also Crisp & Lloyd, 1954). Geijskes (1935) observed that the larvae made tubes along the midrib of dead leaves and fed upon the parenchyma. Without doubt the larvae can also feed on decaying wood, as has been stated for *B. longifurca*.

MICROHABITAT

Tolkamp (1980) and Verdonschot & Lengkeek (2006) found the larvae mainly on dead leaves and other coarse organic matter, and rarely on sand. The larvae also live in debris dams, on submerged wood, between mosses and on living water plants (Lehmann, 1971; Lindegaard et al., 1975; Schmid, 1993; Spänhoff et al., 2000; own data). Gouin (1936) observed that the larvae sometimes live out of water and pupate there.

DENSITIES

B. bifida is never found in high densities. Lindegaard et al. (1975) found 105 larvae per m^2 in the moss carpet of a spring in Denmark; Izvekova et al. (1996) stated up to 250 larvae per m^2 in Russian brooks. The densities may possibly be much higher on wood.

WATER TYPE

Current

The larvae inhabit springs, brooks and streams, and can even be found in hygropetric conditions and in a patch of mud with seepage (Crisp & Lloyd, 1954). They are rarely collected in stagnant water and all records from lakes are probably of larvae or exuviae brought down by brooks (Reiff, 1994). Occurrence of the species has been described as more or less independent of current velocity, but in general the species is much more common in hilly and montane regions than in lowland brooks (Learner et al., 1971; Braukmann, 1984; Laville & Vinçon, 1991; Izvekova et al., 1996; Bitušik, 2000; Janzen, 2003; Limnodata.nl). It is probable that a combination of water movement and the presence of dead leaves is most favourable. Table 7 in Chapter 9 shows the mean probability of finding the larvae at different current velocities. See also the Summary below.

Dimensions

B. bifida is a species of springs, brooks and small streams (Learner et al., 1971; Lehmann, 1971; Thienemann, 1944; Gendron & Laville, 1997; Limnodata.nl). In the river Rhine the larvae are rare (Klink & Moller Pillot, 1982; Caspers, 1991; Becker, 1995).

Shade

Janzen (2003) found a clear correlation with shade in German brooks. In the Netherlands the species has also been found much more often in wooded landscapes. The species often lives abundantly in brooks heavily shaded around the year (own data). It is not clear whether the shade itself is important or mainly the presence of slowly decaying leaves all year round.

Permanence

We found the species sometimes in brooks which dried out completely in summer. It is not known whether eggs or larvae tolerate desiccation or if the presence in such cases is the result of renewed colonising from elsewhere. Schleuter (1985: 94) reared an adult from a wheel track in a wood in Germany. In general the larvae are most common in permanent localities, because the species lives mainly in brooks with strong seepage of groundwater.

pH

Orendt (1999) collected *B. bifida* in brooks with a pH between 4 and 7. Other authors have also stated the presence of larvae in acid water (Verdonschot & Schot, 1987: pH 6.5; Hawczak et al. 2009: pH around 6). Nevertheless, the species is usually absent from acid lowland brooks in Germany (e.g. Janzen, 2003) and in the Netherlands (Limnodata.nl; own data). In general, *B. bifida* is more common in basic conditions. For further comments see also the entry for *Heterotrissocladius marcidus* and section 2.12.

SAPROBITY

In the Netherlands and in other regions most larvae are found in the upper reaches of brooks, where anthropogenic influences are small. As a rule the literature mentions that *B. bifida* is a species of water with little or no pollution (Moog, 1995; Izvekova et al., 1996; Bitušik, 2000). Nevertheless, larvae have been collected in brooks with a BOD > 4 mg O_2/l and ammonium-N content of 1 mg or more (Limnodata.nl; own data). Some of these records seem to be caused by drift transport of larvae. On the other hand the larvae lived abundantly in a patch of woodland mud (with strong seepage) between decaying organic matter (Crisp & Lloyd, 1954). In such cases also the species needs a good supply of oxygen (see table 5 on p. 282).

SUMMARY

Brillia bifida is a common inhabitant of springs and upper reaches of streams with high oxygen content, usually shaded or with much slowly decaying leaves or wood. The larvae tolerate high decomposition levels. They live mainly in brooks with seepage or fast flows, but on sites with slower flow. It is not clear why the species is absent from many brooks, both acid and calcareous.

Brillia longifurca Kieffer, 1921

Brillia flavifrons Langton, 1991: 90, fig. 38i; Langton & Pinder, 2007: 78, fig. 41 C, 136 D (nec Johannsen)

NOMENCLATURE

Few years ago *B. longifurca* has been thought to be a synonym of *B. flavifrons* Johannsen. This is most probably not true and the synonymy is under discussion (Spies & Saether, 2004; see also Saether & Spies, 2010).

DISTRIBUTION IN EUROPE AND THE NETHERLANDS

B. longifurca is distributed throughout Europe (Saether & Spies, 2010). In the Netherlands the species is most common in the central and southern part of Limburg and is present throughout most of the Pleistocene part of the country, but seems to be absent from the provinces of Drenthe and Utrecht and the western part of Overijssel (Limnodata.nl). It has been collected scarcely in the large rivers.

LIFE CYCLE

Emergence can be stated to occur almost the whole year round, but most authors found a peak in spring and in late summer or autumn (Lehmann, 1971; Pinder, 1974; Klink, 1985; own data). See further table 4 on p. 280.

FEEDING

B. longifurca larvae are very often found on submerged wood, grazing on the surface (Cranston, 1982; Spänhoff et al., 2000; own data). Jankovič (1974) found dense populations of the larvae in fresh colonies of *Sphaerotilus natans*. The related American species *B. flavifrons* has been shown to consume decaying conifer needles (Crawford & Rosenberg, 1984).

MICROHABITAT

The larvae are often found on submerged wood, sometimes in large numbers, but they also live on stones and plants (Spänhoff et al., 2000; Klink, 2011; own data). In brooks and rivers the numbers seem to increase sharply when submerged wood is available.

WATER TYPE

Current and dimensions

The larvae are scarce in lowland streams, except in fish ladders, and more common in faster flowing water, as in southern Limburg and central Europe (Lehmann, 1971; Braukmann, 1984; Limnodata.nl; own data). Nevertheless the species more commonly inhabits the lower parts of streams and its scarcity in large rivers seems to be caused by a shortage of submerged wood (Lehmann, 1971; Caspers, 1980; Smit, 1982; Klink & Moller Pillot, 1982; Bitušik, 2000; Móra, 2008; Klink, 2011). In **stagnant water** (mainly lakes) the species is scarce (in the Netherlands even very rare), but more common than *B. bifida* (Brundin, 1949; Reiss, 1968; Otto, 1991; Reiff, 1995). See also table 7 on p. 286.

OR

Shade
Janzen (2003) found a strong correlation with shaded brooks and streams. It is not known if the shade itself is important (in relation to swarming or egg deposition?) or only the availability of submerged wood.

pH
B. longifurca has not been found in acid brooks or streams. Limnodata.nl gives a mean pH of 7.5.

TROPHIC CONDITIONS AND SAPROBITY
Brundin (1949) collected *B. longifurca* in oligotrophic lakes in Sweden. In brooks and streams the species is more abundant in water with organic pollution (Cuijpers & Damoiseaux, 1981; Moog, 1995; Bitušik, 2000). Limnodata.nl gives a number of records with a BOD greater than 5 mg O_2 /l and/or an ammonium-N content higher than 2 mg/l. As can be seen in table 5 in chapter 9, a combination of much decomposing material and good oxygen supply seems to be optimal (compare also Water Type above). Although Jankovi (1974) found dense populations of larvae in fresh colonies of *Sphaerotilus natans*, the presence of submerged wood in particular seems to increase the abundance of the species.

Bryophaenocladius Thienemann, 1934

Eudactylocladius Brundin, 1947: 28–31, figs 54–57 (nec aliis)

The genus *Bryophaenocladius* is very rich in species, many of which are still undescribed. A significant number of the males can be identified using Langton & Pinder (2007) and Brundin (1947). *B. furcatus* is a parthenogenetic species. Some exuviae have been keyed by Langton (1991). Most larvae are undescribed.
The larvae of almost all species are terrestrial in habit. Some of them inhabit very dry habitats, such as moss patches on buildings or in dunes, and benefit from rainy periods. We have indications that such larvae do not survive long periods of inundation. Other species seem to prefer semiaquatic habitats; for example, *B. flexidens* has been collected only along small streams (Lehmann, 1971: 491; Orendt, 2002a; own data). Several authors (e.g. Lehmann, 1971; Stur et al., 2005; Moubayed-Breil, 2007) collected some species known from terrestrial habitats also in springs or even in streams and rivers (Rossaro, 1991: 92; Schmid, 1993: 62–63). Klink (2011) collected larvae from fallen trees in the river Meuse.
Larvae of *B. virgo*, *B. muscicola*, *B. subvernalis* or *B. vernalis* can be common among mosses just above the water level (Strenzke, 1950: 363, 375; Langton, 1991; own data). Such larvae and pupae can sometimes be found below the water surface (see also Hamerlik & Brodersen, 2010).

Camptocladius van der Wulp, 1874

The only species in this genus, *C. stercorarius*, lives mainly in the dung of horses and cows (Strenzke, 1950: 361). In the Netherlands it is very common. We found it rarely also in soils with much organic matter and apparently no dung. It is not known how the species sometimes appears in aquatic samples.

Cardiocladius Kieffer, 1912

SYSTEMATICS AND IDENTIFICATION

The genus *Cardiocladius* is poor in species and related to *Eukiefferiella* (Saether & Halvorsen, 1981). The two western European species are very similar as adult males (see Langton & Pinder, 2007) and as larvae. The characters of the larvae as given by Saunders (1924), Cranston (1982), Moller Pillot (1984), Schmid (1993), Bitušik (2000) and Janecek (2007) are not reliable. The pupae can be identified relatively easy using Langton (1991).

Cardiocladius capucinus (Zetterstedt, 1850)

C. capucinus is more than *C. fuscus* a species of foothill and mountain streams and has not been reported from lowlands (Saunders, 1924; Laville & Vinçon, 1991; Gendron & Laville, 1997, 2000; Michiels, 1999, 2004; Orendt, 2002a). Wilson & Wilson (1984) reported occurrence of the species in the river Rhine; this is probably a misidentification. According to Langton (1991) the species has also been collected in lakes in montane regions.

The data from the Netherlands (Moller Pillot, 1984; Moller Pillot & Buskens, 1990; Limnodata.nl) are based on misidentifications (see under the genus).

Cardiocladius fuscus Kieffer, 1924

DISTRIBUTION IN EUROPE AND THE NETHERLANDS

Most probably the species lives in most of Europe (Saether & Spies, 2010). In the Netherlands the species has been collected at about 20 localities in the provinces of Limburg and Noord-Brabant and in the large rivers, partly as *C. capucinus* (Klink, 1985, 1990; Limnodata.nl; own data). There is one record from the province of Overijssel (Limnodata.nl).

LIFE CYCLE

Adults were caught by Becker (1995) along the river Rhine in all months of the year.

FEEDING

Thienemann (1954) supposed that the larvae feed on larvae of Simuliidae. However, Caspers (1980) found the larvae abundantly in a part of the river Rhine, where no Simuliidae lived. In the guts of the larvae we found only fungi, sand grains and detritus. This is more in agreement with the strong wear of the mental teeth. Klink & bij de Vaate (1994) found diatoms and fine plankton material in the guts of five larvae from the Grensmaas stretch of the river Meuse.

MICROHABITAT

The larvae live on stones and plants in swift currents (Saunders, 1924; Becker, 1995; own data). Klink (1990) found them on artificial substrate in the river Grensmaas in Limburg.

WATER TYPE

C. fuscus has been found in fast-flowing streams and rapids in the upper or lower courses of streams, mainly at current velocities of about 1 m/s and more (Lehmann, 1971; Gendron & Laville, 1992, 1997; Orendt, 2002a; Michiels, 2004). Most of the

Dutch records are from fish ladders and rapids in lowland and foothill streams and ship canals (Limnodata.nl; own data). The species can be common also in large rivers (Klink & Moller Pillot, 1982; Klink, 1985; 1990; Caspers, 1991).
Rossaro (1991) found *C. fuscus* in waters with a rather large range of **temperature** (mean 14.2° C.).

DISPERSAL

Klink (1990) stated the settlement of larvae in the river Grensmaas (Meuse in southern Limburg) in the Netherlands on artificial substrate, apparently after drift of larvae. Garcia & Laville (2001) also collected the species mainly in drift in the lower parts of the river Garonne.
The fact that *C. fuscus* settled rather soon after construction of fish ladders in the province of Noord-Brabant (the Netherlands), far from other populations of the species, is an indication of the dispersal of females over large distances (tens of kilometres).

Chaetocladius Kieffer, 1911

SYSTEMATICS AND IDENTIFICATION

The genus seems to be well defined in the pupal stage, but there are many questions about the systematic position of species only known as adults. The division of the genus into larval groups in Moller Pillot (1984) has no systematic value, but can be used to separate larvae of aquatic species into provisional aggregates. In some cases, a species name of a larval type has been determined (see below). Several other larval types have been collected which could not be identified (Cranston, 1982; Cranston et al., 1983; own data).
Identification of adult males and larvae is treated under the species name. All keys to adult males are incomplete; most aquatic species can be identified as exuviae using Langton (1991). Identification of females is sometimes necessary, especially for species such as *C. vitellinus*, which seems to be parthenogenetic. A list of West Palaearctic species (without *C. femineus*) is given by Langton & Visser (2003).

LIFE CYCLE

Most species are not or hardly found in summer. A diapause in this period seems to be probable; see, for example, under *femineus*, *perennis* and *piger*.

ECOLOGY

Many species of *Chaetocladius* live in springs or in hygropetric environments. The larvae of more aquatic living species such as *C. femineus* and *C. piger* can be found relatively often along the water's edge. This may be why Wilson (1988) found only this genus in a brook severely polluted with zinc.

Chaetocladius acuticornis (Kieffer, 1914)

IDENTIFICATION

The adult is absent from Langton & Pinder (2007). The pupa has been keyed by Thienemann (1936a, 1938, 1944).

DISTRIBUTION IN EUROPE

The species has been recorded in a small number of countries throughout Europe (Saether & Spies, 2010).

ECOLOGY

C. acuticornis lives in springs (Thienemann, 1936a). Lindegaard et al. (1975) found larvae probably belonging to this species in the moss carpet of the Ravnkilde spring in Denmark, mainly in the madicolous zone at the border of the moss carpet. The larvae were collected mainly in November and not at all in spring.

Chaetocladius algericus Moubayed, 1989

This species lives in springs and seepages in the Mediterranean area (Moubayed-Breil, 2008; Saether & Spies, 2010), but has been collected also in a pond in southern England (Ruse, pers. comm.).

Chaetocladius dentiforceps (Edwards, 1929)

SYSTEMATICS AND IDENTIFICATION

The **adult male** can be identified easily because of the characteristic gonostylus (only resembling that of *Zalutschia humphriesiae*), see Langton & Pinder (2007), also figured by Cranston et al. (1989). Identification of the **exuviae** is more difficult, but is possible using Langton (1991). The **larva** has (according to Thienemann, 1938, 1944) two short central mental teeth, often worn off so that only one tooth is left (Moller Pillot, 1984: 40). However, Cranston et al. (1983: 217) figure only one more protruding tooth. Possibly most larvae identified as *C. dentiforceps* agg. belong to *C. laminatus* (see Pankratrova, 1970: 234–235; Moller Pillot, 1984; Schmid, 1993: 68).

DISTRIBUTION IN EUROPE AND THE NETHERLANDS

The species has been recorded throughout Europe, but only in a rather small number of countries (Saether & Spies, 2010). In the Netherlands only larvae of this aggregate have been collected (see under *C. laminatus*), in Limburg, Gelderland and Twente (Moller Pillot & Buskens, 1990: 54; Limnodata.nl). The catch of adult males reported in Klink & Moller Pillot (1996) was based on misidentification.

WATER TYPE

Thienemann (1938) stated occurrence in springs, which has been affirmed by Fittkau & Reiss (1978) and Lindegaard (1995). Langton (1991) gives streams as habitat and Paasivirta (2012) called the species semiterrestrial. The Dutch records of *dentiforceps* agg. (including *laminatus*) are from springs and (usually rather fast-flowing) brooks and streams. All or most larvae were found above the water level. The report of the species by Delettre (1989, also mentioned by Serra-Tosio & Laville, 1991) in an acid temporary pool without seepage influence in Brittany needs to be verified.

Chaetocladius dissipatus (Edwards, 1929)

IDENTIFICATION

The adult male is keyed in Langton & Pinder (2007) and the exuviae in Langton (1991). The larva is unknown.

DISTRIBUTION IN EUROPE

The species has been collected throughout Europe (Saether & Spies, 2010).

ECOLOGY

Paasivirta (2012) found *C. dissipatus* rather commonly in spring areas in Finland. Stur et al. (2005) caught the adults further downstream in a spring brook in Luxembourg. Caspers (1980) caught some males (cf. *dissipatus*) in a small brook near Bonn. This author stated the presence also in the upper parts of the river Rhine, even downstream of Basle (Caspers, 1991). According to Langton (1991) the species lives in streams and lakes in montane regions.

Chaetocladius femineus (Edwards, 1929)

Bryophaenocladius femineus Langton & Pinder, 2007: 82, fig. 44B, 139B
Chaetocladius spec. Herkenbosch Moller Pillot, 1984: 43, fig. VI.9n–p
Chaetocladius spec. Matley Bog British Museum (larva slide coll. Cranston)

SYSTEMATICS AND IDENTIFICATION

Brundin (1947) suggested that the related *C. inconstans* could be placed only when pupa and larva were known. *C. femineus* can be distinguished from that species because of the pubescent tip of the antenna (Edwards, 1950: 156; Brundin, 1947: 29). Albert Dees reared a male pupa, which was used to resolve the systematic position of the species (Dees & Moller Pillot, 2010). A description of the pupa is in preparation. The larva can be identified using Moller Pillot (1984).

DISTRIBUTION IN EUROPE AND THE NETHERLANDS

C. femineus has been recorded from England, Ireland and Belgium (Saether & Spies, 2010; own data) and probably also from Austria (Caspers & Reiss, 1987). We collected one larva in Belarus (near Simonichi). In the Netherlands the species has been found at more than ten localities scattered over the whole Pleistocene region.

LIFE CYCLE

We collected almost all larvae in spring, from February until May. Fahy (1973) reared the species in this period. On two occasions we found a specimen (prepupa, pupa) in October, and Edwards (1950: 156) caught a male in November. Most probably *C. femineus* has the same life cycle as *C. piger*.

WATER TYPE

Several Dutch records are from acid, often temporary upper courses. Fahy & Murray (1972) and Fahy (1973) found the species commonly in small streams in Ireland with much detritus of *Molinia*. At nine localities in the Netherlands, Belgium and Belarus the larvae lived in more or less acid stagnant water such as bog pools, moorland pools and a sand pit, often under the influence of seepage.

Chaetocladius gracilis Brundin, 1956

C. gracilis has been found in spring areas and brooks and streams in northern countries and in the Alps and Pyrenees in France (Laville & Vinçon, 1991; Serra-Tosio & Laville, 1991; Moubayed-Breil, 2008; Saether & Spies, 2010; Paasivirta, 2012).

Chaetocladius grandilobus Brundin, 1956

C. grandilobus has been recorded in Scandinavia and Germany (Saether & Spies, 2010). In Finland the species is rather common throughout the country (Paasivirta, 2012); this author called the species semiterrestrial. In Swedish Lapland the adults were caught near a permanent pool (Brundin, 1956: 127). In Germany adult males were reared from a helocrene of the river Fulda (Lehmann, 1971: 492).

Chaetocladius insolitus Caspers, 1987

The adult male is not keyed in Langton & Pinder (2007). The pupa can be identified using Langton (1991). This species, which is only known from the alpine region and Northern Ireland (Saether & Spies, 2010), is not treated further.

Chaetocladius laminatus Brundin, 1947

Chaetocladius dentiforceps agg. Moller Pillot, 1984: 40, fig. VI.9e, f (pro parte)

IDENTIFICATION

The adult male is absent from Langton & Pinder (2007); the description and figure by Brundin (1947) enables easy identification. However, DNA barcoding has revealed that this species aggregate contains three species (E. Stur, pers. comm.). One of them has been described provisionally as *Chaetocladius* sp. 2 (Ekrem et al., 2010: 401).
The exuviae have been keyed by Langton (1991). The larvae correspond with *C. dentiforceps* agg. in Moller Pillot (1984) and sp. *dentiforceps* in Schmid (1993); see also Pankratova (1970: 234).

DISTRIBUTION IN EUROPE AND THE NETHERLANDS

The species has been recorded throughout Europe, but only in a rather small number of countries; it has not been recorded in the British Isles (Saether & Spies, 2010). Nevertheless, it seems to be more common in continental Europe than the related *C. dentiforceps*. Probably most records of *dentiforceps* agg. in the Netherlands (from southern Limburg, the Veluwe, the Achterhoek and Twente) belong to *laminatus* agg. DNA barcoding revealed, that some larvae from southern Limburg belong to *Chaetocladius* sp. 2 (see above; E. Stur, unpublished).

LIFE CYCLE

Lindegaard et al. (1975) collected adults only from February to May in the Danish spring Ravnkilde. In Germany, in Austria and in the Pyrenees emergence has been stated from January to April (May) and in September/October (until December) (Lehmann, 1971; Ringe, 1974; Laville & Lavandier, 1977; Caspers, 1983).

MICROHABITAT

Lindegaard et al. (1975) collected the larvae at the border as well as in the interior part of the moss carpet in the Danish spring Ravnkilde. They lived mainly in the madicolous zone just above the water surface, where the moss is kept moist by capillary water. We also collected larvae of *Chaetocladius* sp. 2 in springs just above the water level among mosses and dead organic matter.

WATER TYPE

C. laminatus was one of the most common chironomids in the moss carpet in the Danish spring Ravnkilde (Lindegaard et al., 1975). Other authors also caught the adults around springs and spring brooks rich in mosses (Lehmann, 1971; Laville & Lavandier, 1977), but also further downstream (Ringe, 1974; Stur et al., 2005).

Chaetocladius longivirgatus Stur & Spies, 2011

This species, which resembles *C. suecicus*, has been collected in montane and boreal regions in Germany and Norway.

Chaetocladius melaleucus (Meigen, 1818)

Chaetocladius spec. Veluwe Moller Pillot, 1984: 41, fig. VI.9g, i–j

IDENTIFICATION

The adult male has been keyed by Langton & Pinder (2007). A description of pupa and larva has been published by Pinder & Armitage (1985). The larva is described briefly and keyed in Moller Pillot (1984) as *Chaetocladius* spec. Veluwe.

DISTRIBUTION IN EUROPE AND THE NETHERLANDS

The species has been recorded throughout Europe, but only in a rather small number of countries (Saether & Spies, 2010). In the Netherlands the larvae have been collected at eight localities on the Veluwe, in Twente and northern Limburg (Limnodata.nl; own data).

LIFE CYCLE

Ladle et al. (1984) found the larvae in an artificial recirculating stream almost exclusively in May and June. They calculated a mean larval life of 33 days. We collected larvae in the Netherlands in autumn and winter also. Reiff (1994) and Michiels (1999) collected exuviae in Bavaria in February and March.

MICROHABITAT AND WATER TYPE

In the artificial stream built by Ladle et al. (1984) the larvae colonised the gravel bed and completed their cycle there. In the Netherlands some larvae were also found among mosses in waterfalls on weirs (leg. F. Repko) and in a spring (leg. Waterschap Peel en Maasvallei). Most records in the literature are from brooks and streams, but Langton (1991) mentioned also lakes in montane regions (cf. Reiff, 1994) and Paasivirta (2012) called the species in Finland semiterrestrial.

Chaetocladius minutissimus (Goetghebuer, 1942)

Chaetocladius minutus Goetghebuer, 1934

This species lives in spring brooks and is only known from Germany and Switzerland (Saether & Spies, 2010). The larva is keyed in Thienemann (1936a, 1944), the pupa in Thienemann (1936a, 1938, 1944). The adult male is very small and characterised by the very low antennal ratio (0.5–0.7) (Goetghebuer, 1940–50: 57, 61).

Chaetocladius perennis (Meigen, 1830)

Spaniotoma perennis Tomlinson, 1946: 2–15, figs 10, 11, 13, 14

SYSTEMATICS AND IDENTIFICATION

The adult male has a characteristic hypopygium, keyed and figured by Langton & Pinder (2007); see also Brundin (1947, fig. 44). The pupa, keyed by Langton (1991), cannot be distinguished from that of *C. vitellinus* (see there). Reliable differences between the larvae of the *vitellinus* group have not yet been stated (Moller Pillot, 1984: 41–42).

DISTRIBUTION IN EUROPE AND THE NETHERLANDS

C. perennis has been stated throughout Europe. Most probably it is a common species, as has been stated in Finland (Paasivirta, 2012), but because of its semiterrestrial habit it is not often registered in aquatic investigations (see Microhabitat). Records of pupal exuviae of this species can apply also to *C. vitellinus* (see above). The Dutch adult specimens were reared from a sewage filter bed at Doorn in 1980. The majority of about 30 other records of larvae of the *vitellinus* group, in the southern and eastern provinces and the dune region (Limnodata.nl; own data), possibly apply to this species.

LIFE CYCLE

Emergence has been stated mainly in early spring, in early summer and in autumn (e.g. Lehmann, 1971; Ringe, 1974; Caspers, 1983). According to Crisp & Lloyd (1954: 295, 308) the species has a diapause as prepupa in summer.

MATING AND EGG DEPOSITION

The adults swarm, but are also capable of mating in a confined space (without swarming in the air). The eggs are deposited on the wet substrate in a gelatinous mass containing 100 to 400 eggs (Tomlinson, 1946).

MICROHABITAT

Most probably the larvae live mainly in more or less hygropetric conditions, for example along the water's edge, as has been stated for some other species of the genus. Collection of the species may therefore depend on the sampling methods.

WATER TYPE

C. perennis has been collected in and around springs and brooks (Thienemann, 1936a; Lehmann, 1971; Ringe, 1974; Caspers, 1980; Laville & Vinçon, 1991; Orendt, 1999). This also applies to most records of larvae of the *vitellinus* group in the Netherlands. In Sweden and Finland the species seems to live often in terrestrial or semiterrestrial habitats (Brundin, 1947: 23; Paasivirta, 2012). There are several records of populations in sewage filter beds (Lloyd et al., 1940; Tomlinson, 1946; Thienemann, 1950; own data). Orendt (1999) collected some specimens (? exuviae of this species) in acid brooks.

SAPROBITY

The presence in sewage filter beds shows that the larvae are very tolerant of organic pollution. Elsewhere they live in very clear water. The larvae are probably tolerant of pollution in hygropetric conditions because the oxygen supply is very good there.

Chaetocladius piger (Goetghebuer, 1913)

SYSTEMATICS AND IDENTIFICATION

Adult males and pupae can be identified using all existing keys. The larvae are described and keyed to the level of aggregate (see below) in Moller Pillot (1984) and Schmid (1993).

Some authors have used a taxon *Chaetocladius* gr. *piger* (e.g. Thienemann, 1944; Cranston et al., 1983; Moller Pillot, 1984). These groups are not always limited in the same manner and have no systematic value (Cranston et al., 1989). *C. piger* agg., as described by Moller Pillot (1984) possibly includes only one species, but perhaps more as the larvae of several species are still unknown.

DISTRIBUTION IN EUROPE AND THE NETHERLANDS

C. piger is a common species throughout Europe (Saether & Spies, 2010). In the Netherlands the larvae have been collected at many hundreds of localities, mainly in the Pleistocene area of the country and in the dune region, but also more scarcely in all other regions (Limnodata.nl; Steenbergen, 1993; own data). In only six cases have adult males been reared or collected; therefore it cannot be excluded that more than one species is involved.

LIFE CYCLE

Emergence has been reported mainly in March and April (Edwards, 1950; Crisp & Lloyd, 1954; Pinder, 1974; Becker, 1995). In the Netherlands we collected adults and pupae from October until April, most numerously in November and March. Goetghebuer (1914) collected full-grown larvae in May and these larvae only pupated in October. We also reared adults in November from material sampled in July. In summer we found only very few larvae and no pupae or adults, suggesting a diapause in the fourth instar.

FEEDING

Crisp & Lloyd (1954) found the larvae numerously in a mud flat where the available food consisted only of decaying leaves and twigs. We also collected larvae from woodland trenches with decaying leaves. Elsewhere we found larvae with detritus and sometimes many diatoms in the guts.

MICROHABITAT

Tolkamp (1980) and Verdonschot & Lengkeek (2006) collected the larvae most numerously on bottoms with coarse detritus such as leaves and twigs. We often found the larvae at the water's edge, sometimes above the water surface, and also in wet grassland close to or far from a watercourse (Moller Pillot, 2005). Because young larvae were collected rarely in the water body itself they could live more above the water level and it cannot be excluded that the eggs are deposited there. A further argument for the idea that the larvae are often more terrestrial is that aquatic samples frequently contain no larvae (e.g. Lindegaard-Petersen, 1972; Cranston, 1982). The skin of the larva is water-repellent and larvae (also in third instar!) and pupae can often be collected drifting on the surface (Moller Pillot, 2003). More than in any other chironomid species, larval skins and dead larvae are collected in drift samples, which is also characteristic of terrestrial species.

WATER TYPE

In the literature *C. piger* has been recorded from a rather small number of springs, brooks and streams, and rarely from large rivers or from lakes (e.g. Pinder, 1974; Fitt-

kau & Reiss, 1978; Schmid, 1993; Paasivirta, 2012). In the Netherlands, however, it is a very common species, especially in Pleistocene regions in small upper courses of lowland brooks on sandy bottoms, more scarcely also in ditches, and even on clay (Steenbergen, 1993; Limnodata.nl; own data). The larvae seem to live mainly around the water's edge (see Microhabitat) and are often found at sites with seepage, not only along watercourses, but also in grassland. There are only a few records from pools.
The species can be found in permanent as well as in temporary watercourses, possibly due to a summer diapause. It lives both in woodland and in open landscapes.

pH

Most water bodies where *C. piger* has been found have a pH of 7 or higher (Limnodata.nl; own data). However, Moller Pillot (2003) stated that the species can also be numerous in very acid upper courses if all other factors are favourable, including a seepage zone above the water's edge. Therefore it is not clear if pH plays a role at all.

SAPROBITY

The larvae live mainly in dune brooklets, dune fringe waters and upper courses of lowland streams, which are usually hardly or not polluted. However, sometimes larvae are found in severely polluted water (Limnodata.nl; own data). Also, Wilson & Ruse (2005) call the genus tolerant of organic pollution. If it is true that the young larvae in particular live above the water level (see Microhabitat), this fact can play a role. The fatty skin of the larva could also be important. The figures in table 5 on p. 282 refer to the probability of encountering the species, which in this case mainly depends on other factors (e.g. seepage along the water's edge) usually correlated with water quality.

SALINITY

Steenbergen (1993) collected *C. piger* aggregate larvae rather often at chloride contents above 300 mg Cl/l and sometimes at more than 1000 mg Cl/l.

Chaetocladius rusticus (Goetghebuer, 1932)

Soponis (1986) redescribed the type specimen and figured the hypopygium. Identifying the species from this description and figure may be difficult.
The holotype has been caught at Postel, Belgium, near the border with the Netherlands. Nothing is known about its ecology.

Chaetocladius suecicus (Kieffer, 1916)

SYSTEMATICS AND IDENTIFICATION

The adult male has been keyed and figured by Langton & Pinder (2007), see also Brundin (1947, fig. 46). The pupa can be identified using Langton (1991). The larva cannot be distinguished from other species of the *vitellinus* group; however, according to Pankratova (1970: 231) the mental teeth exhibit some small differences between the species. DNA barcoding has revealed that there are two other species resembling *C. suecicus* (see Stur & Spies, 2011).

DISTRIBUTION IN EUROPE

C. suecicus lives throughout Europe and has been found in a rather large number of countries (Saether & Spies, 2010). In Sweden and Finland it seems to be common

(Brundin, 1947; Paasivirta, 2012). In the Netherlands only one male has been collected: near Riethoven on 18 March 1995 (Klink & Moller Pillot, 1996).

LIFE CYCLE

Emergence has been stated mainly in early spring and in autumn, much less in summer (Brundin, 1947; Edwards, 1950; Lehmann, 1971; Michiels, 1999).

ECOLOGY

Moubayed-Breil (2008) collected the adults near eight high mountain streams in the Pyrenees, but not along lowland streams in France. Most other records are also from spring areas or fast-flowing streams (Lehmann, 1971; Michiels, 1999; Paasivirta, 2012). The male specimen from the Netherlands was caught near a lowland stream (Keersop). Most probably the larvae live mainly in wet substrates in more or less hygropetric environments, as stated by Pankratova (1970). See also *C. perennis* and *C. vitellinus*.

Chaetocladius tenuistylus Brundin, 1947

C. tenuistylus has been collected only in the central and northern regions of Sweden (Brundin, 1947) and Finland (Paasivirta, 2012) and in the Netherlands. The Dutch male adult was reared from wet mosses on a wall of a watermill near Haelen (Limburg). Paasivirta called the species semiterrestrial.

Chaetocladius vitellinus (Kieffer, 1908)

SYSTEMATICS AND IDENTIFICATION

C. vitellinus is only known as female, pupa and larva and is therefore absent from keys to adult males. The **female** can be distinguished from other species of the (larval) *vitellinus* group by the yellow colour and the six antennal segments (Goetghebuer, 1940–50: 58, 63). The **exuviae** are not keyed in Langton (1991). Pupa and larva have been described by Potthast (1914: 331–332). This description of the pupa leads in Langton (1991: 152) to *C. perennis*. The difference between these two species in the key to pupae in Pankratova (1970: 231) is incorrect. The **larvae** cannot be distinguished from those of *C. perennis* and *C. suecicus*; it is not known whether the differences in shape and colour of the mental teeth given by Pankratova are reliable. At present these three species form the *vitellinus* group as described by Moller Pillot (1984); this group probably has no systematic value (Cranston et al., 1989).

DISTRIBUTION IN EUROPE

C. vitellinus has been recorded throughout Europe (Saether & Spies, 2010). However, it is not recorded from many countries, probably because the species is absent from keys to adult males and the key to pupal exuviae of Langton (1991).

ECOLOGY

Kieffer & Thienemann (1908: 278) found the larvae between wet leaves on a slope with strong seepage in Rügen (Germany). This description was copied by Potthast (1914: 331–334). Such habitats seem to be characteristic of the whole *vitellinus* group. Most larvae of this group have been found in the Netherlands along brooks or in helocrenes (see under *C. perennis*).

Clunio marinus Haliday, 1855

SYSTEMATICS AND IDENTIFICATION

In Lenz (1950) the genus is treated as a separate subfamily Clunioninae. Recently such a separate status is rarely given to this genus and the aberrant characters are considered as adaptations to marine life, which can take place easily during evolution of the family. Identification of all stages is possible with all existing keys, but the genus is absent from the keys to larvae in Schmid (1993), Bitušik (2000) and Janecek (2007).

DISTRIBUTION IN EUROPE

Clunio marinus lives almost everywhere in Europe if rocky sea coasts are available (Remmert, 1955; Kaiser et al., 2010). There are no records from the Netherlands, but the species lives there without doubt on artificial rocky coasts.

LIFE CYCLE

The species displays a very strong synchronisation of emergence (Neumann, 1976). Adult emergence takes place mainly in winter and early spring. Emergence is immediately followed by swarming, reproduction and oviposition and only takes place around the days of spring tides (lunar or semilunar rhythm) and on these days, only during one of the two daily low tides (diurnal rhythm). Within the species range, the tidal regimes differ considerably from place tot place and therefore also the circadian emergence times of the local populations (Kaiser et al., 2010).

FEEDING

The larvae feed on algae, including blue-green algae, and probably also on organic debris and dead animals (Remmert, 1955; Neumann, 1976; Cranston et al., 1983).

MICROHABITAT

The larvae live on stones near the low-water mark (often among seaweed), locally also on sand (Remmert, 1955; Neumann, 1976; T. Kaiser, pers. comm.).

WATER TYPE AND SALINITY

The species lives only in the sea and in water bodies in open connection with the sea, in the littoral zone, but survives in the brackish Baltic Sea down to 4‰ and elsewhere at 33‰ salinity (Remmert, 1955; Neumann, 1976).

Corynoneura Winnertz, 1846

SYSTEMATICS

In the older literature (e.g. Goetghebuer, 1939) the genera *Corynoneura* and *Thienemanniella* were considered to constitute a separate subfamily Corynoneurinae. In later years these genera were attributed to the tribe Metriocnemini (e.g. in Pinder, 1978). Saether (1977a, 1979a) demonstrated that the group deserves a separate status as tribe Corynoneurini. Saether & Kristoffersen (1996) stated that the exact placement of the *Corynoneura* group cannot as yet be ascertained: it is not clear whether the group is related to the most advanced orthoclads or not. At present this tribe name is rarely used.
Within the genus the relationships are still unclear. Goetghebuer (1939), Schlee (1968) and Hirvenoja & Hirvenoja (1988) came to very different results. Fu et al. (2009) argue that the characters concerning the apodemes, volsella and gonostylus are obviously more significant than other characters, and on this basis made changes to the *scutellata* group, adding the species *lobata*, *celtica* and *coronata*. Here we pay no further attention to these classifications.

IDENTIFICATION

Most **adult males** can be identified using Langton & Pinder (2007); the older key by Pinder (1978) is not usable and has led to mistakes. As a rule, the keys and figures of Schlee (1968) and Hirvenoja & Hirvenoja (1988) can also be recommended, although the former has some mistakes and the latter does not treat all species. In the key by Makarchenko & Makarchenko (2010) the shape of the last flagellomere and the superior volsella are used more often as characters; some European species are absent from this key.
The **exuviae** of almost all species are keyed by Langton (1991); the publications by Schlee (1968) and Hirvenoja & Hirvenoja (1988) give valuable additions. *C. minuta* is absent from these keys and still more species may be distinguished in future.
Some **larvae** can be identified using Cranston (1982) or Moller Pillot (1984). However, because most species are absent from these keys, the identifications have to be used with caution. The keys by Hirvenoja & Hirvenoja (1988) and Brooks et al. (2007) do not give reliable results. The true names of several larval types are still unknown, such as *Corynoneura antennalis* (sensu Thienemann, 1944: 651; Moller Pillot, 1984: 46–47; Schmid, 1993: 72, fig. 37) and *Corynoneura* species A (sensu Cranston, 1982: 60, fig. 23e; Schmid, 1993: 74, fig. 39). These species are not treated here. Probably there are still European species undescribed as adult.
The larvae of this genus in which the antennae are missing can be recognised usually by the reticulation of the head capsule, which is absent in *Thienemanniella* and almost always present in *Corynoneura* (Brooks et al., 2007: 164–165).
More detailed information about identification and nomenclature is given under the species names.

LIFE CYCLE

Emergence of *Corynoneura* species has been stated as occurring all year round (Pinder, 1974; Ferrington, 2000; own data). However, Ferrington stated no emergence of the genus in lotic habitats at water temperatures lower than 7 °C. Many authors have stated a restricted emergence period (e.g. from April to November: Lehmann, 1971; Ringe, 1974; Schleuter, 1985).
The larvae can develop in a very short time. Mackey (1977) calculated a larval development time of 5 days at 15 °C. Lenz (1939a) stated development of larvae within ten days. In rearing experiments of a parthenogenetic strain, Schleuter (1985) stated a development time of 17 to 20 days from egg to adult at 18 °C. *C. lobata* completed one generation within 14 days in rearings in situ in July/August at a mean temperature of 9.8 °C (Nolte, 1991). Some species have up to seven generations in a year (Armitage et al., 1995: 230). Nevertheless, almost all authors have stated a very irregular occurrence of *Corynoneura* species in summer in nature (e.g. Otto, 1991: 87; Reiff, 1994: 246). Such ostensible temporary absence cannot be ascribed to a diapause or dormancy, but is without doubt the result of factors such as food, mortality, etc., as has been suggested in *Cricotopus* and other genera.

SWARMING AND OVIPOSITION

More than most other chironomids, *Corynoneura* males **swarm** in bright sunlight (Darby, 1962). Darby observed swarming close to open water and over a weir box. Crisp & Lloyd (1954) saw swarming of *C. scutellata* in November between holly (*Ilex*) trees. A very aberrant manner of swarming was observed by Schlee (1968: 31), who saw a large number of *C. carriana* males swarming just above the surface of a pool in Germany, during which their legs appeared to hang on the water surface by adhesion. G. Mothes (in litt. to D. Schlee) also saw swarms of this species above a lake shore.
At least one or more strains of *C. scutellata* are **parthenogenetic** (Lenz, 1939a; Langton, 1991). Schleuter (1985: 171–174) reared five generations of parthenogenetic females, probably of this species; the body size of these females diminished from genera-

tion to generation. Lenz (1939a) also mentioned parthenogenesis in *C. celeripes* (see, however, the comments under this species). We reared more than ten females and no males of *C. arctica* from a moorland pool near Tilburg (the Netherlands), suggesting that parthenogenesis is possible in this species as well (cf. Hirvenoja & Hirvenoja, 1988: 224). Caspers (1980) collected about equal numbers of males and females (more than 1000) in *C. lobata*.

Goetghebuer (1932, fide Darby, 1962) observed **oviposition** on the surface of leaves of aquatic plants. Bergers (unpublished) caught many adults of *C. edwardsi* on floating leaves of nymphaeid plants. Nolte (1993) described an egg mass of *C. lobata*, found among moss at the water's edge. Steenbergen (1993) stated a significant preference for a restricted cover with floating water plants (not absent and not too high), suggesting that vegetation structure plays an important role in oviposition.

FEEDING

The larvae graze on plants and stones with quickly moving head or they collect fine particles selectively. They feed on diatoms and disintegrated algae, but also on other microorganisms and detritus (e.g. Lenz, 1939a; Darby, 1962; Kesler, 1981). They seem to perceive food at a greater distance than other chironomids by means of their actively moving antennae (own observations). Abrasion of mental teeth is very often observed. The suggestion of filter feeding (Cranston, 1982: 60) based on the structure of the chaetulae laterales seems to be a misconception in view of the many observations mentioned above.

MICROHABITAT

Larvae of *Corynoneura* walk or run freely on water plants on the upper- and undersides of the leaves, often near the water surface (Darby, 1962; Lenz, 1939a). In lower numbers they are found on wood, stones and dead leaves and on the bottom. Higler (unpublished data) found the larvae of *Corynoneura* both on artificial (plastic) '*Stratiotes*' plants and on true plants. We found large differences in the presence of larvae on different plant species, apparently because the surfaces of the plants presented varying degrees of hindrance to walking. According to Luferov (1971) all larval instars move to the surface layer of the water body when light intensity increases. For the possible role of floating water plants, see under Oviposition above.

In fast-flowing streams the larvae live more on stones, often in small fissures, and have been frequently overlooked because of their small size (Cranston et al., 1983). Pupae stay at sheltered spots in silken tubes, attached to plants and stones (Lenz, 1939a; Darby, 1962; own observations).

OR

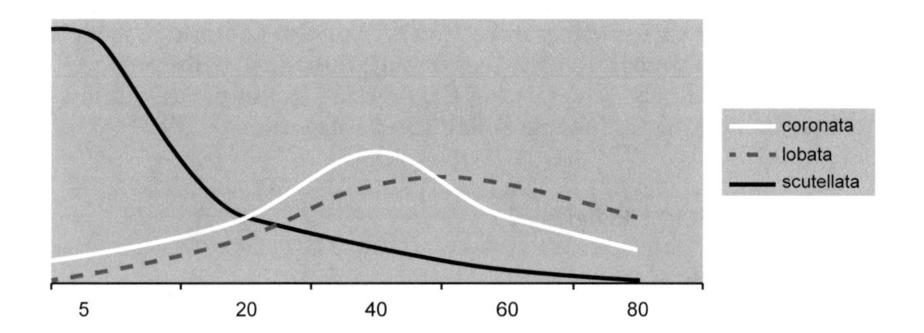

2 *Occurrence of three species of Corynoneura in small streams and ditches with different current velocity (in cm/s)*
(Note: occurrence in lakes is not considered here)

DENSITIES

Adults, exuviae and larvae can often be overlooked because of their small size. Our experience is that the genus is often very abundant, although abundance can vary widely between locations and over time. From the descriptions of methods in reported studies it is not always possible to ascertain whether whether such small animals have been underestimated.

WATER TYPE

The genus as a whole can be found in very different types of water bodies, such as lakes, pools, streams, springs and marshes. The species vary widely in their occurrence at different current velocities (fig. 2). The larvae are scarce in large rivers. Most species seem to live mainly in undeep parts of water bodies; *C. lacustris* could be an exception. Steenbergen (1993) collected the larvae more in lakes than in narrow ditches, but it is not clear whether management (vegetation cutting), vegetation structure or sampling method played a role (see Microhabitat). The abundance of subfossil *Corynoneura/ Thienemanniella* head capsules at 20–30 m depth in a Canadian lake could be due to redeposition of littoral material (Quinlan, unpubl.).
Because many species appear to be polyvoltine and the dispersal capacity can be high (see below), temporary water can be colonised within a short time.

TROPHIC CONDITIONS AND SAPROBITY

Some authors have stated that the genus has a preference for oligotrophic or mesotrophic conditions (e.g. Brodersen et al., 1998; Maasri et al., 2008). However, this is not true for all species and therefore only valid for the investigated water bodies (see under *C. scutellata*).
During the rearing of *Corynoneura* larvae a high oxygen content seemed to be important, but also it appeared to be necessary to remove 'old water' with decomposition products (own observations).

SALINITY

Tourenq (1975: 175) collected *C. carriana* abundantly in marshes in the Camargue with a chloride content of about 10,000 mg/l. In addition to this species only *C. edwardsi* and *C. scutellata* are known from brackish water. Steenbergen (1993) collected larvae of the genus in the Dutch province of Noord-Holland mainly in fresh water, but rather often in water with a chloride content of 300 to 1000 mg/l and rarely at higher chloride concentrations.

DISPERSAL

The larvae may often be transported by drift because they do not live in tubes and are especially common near the water surface. However, reliable information about larval drift is lacking; most young larvae will pass through the meshes of the nets.
Matena (1982) stated mass settlement of *Corynoneura* in fish ponds within a few weeks after flooding, suggesting large scale dispersal of females.

Corynoneura arctica Kieffer, 1923

SYSTEMATICS AND IDENTIFICATION

In the opinion of Hirvenoja & Hirvenoja (1988), *C. arctica* is very closely related to *scutellata*. According to Fu et al. (2009), these two species are not so closely related, but belong to a much larger *scutellata* group. Before 1988, *C. arctica* was not recognised and was identified as *C. scutellata* (Schlee, 1968; Langton, 1984). Adult males and exuviae can be identified using keys published since 1988. The larvae cannot be identified with certainty: the differences in antenna length and head capsule sculpture given by Hirvenoja & Hirvenoja (1988) appear not to be reliable. Nevertheless, identification will often be correct when this key is followed (see also Brooks et al. below).

DISTRIBUTION IN EUROPE AND THE NETHERLANDS

C. arctica has been collected in northern, central and western parts of Europe (Saether & Spies, 2010). In the Netherlands there are about ten unpublished records from all parts of the country.

WATER TYPE AND pH

The species lives mainly in lakes and ponds (Ruse, 2002; Moubayed-Breil, 2008; Paasivirta, 2012; own data), but in Finland has also been collected in streams (Hirvenoja & Hirvenoja, 1988). Ruse collected it only in lakes with low conductivity (less than 250 μS/cm). On our latitude it has been found most often in acid water (even at pH 3), but sometimes also in non-acid water (Moubayed, pers. comm.; Ruse, pers. comm.; own data). In Dutch acid moorland pools it is the most common species of the genus.

TROPHIC CONDITIONS AND SAPROBITY

In contrast to *C. scutellata* the species appears to be scarce in eutrophic conditions and absent from polluted water bodies, at least in the Netherlands, where water temperatures in pools in summer can be high. Paasivirta (pers. comm.) collected this species in Finland mainly in eutrophic lakes. Such differences between Scandinavia and the Netherlands are normal (see par. 2.12 and 2.13). Nothing is known about the interaction between factors like temperature, oxygen and products of decomposition. Brooks et al. (2007: 166) found the *C. arctica* type of larvae most abundant in cool alpine and arctic lakes.

Corynoneura brundini Hirvenoja & Hirvenoja, 1988

SYSTEMATICS AND IDENTIFICATION

The species is very closely related to *C. edwardsi* (Hirvenoja & Hirvenoja, 1988). Identification of adult males is only possible using this article, in which the authors also describe the female and pupa. Identification of pupal exuviae is not very reliable (see also Langton & Visser, 2003). The larva is unknown.

OR

DISTRIBUTION IN EUROPE

C. brundini has been recorded only in Finland and Germany (Saether & Spies, 2010; Steinhart, 1999). In Finland the species has been collected in relatively few provinces in comparison with other species of the genus (Paasivirta, 2012). In other countries it may have been overlooked because it is absent from most keys.

WATER TYPE

The larvae live in lakes and ponds (Hirvenoja & Hirvenoja, 1988). Steinhart (1999) collected large numbers in flooded parts of the floodplain of the river Oder in Germany.

Corynoneura carriana Edwards, 1924

SYSTEMATICS AND IDENTIFICATION

Goetghebuer (1939) and Edwards (1950) placed *C. carriana* in a separate subgenus

Paracorynoneura. The antennae of male and female are quite aberrant from those of other species and the species has several other mostly plesiomorphic characters (Schlee, 1968). Hirvenoja & Hirvenoja (1988) place *C. carriana* together with *C. gynocera* in a *carriana* group, without giving this the status of subgenus.
Because of the characteristic antennae, males and females can be identified easily using all existing keys. Identification of pupae and larvae is hardly possible and seems to be not very reliable (see Hirvenoja & Hirvenoja, 1988; Langton, 1991; Makarchenko & Makarchenko, 2010), unless the antennae in the pupa can be studied.

DISTRIBUTION IN EUROPE AND THE NETHERLANDS

C. carriana has been recorded in many countries throughout Europe (Saether & Spies, 2010). It is noteworthy that the species seems to be scarce in Sweden and absent from the northern provinces of Finland (Schlee, 1968; Paasivirta, 2012). In the Netherlands it is probably common and widespread, but there are only four verified records.

WATER TYPE

Based on identification of adults, the species is a common inhabitant of pools and lakes, even in temporary water (Schlee, 1968; Tourenq, 1975; Hirvenoja & Hirvenoja, 1988; own data). Fahy & Murray (1972) collected it very commonly in a small fast-flowing stream in Ireland.

pH

C. carriana seems to be relatively common in acid water (own data), but also lives in water with pH values > 8 (Tourenq, 1975; Learner et al., 1989).

TROPHIC CONDITIONS AND SAPROBITY

The species has been found often in eutrophic as well as oligotrophic lakes and pools, but seems to be absent from organically polluted water (Schlee, 1968: 31; Tourenq, 1975: 175; Learner et al., 1989; own data).

SALINITY

Tourenq (1975: 175) collected *C. carriana* abundantly in marshes with a chloride content of about 10,000 mg/l. In Finland it has also been found in brackish water (Hirvenoja & Hirvenoja, 1988).

Corynoneura celeripes Winnertz, 1852

NOMENCLATURE AND IDENTIFICATION

There is much confusion about this name. In Schlee (1968) and Fu et al. (2009) the **male** is figured and described as having a short scalpel-like phallapodeme. We assume this is a mistake and follow Langton & Pinder (2007). The figure of the hypopygium in this key corresponds with those in Goetghebuer (1939: fig. 16), Brundin (1949: fig. 205) and Pinder (1978: fig. 125D).
The **exuviae** of the Dutch material differ from the description and figure in Langton (1991: 86, 88–89), because the frontal apotome has no tubercles. The ring organ on the first antennal segment of the **larvae** is absent in our material or lies in the middle of this segment and not at the base as described by Lenz (1939a: 17) and Pankratova (1970: 315–316, fig. 205). In the key by Thienemann (1944: 651) the larva conforms to the description of *C. celeripes* Winnertz-Kieffer (sub 145) and not to *C. celeripes* Winnertz-Goetghebuer (sub 143). Although the species does not belong to the *scutellata* group, the larval mentum cannot be distinguished from that of the species of this group (own data).

Although it has sometimes been suggested that *C. antennalis* could be a synonym of *C. celeripes*, this is not true for the larva described as (cf.) *antennalis* (Thienemann, 1944; Moller Pillot, 1984; Schmid, 1993). See the introduction to the genus.

DISTRIBUTION IN EUROPE AND THE NETHERLANDS

C. celeripes has been collected throughout Europe (Saether & Spies, 2010). In the Netherlands adult males have been reared from two moorland pools near Oisterwijk and Nuland (province of Noord-Brabant). Probably all the data given in Limnodata.nl are based on a mistake.

ECOLOGY

Because of the difficulties in identifying the species all data in the literature have to be used with caution. The species lives mainly or only in stagnant water, although Brundin (1949) writes that it is also known from brooks. This author collected adults at a number of Swedish lakes and reported the occurrence of larvae in bog pools in Lapland. In Finland and the British Isles the species has been found in lakes and pools (Koskenniemi & Paasivirta, 1987; Langton, 1991; Paasivirta, 2012). Lundström et al. (2010) collected the species rather numerously in temporary flooded wetlands in Sweden. We collected larvae and reared adults from a woodland puddle in Belarus and two acid moorland pools in the Netherlands.
Possibly the mention of parthenogenesis in *C. celeripes* (Lenz, 1939a etc.) applies to *C. scutellata* (see Hirvenoja & Hirvenoja, 1988: 222).

Corynoneura celtica Edwards, 1924

SYSTEMATICS AND IDENTIFICATION

Hirvenoja & Hirvenoja (1988) consider *C. celtica* not to belong to the *scutellata* group, but Fu et al. (2009) place the species within this group. The adult male can be identified using Langton & Pinder (2007) and, if their view is correct, it will also be correctly identified using Pinder (1978). However, figs 84–85 in Schlee (1968) apply to quite another species. According to correspondence with N. Reiff, the identity of the material described by different authors is still not sure.
The exuviae are very characteristic because the anal fringe is restricted to the posterior half of the lobes (Langton, 1991: 86). The larva is unknown.

DISTRIBUTION IN EUROPE

C. celtica has been recorded throughout Europe (Saether & Spies, 2010). There are no records from the Netherlands.

ECOLOGY

In most parts of Europe the species seems to live only in brooks and streams (Garcia & Laville, 2001; Orendt, 2002a; Michiels, 2004; Paasivirta, 2012). We have not collected *C. celtica* in the Netherlands and the exuviae only in a foothill brook in Germany (Aachen). Brundin (1949) collected the species rarely along lakes in Sweden.

Corynoneura coronata Edwards, 1924

IDENTIFICATION

For identification of adults and exuviae see the introduction to the genus. Male and female have an S-shaped apical seta on the spur of the hind tibia.

The larvae are keyed in Cranston (1982) and Moller Pillot (1984). So far as is known, *C. coronata* is the only European species with two median mental teeth without interspace. Therefore, all specimens identified as *C. coronata* (agg.) using these keys will belong to this species.

DISTRIBUTION IN EUROPE AND THE NETHERLANDS

C. coronata has been found throughout Europe (Saether & Spies, 2010). In the Netherlands the species has been collected in the whole Pleistocene region (based on identification of larvae and adult males), but it seems to be scarce in the Holocene regions and southern Limburg (Limnodata.nl; own data).

FEEDING

Mackey (1979) found only detritus (mostly fine detritus) in the guts of larvae.

WATER TYPE

Current

Paasivirta (2012) called *C. coronata* a species of brooks and rivers. Schlee (1968), who investigated only lakes and pools, considered it as a rare species. In lowland brooks and streams in the Netherlands it is one of the most common species of the genus (Limnodata.nl; own data). Relatively few articles report it to be an inhabitant of fast-flowing streams (e.g. Orendt, 2002a; Michiels, 2004; Maasri et al., 2008) and we collected the species in fast-flowing streams scarcely. Punti et al. (2009) found it in Spanish streams at much lower altitudes (200–700 m) than *C. lobata* (800–1200 m). *C. coronata* seems to be absent from springs. The relation between current velocity and the occurrence of the species is illustrated in fig. 2 on p. 43.
The species has occasionally been collected in lakes and pools (Schlee, 1968; Langton, 1991; Limnodata.nl).

Dimensions

Although *C. coronata* is rarely collected in large rivers (e.g. not found in the river Rhine by Klink & Moller Pillot (1982) Caspers (1991) and Becker (1995)), the larvae can live there without problems when much vegetation is available (Mackey, 1976a). The species seems to be scarce in small rivers as well: Klink (2010) did not collect the exuviae in the tributaries of the river Seine. We collected the species mainly in upper courses from 1 to 7 m wide.

Permanence

Lundström et al. (2010) collected the adults regularly in small numbers in emergence traps in temporary flooded wetlands in Sweden. We collected larvae and adults sometimes in temporary upper courses in the Netherlands and Belarus.

pH

C. coronata can survive acid conditions, but it seems to be scarce in such environments. Koskenniemi & Paasivirta (1987) collected the species very scarcely in a lake in south-western Finland at pH 5.5–6. Van Kleef (unpublished) sometimes found the exuviae in acid moorland pools in the Netherlands. We collected the larvae only rarely in acid upper courses with a pH of about 6 (Moller Pillot, 2003).

SAPROBITY

As stated for almost all species of the genus, there are no records of *C. coronata* from heavily polluted water. Maasri et al. (2008) collected the species only in a reach of a Mediterranean stream in France free of enrichment. We found it in the Netherlands,

Germany and Belarus in eutrophic, but only slightly organically polluted brooks and streams (see table 5 and fig. 3 on p. 55). However, in such brooks and streams the larvae can be found at muddy sites with intensive decomposition.

Corynoneura edwardsi Brundin, 1949

SYSTEMATICS AND IDENTIFICATION

Hirvenoja & Hirvenoja (1988) separated *C. edwardsi* together with *C. brundini* provisionally from the *scutellata* group, mainly on the basis of the gonostylus. Schlee (1968) described the intraspecific variation of the adult males. Adult males and exuviae can be identified using keys published since 1988. The larvae cannot be identified with certainty.

DISTRIBUTION IN EUROPE AND THE NETHERLANDS

C. edwardsi has been recorded throughout Europe, but in a relatively small number of countries (Saether & Spies, 2010). According to Brooks et al. (2007) the *edwardsi* type of larvae is associated with temperate conditions, but Paasivirta (2012) collected the species even in the most northern provinces of Finland. In the Netherlands the species probably lives in all regions, but there are no data from the northern part of the country.

WATER TYPE

C. edwardsi is a rather common species in lowland streams, lakes and pools (Schlee, 1968; Otto, 1991: 87; Paasivirta, 2012; own data). Steinhart (1998) collected the species mainly in summer on the flooded parts of the floodplain of the river Oder. Reiff (1994) collected the exuviae in Bavarian lakes mainly in summer and autumn. Hirvenoja & Hirvenoja (1988) suggest a preference for larger water bodies. We reared the species sometimes numerously from very slow-flowing streams and canals in the Netherlands and less often from moorland pools. The species has been found scarcely in large rivers (Klink & Moller Pillot, 1982; Langton, 1991; Becker, 1995; see also Peeters, 1988).

pH

In a (? polyhumic) lake in south-western Finland with pH 5.5–6 *C. edwardsi* was the most common species of the genus (Koskenniemi & Paasivirta, 1987). We reared the species scarcely from acid moorland pools and much more frequently from water with a pH above 7.

TROPHIC CONDITIONS AND SAPROBITY

Data given by Schlee (1968) indicate that *C. edwardsi* lives in oligotrophic as well as eutrophic lakes and pools. Data by Reiff (1994) and Dutch data confirm this, but suggest a preference for very eutrophic water (see also Learner et al., 1989; Orendt, 1993). There are no records from water with severe organic pollution.

SALINITY

C. edwardsi lives in fresh and brackish water (Hirvenoja & Hirvenoja, 1988).

Corynoneura fittkaui Schlee, 1968

IDENTIFICATION

Adult males can be identified using Langton & Pinder (2007); good figures can also be found in Makarchenko & Makarchenko (2010: fig. 20–25). In the key by Fu et al. (2009) *C. fittkaui* is placed incorrectly among the species with a long, curved phallapodeme. The exuviae are keyed in Langton (1991) and Makarchenko & Makarchenko (2010). Moderately long spinules may be present on sternite I and II (own data). The larva resembles those of the *scutellata* group (own data).

DISTRIBUTION IN EUROPE AND THE NETHERLANDS

C. fittkaui has been recorded throughout most of Europe, but could be absent from some parts of the Mediterranean area (Saether & Spies, 2010). In the Netherlands it has been collected only at one locality (near Loon op Zand, 2011, unpublished).

WATER TYPE

Current

The larvae are most common in springs and brooks (Schlee, 1968; Orendt, 1999; Stur et al., 2005; Moubayed-Breil, 2008; Paasivirta, 2012). However, they have often been found in lakes and ponds (Pedersen, 1971; Janecek, 1995; Ruse, 2002). We collected the species in the lagg zone of a bog in Belarus and in an acid moorland pool in the Netherlands.

Temperature

Several authors call *C. fittkaui* cold-stenothermic (Schlee, 1968; Langton, 1991; Paasivirta, pers. comm.). However, if this were the case the species would not live in so many stagnant water bodies. In the Dutch moorland pool mentioned above, the temperature rose to more than 20 °C. In such cases the water was acid. Slow decomposition processes or similar factors are probably more important than merely water temperature.

pH

Orendt (1999) collected *C. fittkaui* only in acid brooks with a pH between 3 and 5. In many other cases the species was found in acid water (pH about 5 or not exactly known: Moubayed, pers. comm.; Ruse, pers. comm.; own data). However, Pedersen (1971) collected the species in a small lake on the Faeroe Islands with a pH of about 7 (pers. comm. J. Hansen) and Paasivirta (pers. comm.) found it in Finland not only in acid water. See above under Temperature.

TROPHIC CONDITIONS AND SAPROBITY

Ruse (2002) collected the species only in acid lakes with low conductivity. As far we know it lives only in unpolluted brooks and lakes with low temperature or acid conditions (see under Temperature above).

Corynoneura gratias Schlee, 1968

SYSTEMATICS AND IDENTIFICATION

C. gratias belongs to the *scutellata* group (Hirvenoja & Hirvenoja, 1988; Fu et al., 2009). Only the adult male can be identified without difficulty. The differences in the pupal characters of *gratias* and *lobata* are smaller than given by Langton (1991: 88). The larva probably resembles that of *scutellata* (Hirvenoja & Hirvenoja, 1988: 219, 235).

DISTRIBUTION IN EUROPE AND THE NETHERLANDS

C. gratias has been recorded throughout Europe, but in a relatively small number of countries (Saether & Spies, 2010). In the Netherlands the species has been collected only twice (near Katwijk and Tilburg, unpublished).

WATER TYPE

The species has been found in lakes and ponds (Schlee, 1968; own data) and in lowland and mountain streams (Hirvenoja & Hirvenoja, 1988; Michiels, 2004; Moubayed-Breil, 2008).

Corynoneura lacustris Edwards, 1924

IDENTIFICATION

Adult males can be identified using Schlee (1968) or Langton & Pinder (2007). Using other reference works can lead to mistakes in identification. In the key by Fu et al. (2009), *C. lacustris* is placed incorrectly among the species with a long, curved phallapodeme. We cannot judge the reliability of identifying the exuviae using Langton (1991). Larvae identified as *C. lacustris* using Cranston (1982) may belong to other species also. The short description in Brooks et al. (2007: 165) is not sufficient for identification.

Schlee (1968) found large differences between specimens collected in spring and summer.

DISTRIBUTION IN EUROPE

C. lacustris has been collected throughout Europe (Saether & Spies, 2010). In Finland it is one of the most common and widespread species (Paasivirta, 2012). There are no records from the Netherlands.

WATER TYPE

The species has been collected mainly in lakes (Brundin, 1949; Schlee, 1968; Ruse, 2002), but there is a number of records from rivers and streams in France (Garcia & Laville, 2001; Moubayed-Breil, 2007; Maasri et al., 2008). It is noteworthy that more individuals in Lake Schöhsee emerged from a depth of 6 to 8 m than from an undeep zone.

TROPHIC CONDITIONS AND SAPROBITY

According to data from Sweden and Germany mentioned by Schlee (1968: 24), the species lives mainly in oligotrophic lakes. Ruse (2002) found the exuviae only in small lakes in England and Wales with low conductivity. *C. lacustris* has also been collected in eutrophic lakes in Germany and Austria (Schlee, 1968; Otto, 1991; Wolfram, 1996). Maasri et al. (2008) collected the species only in an unpolluted reach of a stream in France.

Corynoneura lobata Edwards, 1924

IDENTIFICATION

The adult male can be identified using Langton & Pinder (2007); however, the most reliable difference from *C. celeripes* is the sternapodeme (see figs 145–146). There appear to be difficulties with identifying exuviae using Langton (1991). Regarding identification of the larvae, it has not been definitively ascertained whether *lobata* in Cranston (1982) and Moller Pillot (1984) only applies to this species.

DISTRIBUTION IN EUROPE AND THE NETHERLANDS

C. lobata is widely distributed throughout Europe (and the whole northern hemisphere) (Saether & Spies, 2010). In the Netherlands it is a common species in Limburg, the Veluwe, Twente and Winterswijk, but scarcely found in other Pleistocene regions (Limnodata.nl; own data).

LIFE CYCLE

Emergence of *C. lobata* has been stated from March/April or May until November or December (Lehmann, 1971; Ringe, 1974; Drake, 1985; Lindegaard & Mortensen, 1988; Krasheninnikov, 2011). The largest numbers are often found in summer and autumn. In winter we found mainly young larvae, but Schmid (1993a) collected all larval instars during the whole year in Austria.

FEEDING

Schmid (1993a) called the species a detritus feeder.

MICROHABITAT

According to most authors the larvae live mainly on stones, wood, gravel and plants (Cranston, 1982; Nolte, 1989; Verdonschot & Lengkeek, 2006), and more rarely in soft bottoms (Schlee, 1968: 47). However, Schmid (1993a) found large numbers within the gravelly substrate in a stream in Austria, some larvae up to 40 cm deep; the larvae were distributed in a clumped pattern.

WATER TYPE

Current

C. lobata is a characteristic inhabitant of running water (Schlee, 1968; Paasivirta, 2012; Limnodata.nl). The relation between current velocity and the occurrence of the species is illustrated in fig. 2 on p. 43. Punti et al. (2009) found the larvae in Spanish streams at higher altitudes (800–1200 m) and at much lower temperatures than *C. coronata* (200–700 m). Nevertheless, Nolte (1989) stated that the larvae seemed to avoid very lotic zones. In the Netherlands the species has been collected mainly in rather fast-flowing brooks and streams; more rarely there are records from very clear upper courses of slow-flowing lowland brooks (cf. section 2.10). Stur et al. (2005) collected the species in many springs in Luxembourg, but it was very scarce in the moss carpet of the Ravnkilde spring in Denmark (Lindegaard et al., 1975).
Records from stagnant water have to be verified; they could apply also to *C. minuta*.

Dimensions

The species is mainly an inhabitant of springs and upper courses (Schlee, 1968; Lehmann, 1971; Stur et al., 2005). *C. lobata* appears to be absent or very rare in large rivers, and it is scarce in smaller rivers (Caspers, 1991; Becker, 1995; Michiels, 1999; Garcia & Laville, 2001; Klink, 2010). We found it in the Netherlands and Germany sometimes in streams 20 m wide, more often in streams 5 to 10 m wide and most abundantly in streams less than 5 m wide.

Shade

The larvae can be numerous both in closed woodland and in open landscapes (own data).

pH

C. lobata appears to be rather common in acid brooks (Fahy & Murray, 1972; Orendt, 1999; Stur et al., 2005; H. Cuppen, unpublished). The available literature does not allow a more detailed relation to be given. As a rule the temporary acid upper courses of lowland brooks in the Netherlands appear to be an unsuitable habitat for this species (see also section 2.12).

TROPHIC CONDITIONS AND SAPROBITY

Maasri et al. (2008) collected the species only in a reach of a Mediterranean stream in France free of enrichment. Peters et al. (1988: 480) call *C. lobata* characteristic of flowing water of very good quality in the Netherlands. Moog (1995) listed the species for oligosaprobic and β-mesosaprobic water. According to Limnodata.nl, it occurs in brooks and streams with low concentrations orthophosphate and ammonium-N. The data are summarized in fig. 3 on p. 55.

Corynoneura minuscula Brundin, 1949

? *Corynoneura aurora* Makarchenko & Makarchenko, 2010: 356–358, fig. 1–5
Thienemanniella minuscula Saether & Spies, 2010; Paasivirta, 2012 (see below)
nec *Thienemanniella minuscula* Fu et al., 2010: 33, 38

NOMENCLATURE

Fu et al. (2010: 33, 38) have placed *C. minuscula* Brundin, 1949 in *Thienemanniella* without giving clear arguments and without a description. They mention, that the eyes are hairy as in most *Thienemanniella* species. It is not clear if they have seen the type material or not.

As far as is known the eyes of the European material are bare; the adult males cannot be distinguished from *C. aurora* (Krasheninnikov, 2011; Paasivirta, pers. comm.; Spies, pers.comm.; own material from the Netherlands and Belarus). Therefore, in this book we continue to use the name *Corynoneura minuscula* and consider *C. aurora* (under reserve) as a synonym. We assume that all European data on *T. minuscula* and *C. aurora* apply to *C. minuscula*.

IDENTIFICATION

C. minuscula has not been keyed by Langton & Pinder, but can be identified using Brundin (1949: 833), Schlee (1968) or Makarchenko & Makarchenko (2010, as *C. aurora*). Only the last publication contains figures showing all the important characters. The hypopygium resembles that of *C. celeripes*. The sternapodeme and phallapodeme are illustrated in Schlee (1968: fig. 40–41). The pupa and larva are unknown.

DISTRIBUTION IN EUROPE AND THE NETHERLANDS

The species has been found in northern and central Europe (Saether & Spies, 2010). In Finland it has been stated in almost all provinces throughout the country (Paasivirta, 2012). In the Netherlands it has been collected at three localities (Veenendaal, Naardermeer, Weerribben; leg. H. Siepel, unpublished).

ECOLOGY

C. minuscula is a semiterrestrial species living in marshland and trembling bogs. Brundin (1949) caught the adults near lakes in Sweden. Lundström et al. (2010) caught many adults with emergence traps in temporary flooded wetlands in the floodplain of a Swedish river. The species appeared to be very numerous in the lagg zone of a bog in Belarus (own data, unpublished). The three localities in the Netherlands are trembling bogs and marshland in nature reserves.

Lundström et al. (2010: 437) suppose that *C. minuscula* is an opportunistic species, because the investigated flooded wetlands are highly unstable habitats. However, all other records apply to habitats with a very stable water regime.

Corynoneura minuta Winnertz, 1846

According to unpublished investigations by M. Spies, *C. minuta* is not a synonym of *lobata* or a nomen dubium as suggested earlier (e.g. Moller Pillot, 1984: 47). The adult male differs from *C. lobata* in having the antennal rosette much larger and the last segment shorter (Edwards, 1950). Larva and pupa of this species are unknown and this could be the reason behind the difficulties in identifying exuviae of other species. Hirvenoja & Hirvenoja (1988: 214) stated that this species belongs to the subgenus *Eucorynoneura* sensu Goetghebuer (1939: 4), to which the species *coronata, lobata* and *celtica* also belong.

Corynoneura scutellata Winnertz, 1846

IDENTIFICATION

The definition of the *scutellata* group differs considerably between publications. The name *scutellata* agg. has been used for a number of similar species, some of which do not belong to the *scutellata* group. Before 1988 all identifications of *C. scutellata* could apply to other species as well: in the keys by Pinder (1978), Langton (1984) and Cranston (1982) the differences between the species in the genus were insufficiently described for all stages.

Hirvenoja & Hirvenoja (1988) or Langton & Pinder (2007) can be used to identify the **adult male** without difficulty. The S-shaped seta on the tibial spur in the **female** is not unique within the genus as a whole, but can be used within the *scutellata* group. **Exuviae** identified using Langton (1991) will as a rule belong to this species if all the characters have been studied accurately. Hirvenoja & Hirvenoja (1988) and Makarchenko & Makarchenko (2010) place *scutellata* among the species without long spinules on sternite I and II. The characters for identification of **larvae** in Hirvenoja & Hirvenoja (1988) are not sufficient for reliable identification.

For the systematic value of parthenogenetic and other strains, see under Trophic conditions and saprobity.

DISTRIBUTION

C. scutellata is di stributed almost worldwide distributed and has been collected in many European countries (Saether & Spies, 2010). It has been collected throughout the Netherlands.

LIFE CYCLE

The adults emerge mainly from March to November, rarely also in winter (Reiff, 1994; own data). There are several parthenogenetic strains, e.g. under the name *innupta* Edwards (Hirvenoja & Hirvenoja, 1988; Langton, 1991; own data). Schleuter (1985) stated a continual decrease of the body size when rearing such parthenogenetic populations. Schlee (1968: 42) stated a 25% smaller body size and lower antennal ratio in summer.

FEEDING

Kesler (1981) observed intensive grazing on the diatom *Cocconeis placentula* and influence of grazing on the periphyton density. According to Ramcharan & Paterson (1978), the larvae consumed only fine detritus (but it is uncertain whether both authors indeed studied *C. scutellata*, although they used this name).

MICROHABITAT

In many cases the larvae have been found living on stems and leaves of water plants. In polluted water they seem to be confined to floating leaves, for instance of *Nuphar*, *Glyceria*, *Azolla* en *Lemna* (own data). On duckweed the larvae seem to be most numerous when much *Azolla* is present (van Geloven & Thielen, unpublished).

WATER TYPE

Current

In literature *C. scutellata* has been rarely reported as an inhabitant of running water and a part of these records will apply to other species. However, we sometimes reared the species from lowland brooks and small streams in the Netherlands and it seems to be not rare at currents up to 30 cm/s. As can be seen in table 7 (Chapter 9), the larvae live mainly in stagnant water.

Dimensions

The species lives in small puddles, marshland, pools, ditches and large lakes; in lakes it lives mainly in the littoral zone (e.g. Hirvenoja & Hirvenoja, 1988; Reiff, 1994; Ruse, 2002; own data). However, Laville (1971) collected it in a lake in the Pyrenees to a depth of 11 m. In flowing water it has been collected almost only in brooks and small streams, but Hirvenoja & Hirvenoja mentioned also rivers in Finland.

Temperature

Laville (1971) collected the species most numerously in cold lakes in the Pyrenees. In the Netherlands it is mainly found in small water bodies where the temperature can rise to 25 °C. or more. This question will be treated under Trophic conditions and saprobity below.

Permanence

More than any other species of the genus, *C. scutellata* can be found in temporary water such as puddles, marshland, pools and ditches.

pH

Some authors (e.g. Brodin & Gransberg, 1993) report the occurrence of *C. scutellata* in acid water, but in most cases this could apply to other species (e.g. *arctica*). In the Netherlands it lives rarely in acid moorland pools, but Paasivirta (pers. comm.) has reliable records from acid lakes in Finland.

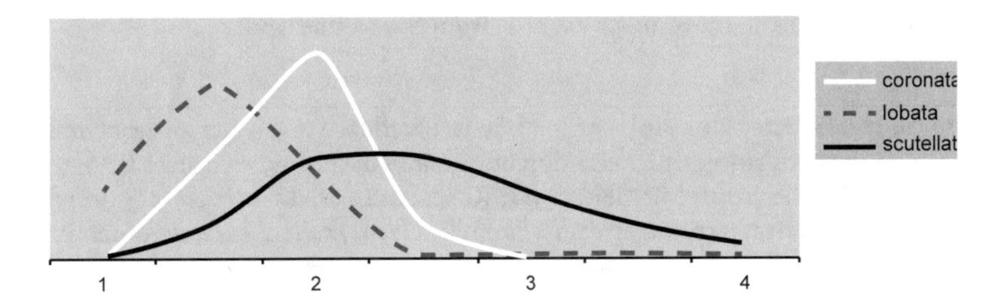

3 *Relative occurrence of three species of Corynoneura in water with different saprobity*

1. *oligosaprobic*
2. *β-mesosaprobic*
3. *α-mesosaprobic*
4. *polysaprobic*

TROPHIC CONDITIONS AND SAPROBITY

Schlee (1968) and Laville (1971) stressed that *C. scutellata* is without doubt a species of oligotrophic, often cold lakes in Europe and doubted the validity of identifications from other water types and other parts of the world. For this reason, Reiff (1994), who found the species (based on identification of exuviae) numerously in eutrophic lakes, supposed her results could be misidentifications. However, Ruse (2002 and pers. comm.) also found the species mainly in lakes with relative high conductivity and Peters et al. (1988: 48) call larvae of the *scutellata* group characteristic of moderately polluted lowland brooks. We collected the species *C. scutellata* in small, warm and hypertrophic or even severely polluted water bodies (see also under Water type / Temperature and under pH). In many cases (not always) these Dutch data concern parthenogenetic populations. We may provisionally conclude that the species *scutellata* as currently conceived is very euryoecious and that it may in fact encompass a number of different species. Figure 3 and the figures in table 5 in Chapter 9 are mainly based on the Dutch data.

SALINITY

C. scutellata has been collected sometimes in large quantities in brackish water (Ruse, 2002; Paasivirta, pers. comm.). Identification in these cases appears to be reliable.

Corynoneurella paludosa Brundin, 1949

Thienemanniella majuscula Langton, 1991: 84 (nec aliis)

SYSTEMATICS AND IDENTIFICATION

Brundin (1949) placed *C. paludosa* in a separate genus *Corynoneurella*. Although many authors (e.g. Langton, 1997; Saether & Spies, 2010) affirm its intermediate position between the genera *Corynoneura* and *Thienemanniella*, the genus is not treated in Coffman et al. (1986) and Cranston et al. (1989).
The **adult male** can be identified using Langton & Pinder (2007); the nearly transparent superior volsella and different views of the inferior volsella can be seen more clearly in the figure by Langton (1997). The **exuviae** are keyed in Langton (1991, as *Thienemanniella majuscula*) and Langton & Visser (2003). However, pearl rows are sometimes present on the wing sheaths (Langton, 1997). The larval antenna resembles that of most *Thienemanniella* species (Michiels, pers. comm.), but the **larva** is still undescribed.

DISTRIBUTION IN EUROPE

The species has been collected in only a few countries in western and northern Europe (Saether & Spies, 2010). There are no records from the Netherlands.

ECOLOGY

Brundin (1949) collected the adults near a lake in Sweden. Most other authors found exuviae or adults in or along streams and rivers, mainly in the lower courses (Langton, 1997; Garcia & Laville, 2001; Michiels, 2004; Ruse: 2012). Michiels (pers. comm.) did not find the species in calcareous streams. Paasivirta (2012) called the species semiterrestrial, but Michiels found the larvae numerously in streams.

Cricotopus v.d.Wulp, 1874

SYSTEMATICS

The genus *Cricotopus* is divided into three subgenera: *Cricotopus*, *Isocladius* and *Nostococladius* (Cranston et al., 1983). Hirvenoja (1973) divides the first two subgenera

into a number of species groups. We mention the subgenus in the head of species of the last two subgenera.
Before 1956 the species of the genus *Cricotopus* and many other genera with hairy eyes in the adult were placed under *Trichocladius* Kieffer. These species are sometimes not related at all; moreover, *Trichocladius* is a nomen dubium (see Brundin, 1956: 113–114; Hirvenoja, 1973: 45; Ashe, 1983: 53).
A **general description of the larva** is given in Hirvenoja (1973) and Cranston et al. (1983). There are no characters to separate all larvae of *Cricotopus* from other genera, such as *Orthocladius*. Most species have a bifid S I on the labrum and lateral setal tufts on the abdominal segments (such setal tufts are rare in other genera). Setal length increases after each moult, but younger instars have relatively longer setae, for example the l_4. The number of setae in an l_4 tuft increases after each moult (Rodova, 1966). The setal tufts can have a function in swimming or resisting predation (see below under Parasitism and predation).

IDENTIFICATION

Adults of almost all European species can be identified using Hirvenoja (1973). Most **exuviae** are keyed in Langton (1991), with some additions and many descriptions in Langton & Visser (2003). Many **larvae** are still unknown. Only a relatively small number of the species can be identified using Cranston (1982), Moller Pillot (1984) or Schmid (1993). We hope to publish a more extended key to the larvae shortly. The first instar of *C. sylvestris* has been described by Rodova (1966). LeSage & Harrison (1980) described egg masses and **eggs** of many species and Nolte (1993) described the egg masses and eggs of four unidentified species of the genus.

LIFE CYCLE

According to Pankratova (1970: 25), each subsequent instar from second to third to fourth lasts for increasingly longer periods in the laboratory. This is not always true in nature and does not apply in the autumn and winter.

Diapause and dormancy
In general, the genus *Cricotopus* seems to have an obligate diapause in late autumn in the second and third instar, and a growth phase in late winter or spring; two or more generations emerge during spring, summer and autumn. For instance, Berg & Hellenthal (1992) found only young larvae of *C. bicinctus*, *triannulatus* and *trifascia* in November and December. Almost all the larvae of *bicinctus* and *sylvestris* we collected in these months (in many different localities) were in second and third instar. However, Tokeshi (1986) found a one-year life cycle for *Cricotopus annulator* in England. This author suggested this species is dormant in the egg stage from summer until November. If this is true, the dormancy can be ended by environmental factors (e.g. temperature), because other authors report two or three generations for this species. Another exception seems to be *C. trifasciatus*, which has a winter diapause in the egg stage.

Number of generations
LeSage & Harrison (1980: 410, 411) believe that all *Cricotopus* species have five generations a year. There is no reason to think that the genus has such a genetically fixed number of generations, see par. 2.4. Also a fixed growth pattern during the year for every species as suggested by Berg & Hellenthal (1992) appears not to apply to other localities and is most probably the result of specific environmental conditions in the investigated stream.

Fluctuations in density
Many authors have reported a large spring generation and one or more smaller genera-

tions in summer, for instance in *C. bicinctus* (Lesage & Harrison, 1980: 397; Tokeshi, 1986a: 498). In other investigations the population density of this species seems to be higher in late summer (e.g. Drake, 1985; Reiff, 1994; own data). Such differences are mainly caused by environmental factors such as food, predation and competition (Tokeshi, 1986a; Matěna, 1989) and sometimes by drift (Gendron & Laville, 2000; Moller Pillot, 2003). The low densities in summer of all *Cricotopus* species in Salem Creek in Canada in two years of investigation reported by Lesage & Harrison (1980) suggest that mortality is especially high in summer, as has been found by Matěna (1989) in *Chironomus plumosus*.

The most important **consequence** of the life cycles for collecting larvae, when mesh sizes of more than 250 μm are used, is that large larvae are hardly or not caught from late October until February or March and that exuviae and adults can be collected from spring to autumn (depending on the species).

OVIPOSITION

Eggs can be deposited in very different ways, possibly even within one species. Nolte (1993) found egg masses among floating algae or attached to different firm substrata, such as reed and (floating) wood. Egg masses of *C. trifasciatus* are always deposited on leaves or stems of plants (see under this species). Darby (1962) observed that eggs of *C. bicinctus* were deposited by a female skimming the water surface. Both linear and bale-shaped egg-masses occur within the genus (Nolte, 1993: figs 14–16, 47e). See also par. 2.5.

SWARMING

LeSage & Harrison (1980) studied the swarming of males of many species. *Cricotopus* swarms were never found during the brighter hours of the day, but all species began dancing two hours before sunset. Interspecific differences were found in the use of markers as foci of swarming and in the location of swarms. Some species swarmed over grassland, others up to a height of 3 m. *C. annulator* swarmed along banks of watercourses, *C. festivellus* up to more than 15 m from the stream.

FEEDING

The larvae of many *Cricotopus* species are scrapers or collectors. Feeding is usually connected with tube building, the larva eating while incorporating particles into the tube (LeSage & Harrison, 1980). Some species have been found to be more specialised. For example, Tokeshi (1986a) stated that in a population of *C. annulator* living on *Myriophyllum spicatum* in an English brook, 80% to 100% of the gut contents consisted of diatoms (more than in any other Orthocladiinae species). However, almost all species appear to be more or less opportunists, and even in *C. annulator* it is possible to find almost only detritus and filamentous algae in the gut (own data). Nevertheless, certain differences between species exist and will be mentioned in the descriptions of the species below.
Only few species are able to digest living plant tissue (*C. brevipalpis, C. obnixus, C. trifasciatus* and *C. tricinctus*). The guts of other species have rarely been found to contain green plant tissue. However, *C. sylvestris* and more rarely *C. bicinctus* have been stated to attack rice seedlings (Ferrarese, 1992; Wang, 2000). Animal food (living or dead) is eaten by accident; the larvae are not active predators; it has been observed especially in *C. sylvestris*.

MICROHABITAT

Second, third and fourth instar larvae of *Cricotopus* have been found in large numbers on **water plants** (Tokeshi & Pinder, 1985; Tokeshi, 1986). Higler (unpublished data) found the larvae of the *sylvestris* group both on artificial (plastic) *Stratiotes* plants and on true plants. Lower numbers are found on the **bottom**, more on detritus than on sandy bottoms, but not beneath the mud (Darby, 1962; Ali & Mulla, 1976; Pinder, 1980). Many larvae can be found also on **wood and stones**. In natural streams fallen trees are possibly the most important microhabitat for *Cricotopus* (see Klink, 2011). According to LeSage & Harrison (1980), cobbles and pebbles densely covered by diatoms and filamentous algae are the best microhabitat for this genus, but larvae may also be found deeper than 25 cm within the substrate (based on a quotation on p. 409). There are large differences in microhabitat preference between species; current velocity plays a role. Species feeding on macrophyte tissue can be found almost only on plants.

Tubes
The larvae spin silken tubes, in which they incorporate less algal material than most other tube-dwelling chironomids (Darby, 1962: 131). Some larvae spin a more triangular tube for pupation. Tube construction has been described by Lesage & Harrison (1980: 388–390). See further par. 2.8.

WATER TYPE

Current
The species of subgenus *Cricotopus* tend to inhabit flowing water, whereas those of subgenus *Isocladius* mainly live in stagnant water. However, there are many exceptions, for instance the species of the *cylindraceus* and *festivellus* groups live mainly in lakes and *C. (Isocladius) trifasciatus* is more common in brooks and streams.
As in many other genera, several species of *Cricotopus* are found in flowing water as well as large stagnant water bodies. In such cases oxygen supply is probably more important than water movement.

Table 1. The relative occurrence of some Cricotopus species in small streams and ditches with different current velocities (Note: occurrence in lakes and rivers is not considered here)

mean current velocity in cm/sec	< 10	10–25	25–50	50–75	75–100
Cricotopus albiforceps	2	3	2.5	1.5	1
Cricotopus bicinctus	+	2	4	3	1
Cricotopus similis	0	0	(+)	2	8
Cricotopus tremulus	0	0	0	3	7
Cricotopus triannulatus	0	1	2	3	4
Cricotopus trifascia	0	0	(+)	3	7
Cricotopus (Isocl.) intersectus	8	2	+	+	0
Cricotopus (Isocl.) sylvestris	4	3	2	1	+
Cricotopus (Isocl.) trifasciatus	2	4	3	1	+

Shade

As has been argued in Moller Pillot (2009: 15), the absence or scarcity of species of chironomids in entirely shaded streams may be due to the behaviour of the adults (flying, swarming) or the scarcity of microhabitat or food. The genus *Cricotopus* seems to be relatively scarce in woodland streams. Ringe (1974: 239) and Singh & Harrison (1984: 247) collected substantially fewer *Cricotopus* in a heavily shaded stream than in a more open stream (see also Lesage & Harrison, 1980: 384). The genus was scarce in the shaded lowland streams investigated by Tolkamp (1980). Klink (2011) collected much fewer larvae on shaded trees along the river Meuse than trees exposed to more sunlight, possibly because of diminished periphyton growth on the shaded trees.
Most investigations in the Netherlands cannot be interpreted easily because the landscape is usually very fragmented and because brooks in evergreen woodland are usually acid. A problem in most of the literature is that studies of human influences are not broken down into factors such as shade, changes in slope and streambed, and eutrophication or pollution. For the genus *Cricotopus* it is necessary to keep in mind that the occurrence of these species may be considerably affected by heavy shading.

pH

The whole genus *Cricotopus* is comparatively rare in very acid water (Friday, 1987; Leuven et al., 1987; Moller Pillot & Buskens, 1990; Orendt, 1999; van Kleef, unpublished). Nevertheless, larvae can be collected in very acid water and in these cases the adults seem to emerge without problems (own data). *C. sylvestris* appears to be the most tolerant species, but Simpson (1983) also found other species at pH 5 or lower. Also Koskenniemi & Paasivirta (1987) found relatively many *Cricotopus* in a Finnish reservoir at pH 5.5–6. The species may tolerate a lower pH in boreal regions, but there is little evidence to support this (see, however, par. 2.12 and Moller Pillot, 2009: 16).
All species can live at pHs higher than 7, but Steenbergen (1993) stated significantly higher occurrence of some species in water at pH 8 or more.
In acid stagnant water *Cricotopus* is usually replaced as the dominant Orthocladiinae genus by *Psectrocladius* (see there).

TROPHIC CONDITIONS AND SAPROBITY

Within the subfamily Orthocladiinae the genus *Cricotopus* appears to be more or less characteristic of eutrophic or even polluted water (e.g. Moog, 1995; Brodersen et al., 1998; Maasri et al., 2008). This applies especially to the subgenus *Isocladius* (e.g. Langdon et al., 2006) and most of all to the species *C. sylvestris* (compare the texts on the different species). An important aspect seems to be that many of these species can feed only on easily degradable material with high nutrient value, which allows them to develop quickly and produce more generations in a year.
A number of species appear to live mainly in eutrophic or polluted water in central Europe, but can be found in the Netherlands mainly in water of better quality (e.g. *C. bicinctus* and *C. festivellus*, see Reiff, 1994; Moog, 1995; Bitušik, 2000). In many ditches in the Netherlands *C. sylvestris* is the only species of the genus, while other species can be found in larger water bodies or in faster flowing water. The reason seems to be that these latter species need a better oxygen supply. The relation between the role of oxygen supply and tolerance of toxic products is not sufficiently known.
Notwithstanding the above, there are considerable differences between species of the genus. For instance, the only relation between trophic conditions and the occurrence of *C. brevipalpis* and *C. trifasciatus*, which live on plants at the water surface, could be the presence of the plant species they feed on (possibly in a lowered condition). *C. triannulatus* seems to need a combination of much oxygen and easily degradable food, while *C. sylvestris* can live in a very oxygen poor environment. In many cases

we do not have enough data to quantify these differences in the tables in Chapter 9, but they are described as much as possible in the text. An example of differences in oxygen demand is given in fig. 4.

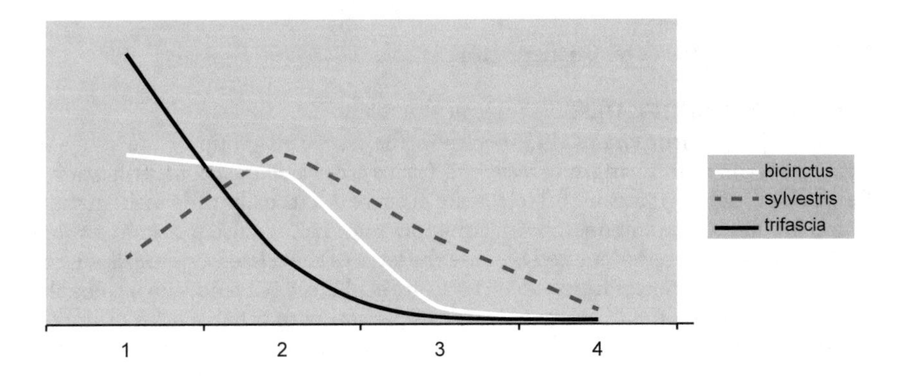

4 Presence of three species of Cricotopus in water with different oxygen regimes.
1 = stable oxygen regime: always above 50% saturation
2 = unstable: minimum between 10% and 50% saturation
3 = low: sometimes (but not longer than a few hours) less than 5% saturation
4 = rotting: in summer almost daily less than 5% saturation for many hours

PARASITISM AND PREDATION

LeSage & Harrison (1980a) stated that a large proportion of *Cricotopus* imagos were infested with mermithid nematode parasites. The level of parasitism differed considerably between species: for example, in one year they found parasites in up to 67% of *C. triannulatus* individuals, but only 5% of *C. bicinctus* individuals. The parasitism may have been responsible for the unbalanced sex ratios stated in many species of the genus. Many larvae were also infested with mites. Nothing is known about infestation with protozoa and viruses. The total infestation by parasites could be an important reason for the low numbers in summer generations mentioned under Life cycle above. Armitage et al. (1995: 428) suggested that predation is not the most important factor affecting numbers of chironomid larvae, although sometimes a substantial influence has been stated. The microhabitat (often between plants) and tube-dwelling behaviour of *Cricotopus* may be an important reason for the relatively low level of predation, as in many cases fish are likely to be the most important predators. Hershey & Dodson (1987) stated that the long abdominal setae of *C. sylvestris* protect the larvae from predation by *Hydra*, but not from the damselfly *Ischnura*. Hershey (1987) found that larvae of *C. sylvestris* spent almost twice as long outside their tube than *C. bicinctus* (which has very short l_4 setae).

DISPERSAL

Cricotopus larvae have frequently been found in new available water bodies (Matěna, 1982; Hamerlik & Brodersen, 2010), but there are no indications that adults of the genus *Cricotopus* fly more than other genera of the Orthocladiinae. Many species (e.g. *C. sylvestris*, *bicinctus* and *trifascia*) often swarm more than 5 m from the water body and at a height of 2 m or more (LeSage & Harrison, 1980), and can therefore easily be transported by air currents. We have regularly collected *Cricotopus* adults above greens and other places in towns far from water, but this may also be due to the fact that the genus is very common almost everywhere in the Netherlands. There are no definite reports of the adults flying less often in woodland (see under Water type).

Cricotopus albiforceps (Kieffer, 1916)

SYSTEMATICS AND IDENTIFICATION

Hirvenoja (1973) incorporates this species in the *festivellus* group of the subgenus *Cricotopus*. The colour pattern of **male and female** are characteristic. Identification of the **exuviae** is not easy; it is useful to closely observe the (usually yellowish) granular conical mound on the antennal sheath (Langton, 1991: 220, see also p. 219 sub *annulator*). The **larva** cannot be identified because the larva of the related *C. festivellus* is still unknown. The differences from *C. vierriensis* are slight (see Hirvenoja, 1973). Possibly the low antennal ratio is characteristic in comparison with other species of the *festivellus* and *cylindraceus* groups (Hirvenoja, 1973; Bitušik, 2000: 63, fig. 22 A–E). However, in contrast to what is stated in Cranston (1982: 66), the first antennal segment of Dutch material is a little more than 50 μm long. In third and second instar the larvae resemble also those of *C. bicinctus*, because the l_4 in younger larvae is relatively longer.

DISTRIBUTION IN EUROPE AND THE NETHERLANDS

The species lives almost everywhere in Europe (Saether & Spies, 2010). In the Netherlands
it seems to be common, but reliable records do not exist for many parts of the country. In the Pleistocene part of the country *C. albiforceps* is much more common than other species of the *festivellus* group.

LIFE CYCLE

The life cycle of *C. albiforceps* conforms to the description in the introduction of the genus. In the Netherlands we found strong indications of a diapause as young larva in winter, and emergence from April to September. Reiff (1994) stated emergence in lakes in Bavaria from early April to the middle of October (see, however, Orendt, 1993: 97); in Sweden and Austria the adults fly from May to August (Brundin, 1949; Hirvenoja, 1973).

MICROHABITAT

The larvae are often found in large numbers among overgrowth of green and blue-green algae on wood and stones, especially in the upper 30 cm of the littoral zone of lakes, canals and rivers (Hirvenoja, 1973: 234; Becker, 1995; Klink, 2011; own data). In lowland brooks and streams the larvae are found near the surface on plants and floating material (own data).

WATER TYPE

Current

C. albiforceps is an inhabitant of flowing and stagnant water (Hirvenoja, 1973). The exuviae or adults have been collected from several fast-flowing streams (Orendt, 2002a; Michiels, 2004), but the species is absent from many other streams with a more or less fast current (e.g. Lehmann, 1971; Ringe, 1974; Moog, 1995; Gendron & Laville, 1997, Orendt, 2002a). In lowland brooks and streams in the Netherlands the species is rather common, but rarely abundant (own data).

Dimensions
In stagnant water *C. albiforceps* appears to be locally abundant in the littoral zone of lakes and canals, but there are few records from pools (e.g. van Kleef, unpublished) or ditches. In flowing water it lives in low numbers in large rivers, such as the Rhine, Meuse and Garonne (Caspers, 1991; Becker, 1995; Garcia & Laville, 2001; Klink, 2011), and also in smaller streams, even sometimes in narrow brooks not wider than 2 m (own data). |Its occurrence in large stagnant water bodies and all types of flowing water is an indication that oxygen supply is the main requirement.

TROPHIC CONDITIONS AND SAPROBITY
In lakes in Bavaria *C. albiforceps* appears to be more or less characteristic of meso-trophic conditions (Orendt, 1993; Reiff, 1994). We have very few data relating to saprobity in streams; in view of the microhabitat, the larvae will have no strong relation to saprobity. Also Moog (1995) does not give figures. The larvae seem to need a rather well oxygenated environment.

Cricotopus algarum (Kieffer, 1911)

nec *Cricotopus algarum* Pankratova, 1970: 196, fig. 116 (= *C. bicinctus*)

C. algarum has been recorded in a number of countries scattered over Europe (Saether & Spies, 2010). Records from eastern Europe identified using Pankratova (1970) apply to *C. bicinctus*. The species seems to be restricted to lakes in highlands. The life cycle has been studied by Schiemer (1968).

Cricotopus annulator Goetghebuer, 1927

SYSTEMATICS AND IDENTIFICATION
C. annulator belongs to the *tremulus* group (Hirvenoja, 1973: 184). The **adult male** cannot be reliably distinguished from *C. curtus*. The **pupae** of these two species can be distinguished if the frontal setae have not been broken off (for the difference from *C. tristis* see under that species). Identification of the **larvae** is still a problem. The width of the median mental tooth in many larvae from the Netherlands and Germany is greater than that given by Cranston (1982) and Schmid (1993). The differences from *C. vierriensis* are still unknown, as is the variability in the width of the median mental tooth in *C. curtus* and *C. tristis*. *C. triannulatus* and *C. bicinctus* can be distinguished from *C. annulator* by their characteristic mandibles, and *C. albiforceps* only by the longer l_4 and the feathered lamellae on the maxilla. Further differences between *annulator* and *bicinctus* are that the dorsal setae on the preanal segment in *annulator* measure about 1/3 of the segment length, in *bicinctus* about 1/5, and that the AR in *annulator* is approx. 1.4 and in *bicinctus* is approx. 1.8.

OR

DISTRIBUTION IN EUROPE AND THE NETHERLANDS
C. annulator is distributed throughout Europe (Saether & Spies, 2010). In the Netherlands there are only reliable records from the eastern and southeastern part of the country, from some streams in Noord-Brabant and from the large rivers.

LIFE CYCLE
Tokeshi (1986) reported a one-year life cycle in England and suggested that *C. annulator* goes through a dormant period in the egg stage from summer until November.

The young larvae remained as second instars until the end of February and the adults emerged from April to June. Elsewhere emergence has been stated from April until autumn, with maxima in April, June or autumn, suggesting two or three generations (but egg dormancy in a part of the population cannot be excluded) (Lehmann, 1971; Hirvenoja, 1973; LeSage & Harison, 1980: 397; Reiff, 1994; Michiels, 1999; Klink, 2011; own data).

FEEDING

From a study of a population of *C. annulator* living on *Myriophyllum spicatum* in an English brook, Tokeshi (1986a) stated that 80% to 100% of the gut contents consisted of diatoms (more than in any other Orthocladiinae species). However, from our data this species appears to be more or less opportunist and it is possible to find almost only detritus and filamentous algae in the gut .

MICROHABITAT

The larvae live on plants, wood and stones, but can also be abundant on sandy or gravelly substrate (Tokeshi, 1986a; Becker, 1995). However, Schmid (1993) writes that they are probably bound to substrate particles with a size > 50 mm.

WATER TYPE

Current

In the Pyrenees *C. annulator* is a euryzonal species occurring at heights from 400 to 2200 m (Laville & Vinçon, 1991). Further downstream Garcia & Laville (2001) caught the pupae and exuviae only in drift samples and not in benthic samples, in contrast to *C. bicinctus*. Elsewhere in Europe the species has been collected in many fast-flowing streams and rivers (e.g. Caspers, 1991; Orendt, 2002a; own data). In lowland brooks we collected the exuviae in some brooks and streams in the eastern part of the Netherlands and the adjacent part of Germany. In brooks in other parts of the Netherlands the species seems to be rare or absent, but this is not or not only caused by current speed. *C. annulator* is also absent from several fast-flowing brooks and streams (e.g. Ringe, 1974). The species has sometimes been collected in lakes (Hirvenoja, 1973).

Dimensions

The species can be found in brooks, streams and rivers from 1 m to more than 200 m wide (Laville & Vinçon, 1991; Becker, 1995; Klink, 2011; own data). It is not known if the larvae live only near the banks of rivers or also at greater depths.

TROPHIC CONDITIONS AND SAPROBITY

The influence of eutrophication on the occurrence of the species is not known, but most probably the larvae are scarce in upper courses poor in calcium. According to Moog (1995) the larvae can be found in organically polluted water to a lesser extent than other species of the genus. However, we sometimes collected larvae and exuviae in streams with rather heavy pollution, if enough oxygen was available. The figures in table 5 on p. 282 are provisional. See also the introduction to the genus for the occurrence of *Cricotopus* in woodland streams.

Cricotopus (Isocladius) arcuatus Hirvenoja, 1973

For the systematics and identification, see under the *laricomalis* group. The adult male is not keyed in Pinder (1978), but can be identified using Hirvenoja (1973) or Langton & Pinder (2007).

The species is known from lakes in northern and north-western Europe (Hirvenoja, 1973; Paasivirta, 2009; Saether & Spies, 2010). In the Netherlands there is only one record: a fen lake near 's Graveland (exuviae det. P. Langton). Nothing is known about the ecology.

Cricotopus bicinctus (Meigen, 1818)

Cricotopus algarum Pankratova, 1970: 196, fig. 116 (nec aliis)
nec *Cricotopus bicinctus* Pankratova, 1970: 199–200, fig. 120

SYSTEMATICS AND IDENTIFICATION

In Hirvenoja (1973) the *bicinctus* group covers three species. Within this group adults and exuviae of *C. bicinctus* can be identified with little difficulty. The larva (also in third and second instar) can be identified most quickly by examining the dark hook-like process proximal of the third inner tooth of the mandible, mentioned and figured only by Schmid (1993: p. 76, 287, fig. 45). This process is totally worn down in only a few individuals.
In East European countries the larva has been identified as *C. algarum* when Pankratova (1970) has been used (see above).

DISTRIBUTION IN EUROPE AND THE NETHERLANDS

C. bicinctus is distributed throughout Europe (Saether & Spies, 2010). In the Netherlands the species occurs in the whole country, although rarely in the province of Zeeland (Limnodata.nl; Krebs, 1981, 1984).

LIFE CYCLE

Cricotopus bicinctus displays the typical life cycle of most species of the genus (see the introduction to the genus). As the autumn progresses the larvae are almost all in second and third instar (Berg & Hellenthal, 1992; own data), indicating a diapause in these instars. Further development usually starts in early spring, but in Germany and the Netherlands adults and exuviae are sometimes collected in December and January and more often in February (Klink & Moller Pillot, 1982; Becker, 1995; own data). There are several generations from March to October or November (Drake, 1985; Tokeshi, 1986a; Reiff, 1994), which is probably dependent on environmental factors (see under the genus). Some generations can be severely reduced in numbers so that the species seems to be almost univoltine (LeSage & Harrison, 1980; Reiff, 1994: figs 107, 108; Gendron & Laville, 2000).
Mackey (1977: 274) found an increase in development at 20 °C, but no increase in growth rate, so that larvae matured and pupated at shorter lengths, in contrast to many other chironomid species.

SWARMING

The males often swarm in small groups of about 20 individuals. Darby (1962) observed that the swarms never mounted high in the air, but remained 2 or 3 feet above the open water or in places protected from the wind by the bank vegetation. He often saw swarms mixed with other chironomid species, e.g. *Tanytarsus*, *Cricotopus sylvestris* and *Parachironomus*. Lesage & Harrison (1980) observed 74 swarms of *bicinctus* and stated a mean height of 2 m above the ground and a distance of 2 to 13 m from the bank of the stream. The swarms were seen dancing over various types of markers, such as bushes, grass tufts and the collector's head.

OVIPOSITION

Darby (1962) observed that a female was slowly skimming the water surface and at points six to eight inches apart quickly dipped the tip of her abdomen into the water. After having apparently oviposited four or five times she settled on a piece of emergent vegetation. After have been taken to the laboratory, this female deposited a last small string of eggs in a rearing cage. (See, however, par. 2.5).

FEEDING

Wiley & Warren (1992) stated that the larvae of *C. bicinctus* seldom completely emerge from the tube while feeding. Tokeshi (1986a: 501) found the guts to contain 50% or more diatoms, but less than in *C. annulator*. However, the larvae are opportunistic feeders like other species of the genus and in some cases the food consists mainly of detritus (Mackey, 1979; Zinchenko et al., 1986). Darby (1962) observed active damaging of the cell walls of filamentous green algae, after which the cell contents were released as food. Animal food is used only by accident (compare *C. sylvestris*). Consumption of living plant tissue and real damaging of rice seedlings, as in some other species of the genus, seems to be rare or absent in *C. bicinctus* (Darby, 1962; Ferrarese, 1992; Wang, 2000).

MICROHABITAT

Darby (1962) found the species in tremendous numbers on pondweed (*Potamogeton*) in ditches and on the upper and lower sides of shelves. The larvae also built their tubes between masses of *Spirogyra*. Most other authors also found the larvae numerously on plants and stones and only in lower numbers in places with only sand and mud (Meuche, 1938; Pinder, 1980; Tokeshi, 1986a; Millet et al., 1987; Moller Pillot & Buskens, 1990; Becker, 1995).
Peeters (1988: 44) found *Cricotopus bicinctus* larvae mainly in the undeep parts of the river Meuse. Relatively large numbers of larvae live here on stones and plants in the wave-break zone (see also Smit, 1982). The most natural microhabitat in large rivers is probably provided by trees (see Klink, 2011).

WATER TYPE

Current

C. bicinctus occurs mainly in flowing water (Hirvenoja, 1973), but it can live in stagnant water if the oxygen supply is good during the day and night and plants or stones are present in undeep water (see under Dimensions). In Finland it is mainly a stagnant water species (Paasivirta, 2012). In streams the larvae are most abundant where the current is slow to rather fast (table 7 on p. 286), but the species is scarce in mountain streams, partly under the influence of other factors (see under Dimensions and Temperature below).

Dimensions

The larvae are very scarce in small **stagnant water** bodies: in the Netherlands they are rarely found in ditches less than 4 m wide and are more abundant in wider ditches and canals (Limnodata.nl). Steenbergen (1993), who studied mainly stagnant water in the western part of the Netherlands, collected *C. bicinctus* rarely in water bodies less than 20 m wide. They are common in lakes, but mainly in the littoral zone. However, Tõlp (1971) collected the larvae up to 25 m deep in large Estonian water bodies. Meuche (1938) found the species fairly often among the epiphytic algae on reed stems and stones in lakes, but it is often absent from lakes which seem to provide a suitable habitat (e.g. Otto, 1991; Orendt, 1993; many lakes of Reiff, 1994, and Ruse, 2002, 2002a).
In **flowing water** *C. bicinctus* is common in brooks, streams and rivers. It is less abun-

dant in narrow upper courses, but it can be found much more abundantly in narrow brooks than in stagnant ditches of the same width (e.g. Limnodata.nl). The scarcity in upper courses may be influenced by low temperatures (see below), high current velocity (see above), poor trophic conditions (see below) or low pH (see below). It is not clear if oviposition (preference for larger water bodies) plays a role too.

Temperature
Rossaro (1991) gives a relatively high mean temperature of 19 °C for *C. bicinctus* in Italian streams. These data will be influenced by other factors, as mentioned under Dimensions above. From investigations by Gendron & Laville (1997) in the Pyrenees, it also appears that the local occurrence of *C. bicinctus* in mountain streams has more to do with locally slower current than with high water temperature, because the species is also present in the spring zone with maximum temperatures of 12 °C.

Permanence
The larvae are rarely collected in temporary brooks, streams or ditches. This is also influenced by other factors, such as current, dimensions, pH and food quality (see above).

pH

Cricotopus bicinctus is absent from very acid water (Steenbergen, 1993; van Kleef, unpublished; own data). Low numbers have sometimes been found at pHs around 6 (Koskenniemi & Paasivirta, 1987; Moller Pillot, 2003). See table 6 in Chapter 9.

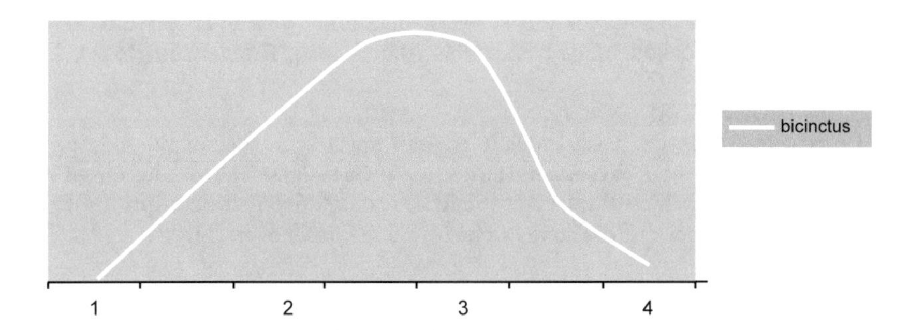

5 *Occurrence of C. bicinctus in lowland brooks of different saprobity*
1 = oligosaprobic, 2 = β-mesosaprobic, 3 = α-mesosaprobic, 4 = polysaprobic

OR

TROPHIC CONDITIONS AND SAPROBITY

In **stagnant water** *C. bicinctus* is collected mainly in mesotrophic to eutrophic lakes and canals (see Water type) (e.g. Reiff, 1994). In the Dutch province of Noord-Holland Steenbergen (1993) found the larvae significantly more abundantly in water bodies with an orthophosphate content of less than 0.05 mg P/l, a chlorophyl-a content of less than 20 μg/l and no floating vegetation. These data seem to be inconsistent with those from flowing water (see below), but this difference is caused by the fact that the larvae need a rather good oxygen supply (Darby, 1962; Moller Pillot & Buskens, 1990). In rice fields most larvae are found in the cooler areas and are only common in moving water (Darby, 1962).

In **brooks and streams** the species is more abundant in water with organic pollution (Cuijpers & Damoiseaux, 1981; Bazerque et al., 1989; Casas & Vilchez-Quero, 1989; Wilson, 1989; Maasri et al., 2008; Milošević et al., 2012). Wilson (1987) found a significant relation with increasing BOD and ammonium-content and Kownacki (1989) collected the species up to a BOD5 of 14–21 mg O_2/l. In all these cases there was no severe hypoxia. In comparison with *C. sylvestris*, the feeding indicates less adaptation to a polluted environment. Figure 5 applies to lowland brooks; in fast-flowing streams the larvae can often be found in more severely polluted water.

SALINITY

C. bicinctus has been collected in brackish water in Finland and Italy (Hirvenoja, 1973; Paasivirta, 2000). In bays in Estonia Tõlp (1971) found *Cricotopus* gr. *algarum* (most probably *bicinctus*) in water with over 3000 mg Cl/l. Krebs (1981, 1984) did not find the species in brackish water in the Dutch province of Zeeland, but Steenbergen stated its occurrence in Noord-Holland in water with more than 300 mg Cl/l. See par. 2.19.

Cricotopus (Isocladius) brevipalpis Kieffer, 1909

DISTRIBUTION IN EUROPE AND THE NETHERLANDS

C. brevipalpis has been collected in many European countries, but seems to be absent from southern Europe (Saether & Spies, 2010). There are relatively few records from the Netherlands and these are scattered over the whole country (Moller Pillot & Buskens, 1990; Limnodata.nl; Vallenduuk, unpublished).

LIFE CYCLE

C. brevipalpis has one to five (possibly seven) generations a year and flies from (May?) June to October (Gripekoven, 1913; Wesenberg-Lund, 1943; Thienemann, 1954).

FEEDING AND MICROHABITAT

The larvae mine the leaves of several *Potamogeton* species and feed on the mesophyll tissue (Gripekoven, 1913; Wesenberg-Lund, 1943; Gaevskaya, 1969). The mines are filled with air from which the larvae obtain oxygen. A detailed description of the mines and the behaviour of the larva can be found in Gripekoven (1913).

WATER TYPE

C. brevipalpis has been collected in ditches, pools, lakes and slow-flowing streams, probably mainly in smaller water bodies or in sheltered places (Hirvenoja, 1973; Limnodata.nl; own data).

Cricotopus caducus Hirvenoja, 1973

C. caducus has been found in a number of countries along the seas of northern, western and southern Europe (Saether & Spies, 2010). Occurrence in the Netherlands is possible, but has never been stated.
The species seems to live in brackish water (Hirvenoja, 1973).

Cricotopus coronatus Hirvenoja, 1973

According to Saether & Spies (2010) the species has been recorded from the Netherlands. Most probably this is based on a mistake. The species has been collected in Scandinavia and northern Russia and is not treated further. See under *cylindraceus* group.

Cricotopus curtus Hirvenoja, 1973

SYSTEMATICS AND IDENTIFICATION

C. curtus resembles some other species of the *tremulus* group, especially *C. annulator*. Only the pupae of these species can be identified reliably, at least if the frontal setae are not broken off. The differences between the larvae of *curtus* and *annulator* as given by Hirvenoja (1973) and Schmid (1993) are only indicative.

DISTRIBUTION IN EUROPE AND THE NETHERLANDS

C. curtus is distributed throughout Europe, but there are no data from many countries (Saether & Spies, 2010). In the Netherlands exuviae have been collected in the large rivers (Klink & Moller Pillot, 1996) and in southern Limburg and Noord-Brabant (own data).

LIFE CYCLE

Emergence has been stated in Germany from April to November (Hirvenoja, 1973; Reiff, 1994). Ringe (1974) stated a high peak in emergence in July and a lower peak in September, and Michiels (1999) found the largest numbers in May and September/October. Reiff (1994) stated large differences in emergence between different lakes or different years.

MICROHABITAT

The microhabitat of the larvae is unknown, but it will be the same as in *C. annulator*. Ertlová (1970) collected the species from artificial substrate.

WATER TYPE

Current

In the Pyrenees *C. curtus* is a foothill species (Laville & Vinçon, 1991; Gendron & Laville, 1997), locally also living higher in the mountain zone, possibly only when the temperature there is not too low. Elsewhere in Europe the species has been collected from waters with a rather fast current (e.g. Orendt, 2002a; Michiels, 2004; own data), even at a current speed of about 2 m/s (Ertlová, 1970, as *alpestris* or *motitator*, see Hirvenoja, 1973: 199). The only record from a lowland stream applies to exuviae underneath a fish ladder in the Beerze, the Netherlands (own data). However, Garcia & Laville (1997) collected the exuviae numerously in a dead branch in the Garonne river in France.

In Bavarian lakes *C. curtus* appears to be rather common (Reiff, 1994). Elsewhere the species seems to be rare in stagnant water.

Dimensions

The species lives in watercourses ranging from narrow upper courses no more than 1 m wide (Ringe, 1974) to large rivers (Klink & Moller Pillot, 1996; Móra, 2008).

Cricotopus cylindraceus group sensu Hirvenoja, 1973

Three species belong to this group: *C. coronatus*, *C. cylindraceus* and *C. patens*. Identification of all stages is possible (but often difficult) using Hirvenoja, 1973: 214–224. The exuviae are also keyed in Langton (1991). Most probably this whole group is rare in the Netherlands and adjacent lowlands. Larvae from this region identified as gr. *cylindraceus / festivellus* using Moller Pillot (1984) belong in the majority of cases to the *festivellus* group (see there).

Cricotopus cylindraceus (Kieffer, 1908)

DISTRIBUTION IN EUROPE AND THE NETHERLANDS

Cricotopus cylindraceus has been reported from a number of countries scattered across Europe (Saether & Spies, 2010). In the Netherlands there are at least three verified records of exuviae: Almere, 1991 (Klink & Mulder, 1992), Vlijmen, 1991 (leg. H. Vallenduuk, unpublished) and Haarsteeg, 1992 (leg. H. Cuppen, det. P. Langton) and two other records from ditches in the province of Noord-Brabant (leg. H. Cuppen). Some other records of this species apply to larvae originally identified as gr. *cylindraceus / festivellus*.

ECOLOGY

Janecek (1995) collected the species in a small eutrophic lake in Austria. The Dutch records are from eutrophic ditches, all of them (or almost all) had strong seepage of groundwater. Hirvenoja (1973) mentioned mainly records from flowing water.

Cricotopus (Isocladius) dobroginus Albu, 1964

Cricotopus dobrogicus Hirvenoja, 1973: 255–257, figs 155–156 (incorrect spelling)

SYSTEMATICS AND IDENTIFICATION

C. dobroginus is the only species in the *dobroginus* group. The adults can be identified using Hirvenoja (1973). The pupa is not known. The larva seems to resemble *C. speciosus* (with reduced first and second lateral mental teeth).

DISTRIBUTION AND ECOLOGY

The type locality of the species is in the delta of the Danube in Rumania. During investigations of the river Danube in 2007, adults of *C. dobroginus* were collected at many localities along the middle and lower course of the river in e.g. Hungary and Slovakia (Janeček, 2010: 21). Janeček supposes that the species has extended its distribution in recent times and can be expected in Austria and Germany too. The larvae probably live on plants in stagnant water bodies in the floodplain.

Cricotopus epihippium (Zetterstedt, 1838)

Trichocladius lacuum Brundin, 1949: 731

SYSTEMATICS AND IDENTIFICATION

This species belongs to the *tibialis* group (Hirvenoja, 1973: 145–152). Exuviae can be identified using Langton (1991). The larva is unknown.

DISTRIBUTION IN EUROPE

C. epihippium has been collected in all parts of Europe, but only in a few countries (Saether & Spies, 2010). Its occurrence in the Netherlands has to be verified.

WATER TYPE

According to Hirvenoja (1973) and Langton (1991) the species is an inhabitant of oligotrophic montane and northern lakes. However, it has been recorded also from streams and other localities elsewhere in Europe (see e.g. Orendt, 2002a). It is not known if these records are reliable.

Cricotopus festivellus group sensu Hirvenoja, 1973

Three species belong to this group: *C. albiforceps*, *C. festivellus* and *C. flavocinctus*. Identification of all stages is possible using Hirvenoja, 1973: 224–235 (but the larva of *C. festivellus* is not known with certainty). The exuviae are also keyed in Langton (1991).

Cricotopus festivellus (Kieffer, 1906)

Trichocladius festivus Meuche, 1938: 480; Brundin, 1949: 730
Cricotopus festivus Kreuzer, 1940: 465–466

SYSTEMATICS AND IDENTIFICATION

See under *festivellus* group above. A larva probably belonging to *C. festivellus* has been described and figured by Schmid (1993). It resembles *C. albiforceps*, but has a slightly wider central mental tooth (30–35µm) and a slightly higher AR (1.7 against 1.5).

DISTRIBUTION IN EUROPE AND THE NETHERLANDS

The species lives almost everywhere in Europe (Saether & Spies, 2010). In the Netherlands there is a number of records from dune lakes and from the provinces Drenthe, Utrecht and Noord-Brabant. Some other records have to be verified, because larvae of the *festivellus* group have sometimes been reported as *C. festivellus*.

LIFE CYCLE

C. festivellus appears to develop rather late in spring; most authors collected exuviae no earlier than May. The spring generation is usually abundant and two or more smaller generations are stated until September/October (LeSage & Harrison, 1980; Kouwets & Davids, 1984; Reiff, 1994; Janecek, 1995). See the introduction to the genus.

MICROHABITAT

Meuche (1938) collected the species from algal overgrowth on reed stems and stones.

WATER TYPE

In Europe *C. festivellus* has been collected scarcely in flowing water (e.g. Lindegaard-Petersen, 1972; Schmid, 1993; Bitušik, 2000; Scheibe, 2002; Orendt, 2002a), although Lesage & Harrison (1980) found the species numerously in a Canadian stream. Most data are from lakes and a few from pools (Kreuzer, 1940; see also Brundin, 1949). In the Netherlands almost all records are from dune lakes and lakes and ditches in the fen peat area, and only very few from moorland pools and lowland streams.

pH

There is only one record from an acid moorland pool (pH 4.65: van Kleef, unpublished). All other data apply to water bodies with a pH of 7 or higher.

TROPHIC CONDITIONS AND SAPROBITY

Reiff (1994: 293) found the species mainly in very eutrophic lakes. Moog (1995) holds *C. festivellus* to be more tolerant of α-mesosaprobic water than most other species of the genus.
However, Brundin (1949) collected the species mainly in oligotrophic and mesotrophic lakes in Sweden, and Steenbergen (1993) collected the whole species group (as gr. *cylindraceus*) significantly more abundantly from water with low phosphate, nitrogen and chlorophyl-a content. Exuviae identified to species level support this.

SALINITY

Most records of the species are from fresh water. In Finland (and possibly also in Belgium) the larvae also live in brackish water with a chloride content between 1000 and 2500 mg Cl/l (Hirvenoja, 1973; Paasivirta, 2000).

Cricotopus flavocinctus (Kieffer, 1924)

SYSTEMATICS AND IDENTIFICATION

See under *festivellus* group.

DISTRIBUTION IN EUROPE AND THE NETHERLANDS

Cricotopus flavocinctus has been recorded in large areas of Europe, with the exception of most parts of the Mediterranean region (Saether & Spies, 2010). In the Netherlands there are at least two reliable records (Kouwets & Davids, 1984; van Kleef, unpublished).

WATER TYPE

Adults and exuviae have been collected in and near lakes, canals, pools and marshes (Hirvenoja, 1973; Tourenq, 1975; Wolfram, 1996; van Kleef, unpublished). The species also sometimes lives in rather fast-flowing water (Janecek, 2000; Scheibe, 2002; Orendt, 2002a).

SALINITY

The larvae live in fresh and brackish water containing up to 4000 mg Cl/l (Hirvenoja, 1973). In the Camargue Tourenq (1975) found *C. flavocinctus* locally common, but rarely if the chloride content was higher than 1000 mg/l.

Cricotopus fuscus (Kieffer, 1909)

SYSTEMATICS AND IDENTIFICATION

C. fuscus and the two similar species (*C. algarum* and *C. pirifer*) form the *fuscus* group (Hirvenoja, 1973). Identification of the three species is difficult in all stages; only the pupa of *algarum* can be easily distinguished because the thoracic horn is absent (in *fuscus* often very small). It is debatable whether *C. pirifer* is a separate species or not (see Reiff, 1994: 103). The larva is absent from the key by Schmid (1993), but it can be identified using Hirvenoja (1973) or Moller Pillot (1984).

DISTRIBUTION IN EUROPE AND THE NETHERLANDS

C. fuscus is distributed throughout Europe, with the exception of some parts of Scandinavia (Saether & Spies, 2010). In the Netherlands it has been collected in the south-eastern part of the country (provinces Gelderland, Noord-Brabant and Limburg). The two other species of the group seem to be absent from the Netherlands and the adjacent lowlands.

LIFE CYCLE

In Bavaria and Austria there are (at least) two or three generations a year, from (February) March until October (November), with large numbers only in spring and autumn (Hirvenoja, 1973; Reiff, 1994). Elsewhere in Germany the highest numbers are stated in late spring (Hirvenoja, 1973; own data). Drake (1982) suggested five generations a year (in England); he also collected few specimens in high summer. See also the introduction to the genus.

FEEDING

We found many filamentous green and blue-green algae, detritus and diatoms in the guts of the larvae (cf. Reiff, 1994). Moog (1995) gives as food for this species more detritus than in other species of the genus. Because of the plasticity of almost all species of *Cricotopus* (see the introduction to the genus) too much importance must not be attached to these data.

MICROHABITAT

The larvae live mainly on wood, stones and plants (Drake, 1982; own data).

WATER TYPE

C. fuscus is only scarcely collected in streams and rivers (Lehmann, 1971; Laville & Vinçon, 1991; Becker, 1995; Garcia & Laville, 2001; Orendt, 2002a; Klink, 2010). Current velocity does not seem to be the most important factor. Rossaro (1982) called the species an inhabitant of springs and fountains. Relatively many records are from spring brooks (Lindegaard, 1995; own data). The species can also be found in lakes (Otto, 1991; Reiff, 1994). In the Netherlands probably all the records from stagnant water apply to water bodies (ponds, ditches, oxbow lakes) fed by groundwater (see also Michiels 1992 in Reiff, 1994).

TROPHIC CONDITIONS AND SAPROBITY

C. fuscus has been collected in mesotrophic and eutrophic lakes (Hirvenoja, 1973; Otto, 1991; Orendt, 1993; Reiff, 1994). There are no available data on tolerance to saprobity; the low indicative weight of the species in Moog (1995) seems to be based on the frequent presence in spring brooks and not on data about tolerance. (See also under Water type).

Cricotopus (Isocladius) intersectus (Staeger, 1839)

Trichocladius dizonias Brundin, 1949: 729
Cricotopus dizonias Pankratova, 1970: 192–193, fig. 113

SYSTEMATICS AND IDENTIFICATION

C. intersectus is currently the only species in the *intersectus* group (Hirvenoja, 1973). The pupa and larva differ only slightly from *C. reversus* and the differences are not always reliable, especially in the larva.

DISTRIBUTION IN EUROPE AND THE NETHERLANDS

Most probably *C. intersectus* inhabits all parts of Europe; the species has been recorded in many countries (Saether & Spies, 2010). In the Netherlands it is common and widespread, but rather scarce in the province of Zeeland (Moller Pillot & Buskens, 1990; Limnodata.nl).

LIFE CYCLE

In Germany and the Netherlands the adults emerge mainly from April or May until October, and sometimes also in February, March and November in small numbers (Klink & Moller Pillot, 1982; Reiff, 1994; Becker, 1995). There are at least two or three, possibly up to five generations (Mundie, 1957; Orendt, 1993; Reiff, 1994). Otto (1991: 87) collected the highest numbers in June, but in many cases the numbers are highest in autumn (Pinder, 1977; Becker, 1995).

FEEDING

Markošová (1979) stated that the larvae are non-selective feeders of periphyton. Huisman & van Breemen (1986) stated a sudden increase in numbers coinciding with an explosive growth of blue-green algae supplied by drift.

MICROHABITAT

The larvae live on plants, wood and stones (Meuche, 1939; Pinder, 1977; Becker, 1995; Klink, 2011), but Huisman & van Breemen (1986) found the larvae numerously on a bottom of sand and clay, sometimes with much detritus. Smit (1982) found more *Cricotopus* larvae on the upper side than on the under side of stones in the river Meuse.

WATER TYPE

Current

The species is rare in brooks and streams, except for streams with very low current velocity, and is often absent in small rivers (e.g. Klink, 2010). A remarkable exception is Tadnoll brook in England, where *C. intersectus* was among the main species on *Ranunculus* (Pinder, 1977). It is a common species in large rivers (see under Dimensions).

Dimensions

In stagnant water *C. intersectus* is an inhabitant of larger water bodies. Steenbergen (1993) collected it very rarely in ditches less than 4 m wide and significantly more often in water bodies more than 10 m wide. In these water bodies the larvae have been collected rarely in the deeper parts and mainly in the littoral zone (Mundie, 1957; Huisman & van Breemen, 1986; A. Kuijpers, unpublished).
In flowing water the species also lives mainly in the littoral zone of large rivers (Peeters, 1988: 44; Becker, 1995). Here, relatively large numbers of larvae live on stones in the wave-break zone (Smit, 1982).

Permanence

There are very few records from temporary water bodies, but settlement is possible if enough water is present during the flying period of the species.

pH

In the Netherlands *C. intersectus* is almost absent from acid water. Steenbergen (1993) collected the species rarely from water with a pH lower than 7.5 and significantly more at pHs above 8. Limnodata.nl gives a range from 6.8 to 9.3 and a mean pH of 8.0. Buskens (unpublished) collected the species scarcely in sand pits with a pH between

6 and 7 and never at pHs lower than 6. Nevertheless, the species lived in a Finnish reservoir at pH 5.5–6 (Koskenniemi & Paasivirta, 1987). It is probably not the pH itself that regulates the occurrence of the species but another factor, for instance digestion of food.

TROPHIC CONDITIONS AND SAPROBITY

C. intersectus is a characteristic inhabitant of **eutrophic lakes**; it was absent from oligotrophic to mesotrophic lakes investigated by Brundin (1949), Orendt (1993) and Reiff (1994). The mean chlorophyll-a content in water bodies of the Dutch province of Noord-Holland amounted to 129 µg/l and the highest densities were found at a content of more than 200µg/l (Steenbergen, 1993). Ruse (2002) found the species mainly in lakes with high conductivity and not in lakes with very low conductivity. The larvae are rarely caught in water with very heavy **organic pollution**, but this can be connected with the need for a good oxygen supply. Dutch data summarised in Limnodata.nl show that the species has been found mainly in water rich in oxygen and never at an oxygen content less than 1 mg/l during daytime. Where the oxygen content in the night can drop to almost zero the larvae are usually absent (Grontmij | Aqua Sense, unpublished; own data). Apart from the availability of algal food (see under Feeding), the influence of organic pollution is still obscure and the figures in table 5 must be used with caution.

SALINITY

In the Netherlands *C. intersectus* is mainly a freshwater species and occurs scarcely in brackish water (Limnodata.nl). The larvae are sometimes collected in water containing 300–1000 mg Cl/l and very rarely above 1000 mg/l (Steenbergen, 1993; Krebs & Moller Pillot, in prep.). The species is very common in brackish water in the Baltic Sea and has been found in water containing more than 2000 mg Cl/l (Paasivirta, 2000).

Cricotopus (Isocladius) laetus Hirvenoja, 1973

SYSTEMATICS AND IDENTIFICATION

The species belongs to the *sylvestris* group. Identification is sometimes difficult. The larvae cannot be identified reliably using Hirvenoja (1973) or Moller Pillot (1984), because the number of setae in the l_4 on segment VII of the larva is not characteristic of this species and the position of the distal setal mark on the antenna is not reliable as a distinguishing character within the group as a whole.

DISTRIBUTION IN EUROPE AND THE NETHERLANDS

C. laetus is known from a small number of countries in the northern, western and eastern regions of Europe (Saether & Spies, 2010). The records from the Netherlands (see Limnodata.nl) have to be verified.

ECOLOGY

The larvae live in brackish water (Hirvenoja, 1973: 268).

Cricotopus (Isocladius) laricomalis group

Two species of the *laricomalis* group have been collected in the Netherlands and adjacent lowlands: *C. laricomalis* and *C. arcuatus*. Both species can be distinguished only as adult and pupal exuviae (see Hirvenoja, 1973; Langton, 1991; Langton & Pinder, 2007).

Cricotopus (Isocladius) laricomalis Edwards, 1932

IDENTIFICATION OF LARVAE

The larvae of *C. laricomalis* (in fourth instar!) can be distinguished from other species of the genus because of the very long tufts of setae on the abdominal segments I to VII (Hirvenoja, 1973: 255). Most probably it is (usually?) a small species. Additional characters can be the dark mandible and (according to Brooks et al., 2007) longitudinal stripes on the mentum. It is not known if the larvae of *arcuatus* resemble those of *laricomalis*.

DISTRIBUTION IN EUROPE AND THE NETHERLANDS

The species is known from a number of countries scattered across Europe (Saether & Spies, 2010). In the Netherlands there is at least one reliable record based on exuviae (Klink & Mulder, 1992).

WATER TYPE

Hirvenoja (1973) saw specimens from lakes in different parts of Europe. According to Oliver & Dillon (1988), in Canada the larvae live mainly in permanent ponds. The Dutch exuviae are from a wide ditch in Flevoland. Rossaro (1991) collected the larvae in Italy in 21 samples from running water, only at relatively high temperatures (mean 20.9 °C.).

SALINITY

Tourenq (1975) caught the species in the Camargue in brackish water containing up to 3000 mg Cl/l.

Cricotopus (Nostococladius) lygropis Edwards, 1929

SYSTEMATICS AND IDENTIFICATION

C. lygropis is the only representative of the subgenus *Nostococladius*, described by Ashe & Murray (1980). The adult male and female have been described by Hirvenoja (1973) and Ashe & Murray (1980). It is the only species of *Cricotopus* with a long anal point (Langton & Pinder, 2007).
The larva has not been keyed by Moller Pillot (1984), Klink & Moller Pillot (2003) or Schmid (1993), but has been described by Ashe & Murray (1980). The abdomen has no lateral brushes and a very short first antennal segment.

DISTRIBUTION IN EUROPE

C. lygropis has been found in a few countries in different parts of Europe (Saether & Spies, 2010). It may be present in the Netherlands.

FEEDING AND MICROHABITAT

The larvae live within colonies of the blue-green alga *Nostoc parmelioides*, feeding on the alga (Ashe & Murray, 1980).

WATER T YPE

The Irish material was collected in the river Flesk. Paasivirta (2012) collected the species in running water in Finland.

Cricotopus (Isocladius) obnixus (Walker, 1856)

? *Cricotopus elegans* Berg, 1950: 88–90, figs 1–5

SYSTEMATICS AND IDENTIFICATION

In Europe *Cricotopus* gr. *obnixus* contains two or three species, of which only *C. obnixus* has been collected outside Finland and Sweden (Hirvenoja, 1973; Saether & Spies, 2010). It has not been ascertained whether the Nearctic *C. elegans* is the same species as *obnixus* or not. The Dutch larvae cannot be distinguished from the descriptions of *C. elegans* in Berg (1950) and Hirvenoja (1973). Also the head length of third and fourth instar larvae of the Dutch specimens is not aberrant from those of the larvae, described by Berg.

DISTRIBUTION IN EUROPE AND THE NETHERLANDS

C. obnixus has been found in few countries and seems to be absent from southern Europe (Saether & Spies, 2010). The Dutch larvae which probably belong to this species (see above) have been collected in different parts of the country, most frequently in the western and northern provinces (Moller Pillot & Buskens, 1990; Limnodata.nl).

LIFE CYCLE

Berg (1950) gives the head capsule lengths of all four larval instars of *C. elegans*. The larvae were found in Michigan (U.S.A.) only from May until October. In the Netherlands fourth instar larvae have been collected from April until September.

FEEDING AND MICROHABITAT

Berg (1950) found the larvae of all instars only mining in floating leaves of several *Potamogeton* species; older larvae mined mainly in stems and petioles of these plants. The larvae fed on fresh tissue of the plants. Learner et al. (1989) reared the adults in England from *Polygonum*, *Myriophyllum* and *Typha*. The adults could be reared also from the silty bottom; this is consistent with Berg's statement that the older larvae often leave the plants. The Dutch larvae were found living on *Potamogeton* species with floating leaves, but also on *P. lucens* and even on narrow-leaved species such as *P. pectinatus* and *P. pusillus*, where mining is impossible and the larvae crept freely on the leaves (see also Schmid, 1993: 77).

WATER TYPE

Hirvenoja (1973) reported the occurrence of *C. obnixus* in lakes and ponds. The Dutch larvae were collected in small lakes, ditches and slow-flowing lowland streams (Peters et al., 1988; Limnodata.nl; own data).

TROPHIC CONDITIONS

A significant proportion of the records of this species apply to very eutrophic to hypertrophic ponds, lakes, ditches and streams. However, in some cases the larvae lived in mesotrophic water (Hirvenoja, 1973; Learner et al., 1989; Limnodata.nl; own data).

SALINITY

Two records from the Netherlands apply to oligohaline pools (Limnodata.nl; own data).

Cricotopus (Isocladius) ornatus (Meigen, 1818)

Eucricotopus atritarsis Thienemann, 1944: 606, 624; Remmert, 1955: 35–36
? nec *Cricotopus ornatus* Gaevskaya, 1969: 91, 92 (cf. Hirvenoja, 1973: 263)

SYSTEMATICS AND IDENTIFICATION

Cricotopus ornatus resembles other species of the *sylvestris* group as adult and as larva (Langton & Pinder, 2007; Moller Pillot, 1984). The number of setae in the l_4 on segment VII of the larva is not characteristic of this species (as supposed in Cranston, 1982, Moller Pillot, 1984 and Oliver & Dillon, 1988). The pupa is very characteristic, but can be confused with *C. laetus* in Scandinavia (Langton, 1991). In the older literature the name is used for light forms of *C. sylvestris* (Hirvenoja, 1973: 263). This is most probably also the case in Gaevskaya (1969): her description of behaviour and feeding is very divergent from other literature.

DISTRIBUTION IN EUROPE AND THE NETHERLANDS

C. ornatus has been found in many countries along the coasts of northern, western and southern Europe, but also in several central European countries (Saether & Spies, 2010). In the Netherlands it is a common species in the delta region and along the coast (Moller Pillot & Buskens, 1990). Records from other parts of the country are doubtful when not identified in the pupal stage (see above).

LIFE CYCLE

In Canada in winter, Swanson & Hammer (1983) collected many third instar larvae and hardly any second instar larvae. The number of third instar larvae increased sharply in September. Their data suggest only one generation a year, emerging mainly in June. In the Netherlands we still collected many third instar larvae in March; pupae were found from April until October. Tourenq (1975) stated emergence in the Camargue during the whole year, with maxima in late spring, but locally in October or in other months.

FEEDING

Most probably the larvae feed on decaying plant material, algae, etc. (see Microhabitat). The larvae described by Gaevskaya (1969: 92) as halfminers in *Nymphaea* probably belong to other species (see above).

MICROHABITAT

Remmert (1955) collected the larvae on floating, decaying *Ulva*, not on *Enteromorpha*. Swanson & Hammer (1983) found *C. ornatus* on macrophyte vegetations (*Potamogeton pectinatus* and *Ruppia maritima*) and submerged shrubs with mats of filamentous algae, and scarcely on mud-detritus substrates. We also found the larvae usually on plants and rarely on the bottom or on stones.

WATER TYPE

C. ornatus lives in lakes, rice fields, creeks, pools in grassland, rock pools and ditches (Remmert, 1955; Hirvenoja, 1973; Tourenq, 1975; Limnodata.nl). Because the species is almost entirely confined to brackish water, records from flowing water are scarce. The larvae live also in temporary pools (Moller Pillot & Buskens, 1990) and even in city fountains in central Europe (Hamerlik & Brodersen, 2010). Records from fast-flowing brooks and streams with fresh water are not reliable.

SALINITY

According to Remmert (1955), in the Baltic Sea the species lives in water containing from 2500 to 8000 mg Cl/l; Tourenq (1975) stated in the Camargue 200 to 6000 mg/l. Krebs & Moller Pillot (in prep.) found *C. ornatus* in the Dutch province of Zeeland most frequently in β-oligohaline water (i.e. between 300 and 1000 mg Cl/l); Steenbergen (1993) gives (in the province of Noord-Holland) many more records from water containing more than 1000 mg Cl/l. In these regions the larvae can also be found in low numbers in fresh water. They appeared to be rather common in city fountains in Copenhagen (Denmark) and Olomouc (Czech Republic) (Hamerlik & Brodersen, 2010).

Cricotopus pallidipes Edwards, 1929

SYSTEMATICS AND IDENTIFICATION

Hirvenoja (1973: 242) placed *C. pallidipes* provisionally in the *bicinctus* group. The larva is unknown; the pupa can be identified using Langton (1991).

DISTRIBUTION IN EUROPE

C. pallidipes is known from a low number of countries in western and eastern Europe (Saether & Spies, 2010). There are no records from the Netherlands.

ECOLOGY

The habitat requirements of the species are still obscure. Langton (1991) mentioned its occurrence in montane and northern lakes and streams, but Tourenq (1975) found it in rice fields and canals in the Camargue. Reiff (1994) collected some exuviae in a lake in Bavaria.

Cricotopus patens Hirvenoja, 1973

SYSTEMATICS AND IDENTIFICATION

C. patens belongs to the *cylindraceus* group. The adult male resembles *C. cylindraceus* and is absent from the key by Langton & Pinder (2007). Identification of the larva is more or less impossible, also because the larva of *C. festivellus* is not known (see Hirvenoja, 1973: 138 and 214). Pupae and exuviae can be identified using Hirvenoja or Langton (1991).

DISTRIBUTION IN EUROPE

According to Saether & Spies (2010) *C. patens* has been collected in Scandinavia and central Europe. A record from the Netherlands has to be verified.

ECOLOGY

The larvae live in stagnant water and have been collected on stony-sandy substrate and among mosses (*Calliergon*) (Hirvenoja, 1973; Palomäki, 1989).

Cricotopus (Isocladius) perniger (Zetterstedt, 1850)

C. perniger is related to *C. reversus*. Identification of the male adult can be problematic (see Hirvenoja, 1973: 310). The pupal exuviae have not been fully described, but can be identified without difficulty (see Langton, 1991: 200 and 206).

The species has been collected mainly in northern lakes, but also lives in central Europe (Hirvenoja, 1973; Saether & Spies, 2010). A record from the Netherlands is doubtful.

Cricotopus (Isocladius) pilitarsis (Zetterstedt, 1850)

SYSTEMATICS AND IDENTIFICATION

C. pilitarsis is a species of the *sylvestris* group. Possibly the species is often not recognized, because the differences from other species of the group as adult and pupa are small. The larva is unknown.

DISTRIBUTION IN EUROPE

C. pilitarsis has been recorded in a number of countries from across the whole of Europe. There are no reliable records from the Netherlands.

ECOLOGY

The species lives in lakes and streams (Hirvenoja, 1973). It has been collected in the lower parts of the Fulda in Germany (Lehmann, 1971) and in lowland brooks (Lindegaard-Petersen, 1972; Drake, 1985). Hirvenoja has also reported its occurrence in brackish water.

Cricotopus pilosellus Brundin, 1956

SYSTEMATICS AND IDENTIFICATION

C. pilosellus and two other arctic species make up a separate group of species within the subgenus *Cricotopus* (Hirvenoja, 1973). The adult males can be identified using Langton & Pinder (2007) and the exuviae using Langton (1991). The larva is still unknown.

DISTRIBUTION IN EUROPE

C. pilosellus has been collected in a number of countries scattered across Europe (Saether & Spies, 2010). Records from the Netherlands have to be verified.

WATER TYPE

The species has been collected mainly in lakes (Reiss, 1968; Hirvenoja, 1973), and in Canada also in rivers (Oliver & Dillon, 1988).

Cricotopus pirifer Hirvenoja, 1973

SYSTEMATICS

C. pirifer is related to *C. fuscus* and may only be an extreme form of this species (Langton, in litt., see Reiff, 1994: 103). The differences between these species are described in Hirvenoja (1973). See also Langton & Visser (2003).

DISTRIBUTION IN EUROPE

C. pirifer has been recorded in very few countries, scattered across Europe. The record from the Netherlands has to be cancelled.

WATER TYPE

Most records are from cold lakes (Hirvenoja, 1973; Laville & Giani, 1974; Reiss, 1984; Reiff, 1994).

Cricotopus polaris Kieffer, 1926

SYSTEMATICS

C. polaris is strongly related to *C. tibialis* (see Hirvenoja, 1973; Langton, 1991). The larva is unknown.

DISTRIBUTION IN EUROPE

C. polaris has been recorded in several countries in northern and central Europe, but seems to be absent from the Mediterranean area (Saether & Spies, 2010). In the western European lowlands it is probably rare; van Kleef (unpublished) collected the exuviae in Denmark.

WATER TYPE

The species inhabits lakes and moorland pools (Hirvenoja, 1973). Koskenniemi & Paasivirta (1987) collected it in a Finnish lake at pH 5.5–6.

C. pulchripes Verrall, 1912

This species, which belongs to the *tremulus* group, cannot be identified in the larval stage. The larva resembles other species of the *tremulus* and *festivellus* groups and the differences in these groups given by Cranston (1982: 66–67) are not reliable. The pupa is very characteristic (Langton, 1991: 210).
C. pulchripes has been recorded in a number of countries scattered across the whole of Europe. There are no reliable records from the Netherlands.
The species is said to be typical for highly oxygenated upland streams with coarse substrate (Cranston, 1982; Wilson, 1988; Hawtin, 1998). It probably also lives in stagnant waters in hilly regions (see Hirvenoja, 1973: 190).

Cricotopus (Isocladius) relucens Hirvenoja, 1973

SYSTEMATICS AND IDENTIFICATION

C. relucens belongs to the *sylvestris* group and is difficult to identify in adult and pupal stage (see Hirvenoja, 1973). The species is absent from the keys in Langton & Pinder (2007) and Langton (1991), but Langton & Visser (2003) suggest that their *C. (Isocladius)* Pe 2 may be the same species. The larva is unknown.

DISTRIBUTION IN EUROPE

C. relucens has been collected in Finland, Germany and Austria (Hirvenoja, 1973; Saether & Spies, 2010). There are no reliable records from the Netherlands. The exuviae of *C. (Isocladius)* Pe 2 Langton have been collected in flowing and stagnant water in England and Wales and in two lakes in Bavaria (Reiff, 1994).

ECOLOGY

Very little is known about the ecology of the species. The material studied by Hirvenoja (1973: 277) had been collected from lakes in Finland and Germany and along the river Fulda.

Cricotopus (Isocladius) reversus Hirvenoja, 1973

SYSTEMATICS AND IDENTIFICATION

C. reversus belongs with *C. perniger* to the *reversus* group of the subgenus *Isocladius*. Identification of adult male and exuviae does not present any problems. The larvae cannot be identified using the existing keys: the setal tufts on abdominal segment VII are not always reliable as different from *C. intersectus*. The larvae may be more bluish than *C. intersectus*.

DISTRIBUTION IN EUROPE AND THE NETHERLANDS

The species lives in large parts of Europe, but is possibly absent from the southwestern countries (Saether & Spies, 2010). Klink (unpublished) caught adult males along the large rivers in the Netherlands.

LIFE CYCLE

In Austria the adults have been collected from May to September (see Hirvenoja, 1973: 308; Janecek, 1995: 293). In Bavarian lakes the exuviae have been collected only in September (Reiff, 1995).

MICROHABITAT

In Austria the larvae live among algae on stones and wood (Hirvenoja, 1973, based on descriptions by Thienemann; Janecek, 2010).

WATER TYPE

C. reversus has been found mainly in lakes (Hirvenoja, 1973; Janecek, 1995; Reiff, 1995). However, Hamerlik & Brodersen (2010) collected the species in all the fountains they investigated in the city of Olomouc (Czech Republic). There are no reliable records from flowing water.

Cricotopus similis Goetghebuer, 1921

Cricotopus Pe 1 Langton, 1991: 210

SYSTEMATICS AND IDENTIFICATION

C. similis is related to *C. trifascia* and may be a small form of this species (Hirvenoja, 1973). Adult male and exuviae can be identified using all keys after 1973. The exuviae are described in Langton & Visser (2003). The larva is still undescribed, but will look like a small *trifascia* larva.

DISTRIBUTION IN EUROPE AND THE NETHERLANDS

C. similis has been found in the greater part of Europe. In the Netherlands it seems to be limited to central and southern Limburg.

LIFE CYCLE

The life cycle will resemble that of *C. trifascia*, i.e. several generations a year with a diapause in winter in second/third instar. However, exact data are not available. We collected exuviae in the Dutch province of Limburg in April, September and October and near Aachen at the end of May.

WATER TYPE

C. similis lives in brooks and streams, but seems to be almost absent from large rivers.

Gendron & Laville (1997) call it a foothill species, only in special cases living in mountain brooks up to 1400 m high, but on average a little higher than *C. trifascia.* It is most abundant at a current velocity of about 1 m/s, but Garcia & Laville (2001) collected the exuviae also at slower currents, possibly because of downstream drift of larvae. In the Netherlands there are no records from brooks and streams with a currentvelocity less than 50 cm/s.

Cricotopus (Isocladius) speciosus Goetghebuer, 1921

SYSTEMATICS AND IDENTIFICATION

The adult male of *C. speciosus* resembles *C. tricinctus*, but the pupa and larva are quite different. The larva has not been described, but, based on much associated material collected by H. Cuppen and D. Tempelman, it can be easily distinguished because of the reduction of the first and second lateral teeth of the mentum. This larva has been figured by Brooks et al. (2007) as *trifasciatus*-type, based on a mistake.

DISTRIBUTION IN EUROPE AND THE NETHERLANDS

C. speciosus has been reported from most West European countries, but there are few records from other parts of Europe (Saether & Spies, 2010). In the Netherlands there are about twelve records, in the majority of cases from the western part of the country. Most of these records are based on identification of larvae, a few on exuviae or an adult male.

ECOLOGY

According to Hirvenoja (1973) and Langton (1991), *C. speciosus* inhabits stagnant and running water. Reiff (1994) collected exuviae in a mesotrophic lake in Bavaria. Most Dutch records are from small lakes, sand pits and ditches, and in one case from a slow-flowing lowland stream. Most of these water bodies were only slightly or not polluted, and some were mesotrophic.

Cricotopus (Isocladius) suspiciosus Hirvenoja, 1973

C. suspiciosus has been recorded from very few countries in different parts of Europe (Saether & Spies, 2010). However, Paasivirta (2012) reported its occurrence only in the northern part of Finland, not in the southern provinces. In view of the overlapping characteristics with related species, a revision of the European material is desirable.

Cricotopus sylvestris group

SYSTEMATICS AND IDENTIFICATION

The majority of species in the subgenus *Isocladius* belong to the *sylvestris* group. Hirvenoja (1973) distinguished twelve species in Europe. Many of them are difficult to identify in all stages and in many cases Gresens et al. (2012) found some overlap in several characters. It is likely that speciation is still ongoing, but hybridisation or phenotypic plasticity cannot be excluded. Identification of the larvae of the group is hardly possible. Many characters, such as the antennal ratio and the number of setae in the setal tuft on segment VII, are variable within the species.

Not all authors separate the species *C. sylvestris* from related species, among which *C. pilitarsis* and *C. suspiciosus*, which sometimes have the same pigmentation as adult and are difficult to distinguish as pupal exuviae.

Cricotopus (Isocladius) sylvestris (Fabricius, 1794)

IDENTIFICATION

See under the *sylvestris* group above. In contrast to what has been written in most keys, the larvae sometimes have a setal tuft on segment VII with 3 to 5 setae.

DISTRIBUTION IN EUROPE AND THE NETHERLANDS

The species has been found almost everywhere in Europe (Saether & Spies, 2010). Within the Netherlands it is a common species in all parts of the country (Limnodata.nl).

LIFE CYCLE

As in most other species of the genus (see the introduction to the genus), *C. sylvestris* has a diapause from November to February in second and third instar. Where the bottom is not penetrable or when it contains no oxygen the larvae live the whole winter among dead or living plants; in other cases they have to be sought in the bottom, sometimes in deeper layers (LeSage & Harrison, 1980; Menzie, 1980; own data).
Usually it is not possible to determine the number of generations, because development is dependent on environmental factors and whole generations can be eliminated. Wotton et al. (1992) reported that in very dynamic environments with abundant food (sand filter beds) mass emergence occurred as early as 18–20 days after egg deposition. *C. sylvestris* probably as a rule has three to five generations a year (Pankratova, 1970; LeSage & Harrison, 1980). Nevertheless, Reiff (1994: figs 119, 120) stated in some lakes in Bavaria only one or two generations. Several authors (e.g. Mundie, 1957; Lehmann, 1971) stated larger numbers in late summer or autumn, but in some Bavarian lakes the highest numbers are found in spring or early summer (Reiff, loc cit.). In a gravel pit in England Titmus (1979) found only one large generation in May/June and hardly any emergence later in summer.
All larval instars are described in detail by Rodova (1966).

OVIPOSITION

See par. 2.5 and under the genus. Usually the egg mass is attached to a plant. Sometimes many females deposit their egg masses together, so that one large mass is formed (Pankratova, 1970: 24). The egg mass is string-shaped; a small portion is attached to a substratum and the main portion floats in the water (Nolte, 1993: 64, fig. 47e). The number of eggs in one egg mass can vary considerably, from 160 to 1300, eggs depending on many factors (Nolte, 1993: 42; see also par. 2.5). Balushkina (1987) gives a mean of 200.

SWARMING

Darby (1962) saw a large cloud of males over a ditch and over the open water of a rice field, not rising high in the air. LeSage & Harrison (1980) saw a swarm about 7 m from the stream at a height of more than 1 m.

FEEDING

The larvae usually gather food around their tube. Rodova & Sorokin (1965) stated that they often also consume material they cannot use as food. These authors further mentioned that different studies give different results on the food of *C. sylvestris*. Later publications give the same picture. Sorokin (1968) stated that the larvae are phytophagous and actively feed on blue-green algae and diatoms. Tarkowska-Kukuryk & Mieczan (2008) found that filamentous algae and diatoms were ignored and the food consisted mostly of detritus and protozoa. Wang (2000) reported *C. sylvestris* as a rice

pest that damages the roots and leaves of rice seedlings. As a rule, the larvae rarely or never gnaw at living water plants, but often attack young rice seedlings and can damage the vegetative apex, roots and leaves (Darby, 1962; Ferrarese, 1992; Wang, 2000; Vala et al., 2000).
Without doubt the larvae are opportunists and consume the available food. However, living green algae are vigorously consumed but are not ingested to a significant extent (Sorokin, 1968). Living or dead animals are often consumed when encountered, but the larvae are not active predators (Rodova & Sorokin, 1965; Mackey, 1979; Izvekova, 2000; own data). Small larvae consume larger amounts of detritus than larger larvae (Tarkowska-Kukuryk & Mieczan, 2008).
It is not clear if the differences in feeding are based on speciation (genetic variability) or on plasticity of the species (see under Systematics of the *sylvestris* group above).

MICROHABITAT

The larvae live mainly on plants (Menzie, 1980), wood (Klink, 2011) or stones (Becker, 1995), but they can be numerous among masses of filamentous algae such as *Spirogyra* and *Cladophora* (Mundie, 1957; Darby, 1962). There are striking differences between different plant species (Learner et al., 1989). In summer the larvae are very scarce in the sediment (unless no plants or stones are available), but from November to February they can be found there in large numbers (Menzie, 1980; own data).

DENSITIES

Many authors give densities of larvae and these differ significantly according to the time of year. Balushkina (1987: 152) found 40 larvae/m^2 at one bank of the river Luga in Russia and 14,000 larvae/m^2 near the other bank. Menzie (1980) studied *C. sylvestris* on *Myriophyllum* plants and found from 2000 to 50,000 larva/m^2 bottom area in summer. Tarkowska-Kukuryk & Mieczan (2008) found 57,000 large larvae/m^2 on the surface of reeds in June and 7000 in August.

WATER TYPE

Current and dimensions

As can be seen in table 7 (p. 286), *C. sylvestris* is a common species in stagnant pools and ditches and slow-flowing brooks and streams, but is scarcely found in fast-flowing streams (Ertlova, 1970; Michiels, 1999; Orendt, 2002a; Limnodata.nl). The species seems to be present mainly in polluted water, especially at faster currents (see below under Saprobity). The larvae are common in large rivers and almost everywhere an increase in numbers downstream can be observed (Lehmann, 1971; Becker, 1995). The larvae are easily carried off by fast currents (Gendron & Laville, 2000). This was observed by Moller Pillot (2003) in a small lowland brook when a faster current was accompanied by acidification and loss of microhabitat and food. In this case, *C. sylvestris* seemed to be a habitat-shifting species, which was not continually present in the brook.

Temperature

Rossaro (1991) found *C. sylvestris* at a relative high mean water temperature of 20 °C and stated a restricted temperature range. From our own investigations, it appears that the larvae can be abundant in small water bodies where the temperature during the day exceeds 30 °C.

Permanence

In temporary water bodies suitable for the species, settlement can be observed very soon when water is present. The most important factor is the presence of flying females at that time and place (Wotton et al., 1992; Moller Pillot, 2003; Hamerlik & Brodersen, 2010; own data).

pH

At pHs lower than 5 *C. sylvestris* is rare and never abundant (Leuven et al., 1987; Moller Pillot, 2003; van Kleef, unpublished; own data). With increasing pH the presence increases (see table 6); Steenbergen (1993) found even significantly more larvae in water at pHs higher than 7.5 and a calcium content higher than 50 mg/l. Most probably this correlation with pH is mainly dependent on the availability of food.

TROPHIC CONDITIONS AND SAPROBITY

The larvae live in ditches, lakes and streams which are mesotrophic or slightly eutrophic to hypertrophic or even severely organically polluted. They are most common at higher trophic levels or in strongly polluted streams, as summarised in table 5 on p. 283. In contrast to many other chironomid species, *C. sylvestris* is also a relatively common species in polyhumic lakes and pools if the pH is not too low (Saether, 1979; own data).

This is illustrated by some examples. In Bavaria Reiff (1994) collected *C. sylvestris* mainly in the eutrophic **lakes**. Buskens & Verwijmeren (1989) found the larvae mainly in sand pits of the highest trophic level. Steenbergen (1993) collected the larvae, often in large numbers, in water with very high phosphate and chlorophyl-a content. He found the larvae in almost the same numbers in water with less than 40% oxygen saturation (during the day). In very eutrophic **pools and ditches** in the Netherlands it can be the only species of the subfamily. In some ditches with very severe organic pollution and almost without oxygen during summer nights we still found numerous larvae, mainly on plants near the water surface. Morris & Brooker (1980) did not find larvae of the *sylvestris* group in the nutrient-poor upland reaches of the **river** Wye in Wales, but they were present everywhere in the nutrient-rich lowland areas (see, however, also under Water type). Bazerque et al. (1989) collected the exuviae in the river Somme in France only in polluted stretches downstream of the town of Saint-Quentin. The species was present in low numbers after severe pollution with organic matter and heavy metals (Zn^-), but the numbers increased downstream as the water quality improved. In the river Trent exuviae of *C. sylvestris* were only abundant at more strongly polluted stations (BOD 9–13, ammonium-N 3–5 mg/l), especially where heavy metal pollution was also evident (Wilson, 1987). This author suggests that the presence of the species is strongly linked to algal growth.

On the other hand, the species can be abundant in ditches, lowland streams and city fountains without organic pollution. In some cases the **dynamic environment** with explosive algal growth could play a role.

SALINITY

Krebs (1981, 1984) collected *C. sylvestris* in the Dutch province of Zeeland often in slightly brackish water, but scarcely in water containing more than 1000 mg Cl/l and never above 2500 mg Cl/l. In the province of Noord-Holland the species is still common at concentrations above 1000 mg Cl/l (Steenbergen, 1993).

Elsewhere, even much higher concentrations have been stated: in the river Jordan in Israel, 3000 mg/l (Ortal & Por, 1978); in the Camargue not rarely between 5000 and 8000 mg/l (Tourenq, 1975); in the Baltic Sea almost everywhere and still abundantly above 3000 mg/l (Paasivirta, 2000); and in northern Germany up to 4000 mg/l (Remmert, 1955, as var. *ornatus*).

SUMMARY

In summer *C. sylvestris* is able to build up large populations in a short time where easily digestible food is available in large quantities, probably only where it is sheltered from fast currents or wave action. It is highly tolerant of toxic chemicals and anoxia.

Cricotopus tibialis (Meigen, 1804)

nec *Trichocladius tibialis* Brundin, 1949: 732 (see Hirvenoja, 1973: 308)

SYSTEMATICS AND IDENTIFICATION

Within the subgenus *Cricotopus*, the *tibialis* group (with six species in Europe, three in the western European lowland) is closely related to the *pilosellus* group (Hirvenoja, 1973: 56, 61). The adult male of *C. tibialis* is not keyed in Pinder (1978), but can be identified using Langton & Pinder (2007). The larvae are absent from Schmid (1993) and cannot be distinguished from other species of the group. They also resemble the larvae of *C. fuscus* (see Cranston, 1982; Moller Pillot, 1984). See also under Water type. Many records of *tibialis* from before 1973 apply to other species, see Hirvenoja (1973: 308).

DISTRIBUTION IN EUROPE AND THE NETHERLANDS

C. tibialis is known from almost the whole of Europe (Saether & Spies, 2010). The species seems to be especially common in northern countries. In the Netherlands it is possibly the only species of the *tibialis* group. It has been collected mainly in the province of Limburg, but also locally elsewhere in the southeastern part of the country (Limnodata.nl; own data). The Dutch records are based on identification of larvae and exuviae.

WATER TYPE

In northern countries *C. tibialis* can be found in springs, brooks, rivers, ponds and lakes, but elsewhere it seems to be almost confined to flowing water (Hirvenoja, 1973: 163; Lindegaard & Jónasson, 1979; Oliver & Dillon, 1988). Lehmann (1971) and Lindegaard-Petersen (1972) collected the species in spring brooks in Germany and Denmark. As far as is known, all records from the Netherlands apply to spring brooks or brooks and ditches fed by groundwater (own data). In a few cases some specimens of *C. tibialis* have been collected in central European rivers (Móra, 2008) or lakes (Reiff, 1994).

Most probably all records of larvae of gr. *tibialis* in spring brooks in or nearby the Netherlands apply to the species *tibialis*, because the two other species of this group live almost only in lakes and pools (see under *epihippium* and *polaris*). The same is true for *C. pilosellus*.

Cricotopus tremulus (Linnaeus, 1758)

IDENTIFICATION

The adult males and females of the *tremulus* group are not always easy to identify (compare Hirvenoja, 1973, and Langton & Pinder, 2007). Identification of the exuviae using Langton (1991) presents no problems. The larva of C. tremulus has no setal tufts on the abdominal segments and this character seems to be quite specific (Hirvenoja, 1973: 185; Cranston, 1982; Schmid, 1993). The l4 is long and thick. The larva is also conspicuous within the genus *Cricotopus* because of the relatively narrow central mental tooth which does not protrude, or hardly protrudes, before the first lateral teeth, and the small second lateral teeth which have grown together with the first lateral teeth (Schmid, 1993: fig. 50). The head capsule of the Dutch material is yellowish, sometimes yellow-brown (own data).

DISTRIBUTION IN EUROPE AND THE NETHERLANDS

C. tremulus has been found in most European countries (Saether & Spies, 2010). In the Netherlands exuviae have been collected scarcely in the Rhine and in the Meuse in Limburg (Klink & Moller Pillot, 1982; Klink, 1985, 1991). In brooks and streams in Southern Limburg the species is scarce also (Limnodata.nl; own data). Other records have to be verified.

LIFE CYCLE

In Germany emergence takes place from April until October in two or more generations (Lehmann, 1971; Ringe, 1974; Michiels, 1999).

MICROHABITAT

The larvae are found on stones and among mosses (Lehmann, 1971; Hirvenoja, 1973; Schmid, 1993).

WATER TYPE

Current

C. tremulus is an inhabitant of running water has been rarely collected in lakes (Hirvenoja, 1973; Langton, 1991; Reiff, 1994). Gendron & Laville (1997) collected the exuviae mainly in the montane upper stretches of the river Aude in the Pyrenees, but in low numbers also in the (still rather fast-flowing) lower stretches. However, Rossaro (1991) found the species at a relatively high mean temperature in Italian streams (18 °C), suggesting that it is not a typical species for montane spring brooks. Nevertheless, *C. tremulus* has been collected in many fast-flowing streams or on stones in fast currents (Lehmann, 1971; Ringe, 1974; Orendt, 2002a; Janzen, 2003). It is absent from slow-flowing lowland brooks.

Dimensions

The larvae live in narrow brooks and small streams (Ringe, 1974; Gendron & Laville, 1997; Orendt, 2002a; own data). In large rivers the species is scarce, although locally abundant, mainly in montane areas (Klink & Moller Pillot, 1982; Wilson & Wilson, 1984; Klink, 1985; Caspers, 1991; Becker, 1995).

Temperature

Rossaro (1991) found *C. tremulus* at a mean temperature of 18° C. The species is not cold-stenothermic as sometimes supposed (e.g. Becker, 1995).

SAPROBITY

C. tremulus has been found in many brooks and streams without organic pollution, but Maasri et al. (2008) collected the species in an organically polluted stream. Moog (1995) considers the species to be rather tolerant and Bitušik (2000) called it an indicator of organic pollution.

Cricotopus triannulatus (Macquart, 1826)

SYSTEMATICS AND IDENTIFICATION

C. triannulatus belongs to the *tremulus* group (Hirvenoja, 1973: 184). Identification of the adult male usually presents few problems. The pupa can be identified only when the frontal setae and the lateral setae of segment VIII are clearly visible (Hirvenoja, 1973; Langton, 1991). The larva is not keyed in Schmid (1993) and identification using Cranston (1982) is not reliable. *C. triannulatus* can probably be distinguished from *annulator*, *curtus* and *vierriensis* by the mandible, which is obviously crenulated at the outer edge and darkened in the whole distal half (Hirvenoja, 1973: fig. 127). See also under *C. annulator*.

DISTRIBUTION IN EUROPE AND THE NETHERLANDS

The species has been found in all parts of Europe (Saether & Spies, 2010). In the Netherlands it has been reported from southern Limburg, almost the whole Pleistocene area and the large rivers (Limnodata.nl). Other data have to be verified.

LIFE CYCLE

C. triannulatus has two or more (up to five?) generations a year and a diapause as juvenile larvae in late autumn. Berg & Hellenthal (1992) stated fast growth of the larvae from January to March. In the Netherlands adults are very scarce in February and March; high numbers of exuviae and adults have been collected from May to September (Klink, 1985, 1990; own data; cf. Lehmann, 1971; Becker, 1995). LeSage & Harison (1980) found large numbers in spring and autumn and only very low numbers in summer. They stated emergence at water temperatures from 7 to 27 °C. Becker (1995) also caught some adults in winter.

SWARMING

C. triannulatus usually swarms very near to the stream, about 1 m above the ground (LeSage & Harrison, 1980).

FEEDING

Klink & bij de Vaate (1994) stated that the larvae in the river Grensmaas (Meuse in southern Limburg) did not consume filamentous algae, but only diatoms, benthic algae and other fine material. This in contrast to *C. trifascia*.

MICROHABITAT

The best microhabitat for the larvae is provided by riffles of cobbles and pebbles densely covered by diatoms and filamentous algae (LeSage & Harrison, 1980). These authors stated that the species is easily adapted to other microhabitats. Becker (1995) found the species less numerously on sandy substrate in the river Rhine. In the Netherlands the larvae are much more common on stones than on plants and they are very scarcely collected from the bottom, but these differences are influenced by the different current velocities between these places.

WATER TYPE

C. triannulatus is known from flowing water and lakes (Hirvenoja, 1973), but it seems to be rare in lakes, except in the alpine and boreal region. In large rivers the species can be very abundant; in brooks and small streams it is rather common in fast-flowing water (Klink, 1985; Becker, 1995, Orendt, 2002a; Michiels, 2004). In slowly flowing lowland brooks the species is scarce (fig. 6), but it can be numerous, especially on stones in fish ladders (own data). Records from water authorities in the Netherlands may refer in part to related species (mainly *C. annulator* and *C. curtus*).

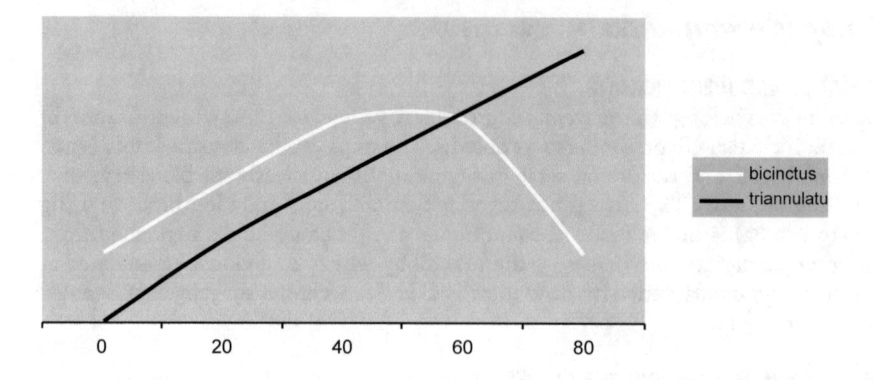

6 *Occurrence of C. bicinctus and C. triannulatus in streams with different current velocity, in cm/s*

SAPROBITY

Several authors call *C. triannulatus* a typical species of organically polluted streams (LeSage & Harrison, 1980; Klink, 1985; Bitušik, 2000; Tang et al., 2010; Milošević et al., 2012). Tang et al. found the species dominating at a BOD of 7–10 mg O_2/l. The larvae also live, in lower numbers, in brooks and streams with no or very little organic pollution (Caspers, 1980; Michiels, 2004; own data). They require a habitat with sufficient oxygen supply. See table 5 in Chapter 9. In lakes the species seems to be characteristic of less eutrophic conditions (Orendt, 1993: table XII.18).

SALINITY

Paasivirta (2000) collected *C. triannulatus* sometimes abundantly in brackish water in the Baltic Sea.

PARASITISM

LeSage & Harrison (1980a) observed a relatively high infestation of mermithids in *C. triannulatus* (in up to 67% of the larvae in one year).

Cricotopus (Isocladius) tricinctus (Meigen, 1818)

SYSTEMATICS AND IDENTIFICATION

C. tricinctus belongs to the *sylvestris* group. Usually the adults can be easily identified from the colour patterns of thorax, abdomen and legs (Hirvenoja, 1973; Langton & Pinder, 2007). The exuviae can be distinguished from other species using Langton (1991). The differences from other species are small; Hirvenoja (1973) gives more variation in the length of the thoracic horn (up to 260 μm). The larva cannot be identified reliably. The l_4 on segment VII of the Dutch larvae often has 3–5 setae.

DISTRIBUTION IN EUROPE AND THE NETHERLANDS

C. tricinctus is distributed throughout Europe (Saether & Spies, 2010). In the Netherlands the species has been collected at several localities, mainly in the provinces of Zeeland, Noord- Holland and Noord-Brabant (Krebs, 1981, 1984; Moller Pillot, 2003; Limnodata.nl), but is probably present in the whole country.

FEEDING
The larvae feed on waterplant tissue (Hirvenoja, 1973; own data).

MICROHABITAT
The larvae live on water plants such as *Potamogeton*, *Callitriche*, *Ranunculus* and *Glyceria* (Hirvenoja, 1973; Tourenq, 1975; own data). In some cases they are halfminers, in other cases they live on the surface of stems and leaves, possibly depending on the plant species.

WATER TYPE
Current
The larvae inhabit stagnant and slow-flowing water (Hirvenoja, 1973; Moller Pillot, 2003; Limnodata.nl) and are very rarely found in faster-flowing brooks or streams (Lehmann, 1971; Scheibe, 2002).

Dimensions
Many records are from ditches and small streams from 1 to 7 m wide with much vegetation (Tourenq, 1975; Krebs, 1981, 1984; Moller Pillot, 2003). Other authors found *C. tricinctus* scarcely in lakes (Macan, 1949; Koskenniemi & Paasivirta, 1987).

Permanence
The species sometimes settles in temporary brooks and ditches and in rice fields (Tourenq, 1975; Vala et al., 2000; Moller Pillot, 2003).

pH
There are some records from acid water with a pH between 5 and 6 (Koskenniemi & Paasivirta, 1987; Moller Pillot, 2003). Krebs (1984) found the species still at pH 9.

TROPHIC CONDITIONS
Most records are from eutrophic to hypertrophic water. Nothing is known about toleration of low oxygen contents. However, the larvae live often near to the water surface and will be not very dependent on the oxygen content of the deeper parts of the water body.

OR

SALINITY
Tourenq (1975) collected the species in the Camargue only in fresh water. Krebs (1984) also collected the species once in a slightly brackish ditch (about 1000 mg Cl/l).

Cricotopus trifascia Edwards, 1929

SYSTEMATICS AND IDENTIFICATION
Within the subgenus *Cricotopus* the two species of the *trifascia* group can be easily distinguished as adult male and pupa. However, in the past specimens of *C. similis* can have been identified as *C. trifascia*. Presently both species cannot be distinguished as a larva. *C. similis* is undoubtedly much smaller than *C. trifascia*, but the differences between both species in third and fourth instar are unknown.

DISTRIBUTION IN EUROPE AND THE NETHERLANDS
C. trifascia has been collected in all parts of Europe (Saether & Spies, 2010). In the Netherlands the species is rather common in southern Limburg and has been collected also in central Limburg (Limnodata.nl). Other records have to be verified.

LIFE CYCLE

We could find hardly any larvae in winter and until March only larvae in third instar. LeSage & Harrison (1980: 392) suggested that the larvae are in second instar from mid November and migrate deeper into the substrate in winter. See also under the genus. In the Netherlands and adjacent areas in Germany exuviae and adults have been collected from (March) April until September (Becker, 1995; own data). The autumn generation can be much larger, as has been stated by Michiels (1999); Pinder (1980) found only larvae in September. Reiff (1994) stated only one generation in July in the Chiemsee in Bavaria.

SWARMING AND OVIPOSITION

Swarming males were seen by Lesage & Harrison (1980) at a distance of 5 to 13 m from the stream, at a height of 1 to 3 m; they used a tree branch as marker. Oviposition will take place during twilight, mainly at dusk.
Most probably the ribbon-shaped egg masses adhere to the substrate; LeSage & Harrison (1980) reported a mean egg mass length of 21.5 mm. The eggs are larger than in most other species of the genus; LeSage & Harrison (1980) stated a mean length of 218 µm. These authors also found a very high mean number of eggs in one egg mass: 526 (compared with *bicinctus*: 291; *sylvestris*: 184).

FEEDING

Klink & bij de Vaate (1994) found the guts of six larvae to contain filamentous algae and the diatoms and other benthic algae living on them.

MICROHABITAT

C. trifascia has been collected mainly on stones, gravel and other hard substrates; in other sediments and on plants the larvae seem to be less numerous (Lehmann, 1971; Pinder, 1980; Klink, 1991; Schmid, 1993; own data).

WATER TYPE

Current

C. trifascia is strongly rheophilous and rarely present in large lakes, such as the Chiemsee in Bavaria (Reiff, 1994). Although Gendron & Laville (1997) call it a plain species, it is mainly found in foothill streams, in riffles in lowland streams, and in mountain brooks up to an altitude of more than 1000 m (Lehmann, 1971; LeSage & Harrison, 1980; Casas & Vilchez-Quero, 1989; Janzen, 2003). In the Netherlands the species is absent from the typical lowland brooks and can be found only in the hillier parts of the province of Limburg.

Dimensions

The larvae live in narrow brooks, but also in larger streams, possibly dependent on the presence of suitable substrate in areas of fast-flowing water. In large rivers they are usually scarce (Becker, 1994; Klink, 2010).

TROPHIC CONDITIONS AND SAPROBITY

C. trifascia appears to require a rather high oxygen supply. It is strongly rheophilous (see above) and the larvae die when reared in a Petri dish without the addition of oxygen, in contrast to *C. bicinctus* and many other species (own data). However, an oxygen content of 5.6 mg/l appeared sufficient in the river Trent (Wilson, 1987).
On the other hand, the larvae tolerate rather severe organic pollution (BOD of at least 10 mg O_2/l), as has been stated by Lesage & Harrison (1980), Laville & Viaud-Chauvet (1985) and Wilson (1987). The species also can be found in hardly polluted water. In

Dutch streams in southern Limburg the larvae seemed to tolerate substantially less pollution than *C. bicinctus*. Milošević et al. (2012) stated in streams in Serbia, that the larvae can live in severely polluted water, but only at low temperatures. As has been stressed in section 2.10 factors such as pollution, oxygen, current velocity and temperature have to be looked at together.

SALINITY

According to Thienemann (1954: 592) the species can also live in brackish water. Paasivirta, (2000) stated tolerance of salinities higher than 2000 mg Cl/l in the Baltic Sea.

Cricotopus (Isocladius) trifasciatus (Meigen, 1818)

SYSTEMATICS AND IDENTIFICATION

C. trifasciatus belongs to the *sylvestris* group (Hirvenoja, 1973). The species name is debatable, but can be maintained (Spies & Saether, 2004). The adult male and female can be identified mainly on the base of the abdominal colour pattern (Hirvenoja, 1973; Langton & Pinder, 2007). Gresens et al. (2012) stated that it is not always possible to distinguish *C. sylvestris* and *C. trifasciatus* by barcoding data. These authors suggested that these species together could constitute a single species. However, there are considerable differences in biology and ecology as can be seen below. Identification of the exuviae can present problems in some specimens. Identification of larvae within the *sylvestris* group is hardly possible, also because the larvae of some species are still unknown.

The larvae figured by Brooks et al. (2007) as *C. trifasciatus* belong to *C. speciosus* (see there).

DISTRIBUTION IN EUROPE AND THE NETHERLANDS

C. trifasciatus has been collected in all parts of Europe (Saether & Spies, 2010). In the Netherlands it seems to live throughout almost the whole country, although some of the data presented in Limnodata.nl may apply to related species. Krebs (1981, 1984) did not find the species in the province of Zeeland.

LIFE CYCLE

Pupae and adults have been observed from April until October (Kettisch, 1936–1938; Schleuter, 1985; own data). At relatively high temperatures development of one generation lasts about one month (Kettisch, ibid.). According to Kettisch, the species has seven larval instars, and she gives the head width of each of these instars. If this is correct, it is an exception in the familyChironomidae.

No larvae can be found in winter. Eggs laid in the autumn developed only when the temperature rose after a cold period (Kettisch, 1936–1938: 200). Young larvae develop only at temperatures above 11 °C. Kettisch observed (in the laboratory) that larvae and pupae which did not emerge in the autumn died in winter.

MATING AND OVIPOSITION

Nothing is known about the swarming of this species. Mating also occurs in the laboratory. Kettisch (1936–1938) observed that the eggs were deposited on floating leaves, preferably in places where the leaves are damaged, for instance in old mines; this protects the eggs against desiccation. In other cases, the eggs were deposited just below the surface of the water. In autumn the eggs were deposited on stems deeper below the surface.

The oviposition lasts 20 to 30 minutes. One egg mass contains 200–300 eggs. The female dies within a few hours after oviposition.

FEEDING

The half-grown and older larvae feed on living green plant tissue of *Potamogeton*, *Nymphoides*, *Callitriche*, etc. (Kettisch, 1936–1938; Berg, 1950; own data). Larvae fed on green leaves of terrestrial plants in the laboratory developed without problems into adults (own observations). Owing to this the larvae can be a pest in rice fields (Ferrarese, 1992; Vala et al., 2000).
Surprisingly, the mouthparts of the larvae of *C. sylvestris* and *C. trifasciatus* display no differences, although they use completely different food. Apparently, the differences lie in the products of the salivary glands or in symbiosis with bacteria.

MICROHABITAT

The larvae live on the floating leaves of water plants such as *Nymphoides peltata* and *Potamogeton*, but also below the surface on other water plants, such as *Callitriche*, *Veronica*, etc. Some larvae live in open mines, only covered by secretions and excrement; others creep freely on the plants (Kettisch, 1936–1938; Lindegaard-Petersen, 1972; Tourenq, 1975; own data). According to Tourenq (1975), the larvae live mainly on *Phragmites* and *Typha*, but no other author has observed this. Van der Velde (unpublished) never found the larvae on *Nymphaea*, but Hirvenoja (1973) mentioned data in the literature concerning this genus. The larvae are mainly found along the edge of the vegetation (Frigge & Olde Loohuis, unpublished).

WATER TYPE

Current

C. trifasciatus is a common species in moderately fast and slowly flowing brooks and streams (Lindegaard-Petersen, 1972; Pinder, 1974, 1977; own data). Gendron & Laville (1997, 2000) call it a plain species, but collected it regularly in small numbers in the river Aude in de Pyrenees (current velocity about 1 m/s). The species seems to be absent from most fast running streams in Bavaria (Orendt, 2002a). In the Netherlands *C. trifasciatus* seems to live mainly in flowing water, but in Finland Paasivirta (2012) called it more an inhabitant of stagnant water.

Dimensions

The larvae are collected in small pools, marshes and ditches, but also in larger ponds and small lakes. They are rare in large lakes and seem to be almost absent from narrow ditches with dense vegetation. In flowing water the larvae live mainly in brooks and small streams, rarely in large rivers (Lindegaard-Petersen, 1972; Pinder, 1974, 1977; Tourenq, 1975; Schleuter, 1985; Janecek, 2000).

Permanence

Most records are from permanent water bodies. Tourenq (1975) found the larvae also in marshes and rice fields, and Schleuter (1985) collected the species in temporary as well as in permanent pools.

TROPHIC CONDITIONS AND SAPROBITY

C. trifasciatus has been reported mainly from eutrophic water and seems to be absent from oligotrophic moorland pools. Moog (1995) gives the species even for polysaprobic conditions and Wilson & Ruse (2005) mentioned that the whole *sylvestris* group is resistant to many forms of pollution. We collected the larvae in very clear upper courses of brooks as well as in a stream with severe organic pollution (Jeker, near

Maastricht). However, we do not have enough exact information to estimate the relation with saprobity. Most of our data apply to water bodies without serious organic pollution. The presence of the species seems to be mainly related to the availability of preferred plant species and other factors than to trophic or saprobic aspects.

SALINITY

Tourenq (1975) found the species in the Camargue in water containing 3000 mg Cl/l, rarely up to 6000 mg/l. Ruse (2002) collected the exuviae in the brackish Martham Broad. Krebs (1981, 1984) did not find the species in brackish water in the Dutch province of Zeeland.

Cricotopus tristis Hirvenoja, 1973

SYSTEMATICS AND IDENTIFICATION

C. tristis is a species of the *tremulus* group, closely related to *C. annulator* and *C. curtus*. The adult male is well characterised (Hirvenoja, 1973; Langton & Pinder, 2007). The pupa can be distinguished from that of *C. annulator* only because pedes spurii are present on segment III. The description of the larva by Oliver & Dillon (1988) gives no clear differences from the larvae of other species of the group, only that the head is darker.

DISTRIBUTION IN EUROPE

Adult males have been collected in northern Scandinavia and Canada. If the pupal character mentioned above is sufficient to determine that a specimen belongs to *C. tristis*, the species has been recorded also from Germany (Orendt, 2002a; Michiels, 2004), Hungary (Morá, 2008) and France (Klink, 2010). In the Nearctic, *C. tristis* does not occur south of the tree line (Oliver & Dillon, 1988). Paasivirta (2012) stated its occurrence only in the most northern provinces of Finland.

WATER TYPE

Adult males have been collected from northern streams and lakes (Hirvenoja, 1973). Exuviae (see above) have been recorded from some fast-flowing streams in Germany (Orendt, 2002a; Michiels, 2004), from the river Tisza in Hungary (Morá, 2008) and from several tributaries of the Seine in France (Klink, 2010).

Cricotopus vierriensis Goetghebuer, 1935

SYSTEMATICS AND IDENTIFICATION

C. vierriensis belongs to the *bicinctus* group (Hirvenoja, 1973). Male adults and exuviae are easy to identify because of the characteristic hypopygium and thoracic horn (Langton & Pinder, 2007; Langton, 1991). The larva has been described by Hirvenoja (1973: 242), but the differences from *C. albiforceps* (and some other species?) are slight, with the exception of the maxilla.

DISTRIBUTION IN EUROPE AND THE NETHERLANDS

C. vierriensis seems to be absent from the northern part of Europe, but has been collected in many countries in the other regions (Saether & Spies, 2010). The species has been reported from the Netherlands, but most records are based on identification of larvae and therefore unreliable. Exuviae are very rarely collected in small streams in the southeastern part of the country (own data).

WATER TYPE

As appears from the text below the factors responsible for the presence or absence of this species remain puzzling. Possibly northwest Europe lies at the margin of the distribution area. Most publications are from more southern regions.

Current

C. vierriensis lives in lakes and streams (Hirvenoja, 1973). The species has been found in several fast-flowing streams in Europe, but not in typical lowland streams with a current velocity lower than 50 cm/s. Gendron & Laville (1992, 1997, 2000) collected the exuviae in large numbers in the river Aude in the Pyrenees at a current velocity of 1 m/s. After a catastrophic spate the numbers of this species diminished much more than those of the typical montane species. The authors call *C. vierriensis* a plain species, living only in the lower parts of the river. However, Casas & Vilchez-Quero (1989) collected the species in the Spanish Sierra Nevada mainly at a height of 760–900 m and sometimes higher.

The larvae identified as *C. vierriensis* by Simpson (1983) in a small acid brook in New York state (current speed 10–20 cm/s) probably belong to another species.

Dimensions

The species seems to be most common in small rivers and streams (Gendron & Laville, ibid.; Klink, 2010; Michiels, 1999, 2004). It may be absent from brooks less than 5 m wide. Garcia & Laville (2001) collected the exuviae in the Garonne, where this river is about 100 m wide. It is not clear why it has not been collected in the river Rhine (Klink & Moller Pillot, 1982; Wilson & Wilson, 1984; Caspers, 1991; Becker, 1995). The species may prefer smaller rivers, but it is also absent from the whole river Fulda (Lehmann, 1971) and from many fast-flowing middle-sized streams in Bavaria (Orendt, 2002a). See the introduction of this section.

Temperature

Rossaro (1991) found *C. vierriensis* in streams in Italy at a (for montane species) relatively high main temperature of 16.5 °C. Ortal & Por (1978) collected the species in the river Jordan in Israel at temperatures of 26–28 °C.

SAPROBITY

Ortal & Por (1978) collected the species in the organically polluted river Jordan with BOD5 up to 10 mg O_2/l and high temperatures (26–28 °C).

Diplocladius cultriger Kieffer, 1908

SYSTEMATICS

Saether (1977a, 1979a) demonstrated that *Diplocladius* takes a plesiomorphic position within Orthocladiinae and placed the genus provisionally in a separate tribe, Diplocladiini. There is only one European species.

DISTRIBUTION IN EUROPE AND THE NETHERLANDS

Diplocladius cultriger has been collected in most parts of Europe, but seems to be scarce in the British Isles and absent from parts of the Mediterranean area (Cranston, 1982; Saether & Spies, 2010). In the Netherlands it is a common species in the Pleistocene and dune region, scarcer in southern Limburg and rare elsewhere (Moller Pillot & Buskens, 1990; Steenbergen, 1993; Limnodata.nl).

LIFE CYCLE

In the Netherlands pupae and adults have been collected in November and from March to May. Krasheninnikov (2011) collected adult males in the Middle Urals in Russia in the same periods. In northern regions adults have been found in late winter walking on snow beside open water bodies of flowing water (Cranston et al., 1989). Most probably the species has a summer diapause in second instar (see Moller Pillot, 1984). Older larvae have been found in the Netherlands and Denmark from October to May, with maximum numbers in March (Lindegaard-Petersen, 1972; own data). Lindegaard & Mortensen (1988) collected third and fourth instar larvae in another Danish lowland stream also in August.

MICROHABITAT

Several authors collected the larvae from a silty bottom (e.g. Shilova, 1976). Tolkamp (1980) stated a significant over-representation in two lowland streams on a bottom of coarse detritus (often in leaf packs). Verdonschot & Lengkeek (2006) found the larvae relatively often on wood. We collected the larvae in low numbers also from stones. The larvae have been found rarely above the waterline, but this microhabitat has been hardly investigated. Nolte (1991) stated that the larvae in moss cushions on stones in a German brook were restricted to mosses with permanently emersed areas.

WATER TYPE

Current

Adults and exuviae are rarely found near or in lakes (Brundin, 1949; Shilova, 1976; Reiff, 1994). The larvae have sometimes been collected in stagnant pools and ditches (Cranston, 1982; Limnodata.nl), but almost all records are from flowing water, often very slow-flowing springs, ditches or streams. The species is often absent from fast running brooks and streams and rare in mountain regions (e.g. Lehmann, 1971; Ringe, 1974; Laville & Vinçon, 1991; Orendt, 2002a; Stur et al., 2005; own data).

Dimensions

The larvae are most common in narrow brooks, springs and ditches, sometimes not more than 1 m wide (e.g. Verdonschot et al., 1992). However, they can be found (sometimes rather numerously) in larger brooks and streams (Garcia & Laville, 2001; own data). Moller Pillot (2003) collected relatively more larvae in drift samples. Most of the larvae and exuviae found in larger brooks and streams may have been transferred by drift.

Permanence

The species is especially common in temporary upper courses (Verdonschot et al., 1992; own data). It is possible that the larvae need a period with a dry environment in summer (see Life cycle, Microhabitat and Dimensions).

pH

As a rule *Diplocladius cultriger* is absent from acid brooks (Orendt, 1999; Limnodata.nl). However, we sometimes found larvae in brooks with a pH around 6 (e.g. Moller Pillot, 2003). In the Netherlands there are hardly records from water with pH > 8 (Steenbergen, 1993; Limnodata.nl), but Reiff (1994) found the exuviae in a lake in Bavaria at pH 8.3.

SAPROBITY

Most data are from eutrophic, but unpolluted or only very slightly polluted watercourses. Usually the oxygen content (October to April!) is well above 5 mg/l and the

BOD less than 3 mg O_2/l. However, in several cases rather severe organic pollution has been stated and the ammonium-N content can be 2 mg/l (Steenbergen, 1993; Limnodata.nl; own data). Because such records are from lower courses and chemical data of upper courses are scarce, nothing can be said about the demands made during the whole life cycle.

SALINITY

We collected the larvae sometimes from a slightly brackish watercourse in the province of Noord-Brabant. Paasivirta (2000) reported the occurrence at one place in slightly brackish water in the Baltic Sea.

Dratnalia potamophylaxi (Fittkau & Lellák, 1971)

SYSTEMATICS AND IDENTIFICATION

Saether & Halvorsen (1981) separated the new genus *Dratnalia* from *Eukiefferiella*. They described the male and female imago, pupa and larva. Adults and larvae are absent from most keys, but the adults can be identified using Cranston et al. (1989) and the larvae using Cranston et al. (1983) and Bitušik (2000). Many beautiful photos of the larva have been published by Schiffels (2009).

DISTRIBUTION IN EUROPE

The species has been collected in a small number of countries scattered over the whole of Europe (Saether & Spies, 2010). It is not known from the Netherlands, but is to be expected in Limburg.

FEEDING AND MICROHABITAT

D. potamophylaxi larvae live within the stone cases of the Trichoptera genera *Potamophylax*, *Halesus* and *Lasiocephala*. They are fixed to the abdomen or thorax of the host with their claws of the posterior parapods. In many cases the cuticle of the host is damaged. Most probably organic particles, bacteria and ciliated Protozoa are important as food, but possibly in some cases the chironomid feeds on their host's hemolymph (Schiffels, 2009; see also Schnell, 1991).

WATER TYPE

The larvae live in brooks and streams, where the Trichoptera hosts occur.

Epoicocladius ephemerae (Kieffer, 1924)

Epoicocladius flavens Cranston, 1982: 70–71, fig. 26; Moller Pillot, 1984: 69

Beautiful photos of the larva can be found in Schiffels (2009).

DISTRIBUTION IN EUROPE AND THE NETHERLANDS

E. ephemerae is distributed throughout Europe (Saether and Spies, 2010). In the Netherlands
there are many records from the Veluwe, the most eastern part of the provinces of Gelderland and Overijssel and the southeastern part of Noord-Brabant. The species seems to be rare in southern Limburg (Limnodata.nl; own data).

LIFE CYCLE

Emergence has been stated from May to early September, usually in one (sometimes two?) generations a year (Lindegaard-Petersen, 1972; Svensson, 1976, 1979; Tokeshi, 1986b). The last author stated hibernation in second to fourth instar.

FEEDING AND MICROHABITAT

The larvae and pupae live only on the nymphs of *Ephemera* and feed on algae and detritus, supplied by the respiration movements of the nymph (Svensson, 1976; Tokeshi, 1986b). Tolkamp (1980: 144) found a preference for a substrate of sand with coarse and fine detritus, like stated for the host species. Nevertheless, we also found the species on a stonier substrate, when *Ephemera* was present.

DENSITY

Tokeshi (1986b) reported a very stable density of about 1000 larvae/m² in a river in England during the whole year. In the Netherlands we usually found very low densities, probably because of the low density of *Ephemera* larvae.

SAPROBITY

The larvae are found mainly in unpolluted or very slightly polluted brooks and streams. Moog (1995) called the species characteristic of oligosaprobic and β-mesosaprobic water. Because a bottom in which water flows through the sandy substrate is more important for *Ephemera* larvae than a high oxygen content, this possibly also applies to the chironomid. However, the *Epoicocladius* larvae may need more oxygen. In a stretch of the Hierdense Beek in the Netherlands the larvae were absent after some pollution and deposition of organic matter (Cuppen, 2006).

PREDATION

Tokeshi (1986b) suggested that the enhanced chance of survival, resulting principally from the provision of a secure site on hosts of appropriate size all through the year, is the clearest benefit accruing to *E. ephemerae* from the commensalism.

Eukiefferiella Thienemann, 1926

OR

Eukiefferiella Lehmann, 1972: 347–405 (pro parte); Moller Pillot, 1984: 70–79 (pro parte)

SYSTEMATICS

The genus *Eukiefferiella* as used by most authors in the 20th century has been divided by Saether & Halvorsen (1981) into *Tvetenia*, *Dratnalia* and a restricted *Eukiefferiella*. These three genera together with *Tokunagaia* and *Cardiocladius* form the *Cardiocladius* group or Cardiocladiini (Saether, 1979a; Saether & Halvorsen, 1981).
Many groups have been distinguished in this genus (see Lehmann, 1972; Coffman et al., 1986). As yet there is no consensus about this classification (G. Halvorsen pers. comm.).

IDENTIFICATION

Lehmann (1972) contains keys to adult males and pupae, together with extensive descriptions and figures of all the European species. Adult males of most species can also be identified using Langton & Pinder (2007), in which a few European species are absent, e.g. *E. lobifera* and *E. similis*. Exuviae of nearly all European species can be identified using Langton (1991). Older keys to adults or exuviae are less complete.

To distinguish the larvae from those of other genera, especially in second and third instar, it is useful to examine the ventromental plates and the insertion of the setae submenti (see fig. 7). The keys to the species by Cranston (1982) and Moller Pillot (1984) are incomplete. The best key to the larvae is that of Schmid (1993), but some species are still unknown as larva. Schmid gives also descriptions and good figures.

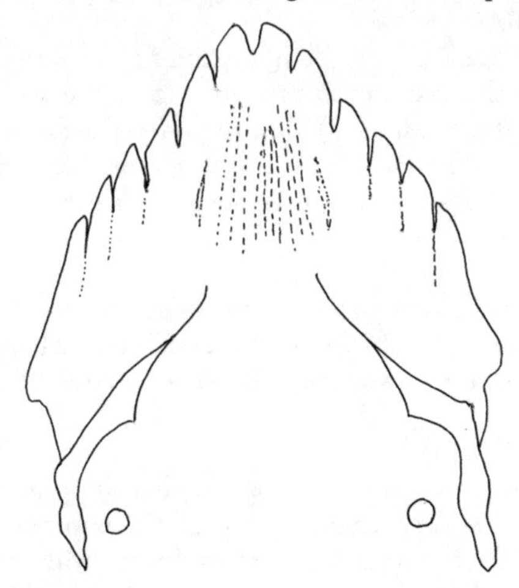

7 Eukiefferiella claripennis: mentum with ventromental plates and setae submenti

LIFE CYCLE

In the western European lowland fourth instar larvae of *Eukiefferiella* are usually found most numerously in spring. In this period the main food organisms (especially diatoms) are most numerous (see e.g. Brennan et al., 1978; Tokeshi, 1986a; Berg & Hellenthal, 1992). Storey (1987) showed that larvae of *E. ilkleyensis* survived better on a 'spring diet' than on a 'winter diet' (both consisting of epiphytic algae and bacteria collected in these seasons from aquatic plants).
Cranston et al. (1989) supposed that most species are univoltine. However, most species have several generations a year, but in many cases summer and autumn generations are small (e.g. Singh & Harrison, 1984, Tokeshi, 1986; own data). Possibly the larvae have a diapause in autumn or early winter, but also a dormancy period in the egg stage in summer (under influence of high temperatures?) seems to be possible. Deviating patterns of emergence are not rare, even within one species (Pinder, 1980; Nolte, 1991; Reiff, 1994). Hannesdóttir et al. (2012) stated more generations in brooks with higher water temperatures (see under *E. claripennis* and *E. minor*).
Development time is dependent on temperature and food conditions (Storey, 1987). The shorter developmental period given for *E. claripennis* in comparison with some other species of the genus (table 4 on p. 280) is probably caused by the usually higher temperatures in brooks with lower current velocities (cf. Rossaro, 1991).
The available data indicate that young larvae live in the same microhabitat as older larvae (Storey, 1987; own data).

FEEDING

From observations of larvae of different species, the main food of *Eukiefferiella* larvae consists of 'Aufwuchs': the flora of all types of algae, bacteria and fungi and associated detritus adhering to the substrate (Tokeshi, 1986a; Storey, 1987; own data). Storey mentions (based on Williams, 1981) that first instar larvae of *E. ilkleyensis* select diatoms from the environment. Survival of larvae was better on the 'spring diet' rich in diatoms than on the 'winter diet' containing relatively more detritus. Diatoms have a higher nutritional value (C:N ratio) than detritus and are possibly more easily ingested because of their size. The role of macrophyte tissue in the food is still unclear (see under *E. ilkleyensis*).

Tokeshi (1986a) did not find food selection in older larvae of *E. ilkleyensis*. We also observed no food selection in older larvae of *E. claripennis*; probably there are differences between different instars and between species. Differences between inhabitants of small upper courses and large lower reaches of streams are also probable (Tokeshi & Pinder, 1985).

In some cases we saw larvae of *E. claripennis* feeding on a dead pupa or living young larvae of other chironomids, but this behaviour is probably quite exceptional.

MICROHABITAT

The larvae are common on stones, but scarcely on smooth stones. Most larvae are found among mosses or algae or on stones and wood with silt crusts (e.g. Zavřel, 1939: 25). They move freely or live in tubes, which are often leaved. On soft sediments the larvae are almost always very scarce (Pinder, 1980; own data), but Holzer (1980) collected them regularly (in low numbers) up to 30 cm deep in the sediment. Many species also live on plants, e.g. *Potamogeton* and *Ranunculus* (Tokeshi & Pinder, 1985; Storey, 1987) and grasses (own data).

Because of their microhabitat choice, the places where the larvae live often have a higher current velocity and a higher oxygen content than those of bottom dwellers.

WATER TYPE

Current

The most typical habitat for *Eukiefferiella* is the mountain brook, where a number of species of the genus can often be found (Thienemann, 1936: 50). In lowland brooks the larvae are significantly more abundant in riffles than in pools (Janzen, 2003). They are relatively scarce in springs, and in the western European lowland they are absent from lakes (Cranston, 1982). In northern and alpine regions their occurrence in the littoral zone of lakes has been stated by e.g. Brundin (1949), Reiss (1968), Pankratova (1970: 27, 160) and Reiff (1994). In Bavarian lakes Reiff (1994) found no less than eight species, but most of them very scarcely. In the Netherlands there are only records from running water.

Dimensions

The larvae are less common in large rivers (e.g. Caspers, 1980, 1991; Becker, 1995). Here the larvae seem to be even more dependent on the presence of stones or other firm substrate on places with fast current. Klink (2011) found the genus on trees in the river Meuse.

Shade

The genus seems to be scarce in closed woodland without clearings (own data; see also Caspers, 1980). Possibly the adults do not swarm in woods or the amount of food is limited. More detailed studies are needed to ascertain this.

Eukiefferiella ancyla Svensson, 1986

IDENTIFICATION

Eukiefferiella ancyla is absent from the key by Pinder (1978). Adult males can be identified using Langton & Pinder (2007) and the pupa using Langton (1991). The larva is absent from most keys; it has been described by Svensson (1986) (see also the colour photographs in Schiffels, 2009).

DISTRIBUTION IN EUROPE

The species is only known from a number of countries in western Europe (Saether & Spies, 2010), but most probably it will also be found in other regions. Until now there are no records from the Netherlands.

MICROHABITAT

The larvae live only in silken tubes in the mantle cavity of the limpet *Ancylus fluviatilis* (Svensson, 1986; Schiffels, 2009).

WATER TYPE

The species has been found only in more or less fast-flowing streams (Svensson, 1986; Michiels, 2004; Schiffels, 2009).

Eukiefferiella brehmi Gouin, 1943

IDENTIFICATION

Adults and exuviae can be identified using Lehmann (1972) and Langton (1991) respectively. The larvae are unknown.

DISTRIBUTION IN EUROPE

E. brehmi has been found in streams in several countries in different parts of Europe, e.g. from Germany (Saether & Spies, 2010). There are no records from lowland streams. The species is not treated here.

Eukiefferiella brevicalcar Kieffer, 1911

IDENTIFICATION

Identification of adult males does not present special problems when using the usual keys. The related *E. tirolensis* is absent from the key by Pinder (1978), but this species cannot be expected in lowland streams. Distinguishing the exuviae of *E. brevicalcar* and *E. lobifera* using Lehmann (1972) or Langton (1991) can be more difficult than the keys suggest. The larva is mentioned in Moller Pillot (1984) as *brevicalcar* agg. However, the species *E. lobifera* has a mentum with two median teeth (if not worn too much) and does not resemble *E. brevicalcar*. In lowland streams *E. brevicalcar* is the only species of this aggregate (see Schmid, 1993: 90).

DISTRIBUTION IN EUROPE AND THE NETHERLANDS

The species is known from nearly the whole European mainland and from the British Isles (Saether & Spies, 2010). In the Netherlands it seems to be restricted to the most eastern part of the country, the Veluwe and the large rivers (Moller Pillot & Buskens, 1990). *E. brevicalcar* is very scarce in the province of Noord-Brabant (GWL, unpublished; own data).

LIFE CYCLE

Data in Lindegaard & Mortensen (1988) suggest three generations a year, emerging from April to October. The second and third generations were much smaller than the first generation. Singh & Harrison (1984) obtained similar results from a southern Ontario stream. In Austria Caspers (1983) collected adults from May until November. In the Netherlands some larvae were in fourth instar in December and one pupa was found in that month. Emergences in March are scarce (own data).

Most probably there is a diapause in early winter as young larvae, but Lindegaard & Mortensen do not exclude delayed hatching of eggs during winter.

FEEDING

We examined the gut contents of some larvae and found mainly detritus and diatoms. Feeding seems to be the same as in other species of the genus (see under the genus).

MICROHABITAT

The larvae live among mosses on stones and on plants like *Sparganium*, mainly near the water surface and sometimes also hygropetric (Lindegaard-Petersen, 1972; Nolte, 1989; own data). In these microhabitats they can endure a relatively strong current. They can be abundant also in the upper layers of the sediment of brooks and streams (Schmid, 1993).

WATER TYPE

The larvae have been collected only in running water, especially with a current velocity between 0.5 and 1.0 m/s. They are most common in small brooks and streams 1–10 m wide; in large rivers the larvae are rather rare (Lehmann, 1971; Becker, 1995; own data). The species is not rare in springs and can appear incidentally in temporary springs and upper courses (Lindegaard, 1995; own data).

Rossaro (1991) collected *E. brevicalcar* (in 11 samples) at a very low mean temperature (6.9 °C).

pH

Orendt (1999) collected *E. brevicalcar* in streams where the pH sometimes dropped to about 5.5 in the course of the year. Hawczak et al. (2009) collected the larvae rather numerously at pH 5.73 in Oakers Stream, a small brook in southern England. We have hardly data about its presence in acid water. Most of our data are from streams with a pH of 7 or more.

TROPHIC CONDITIONS AND SAPROBITY

We collected *E. brevicalcar* almost exclusively in very clean, mesotrophic to eutrophic water. Moog (1995) called the species characteristic for oligosaprobic to β-mesosaprobic conditions. Izvekova et al. (1996) confirm this, but state that the larvae can survive among the vegetation of more polluted streams with high oxygen content.

E. claripennis

Eukiefferiella hospita Brundin, 1949: 702; Pankratova, 1970: 27, 159–160, fig. 90

IDENTIFICATION

For the identification of adult males and exuviae, see under the genus. Larvae of the group *claripennis* sensu Lehmann (1972) cannot be identified correctly using Cranston (1982) or Moller Pillot (1984). The differences between *E. lobifera*, *E. fuldensis* and *E.*

claripennis are given by Schmid (1993). *E. alpestris* is a nomen nudum (see Lehmann, 1972: 348). The larvae of some species are still unknown.

DISTRIBUTION IN EUROPE AND THE NETHERLANDS

The species is known from nearly the whole European mainland, the British Isles and Iceland (Saether & Spies, 2010). In the Netherlands it is common in the southeastern and eastern part of the country, but scarce in Drenthe. There are also some records from the large rivers and the dune region. In other parts of the Holocene region the species is very rare (e.g. Klink & Mulder, 1992).

LIFE CYCLE

The life cycle of *E. claripennis* seems to be different in different circumstances. In western Europe there are often three generations a year, the largest numbers of fourth instar larvae occurring between February and June (Singh & Harrison, 1984; Pinder, 1980; Prat et al., 1983; own data). In the British Isles the first generation often emerges in winter (Fahy, 1973; Wilson, 1977; Drake, 1982, 1985). Hannesdóttir et al. (2012) stated one generation a year in a cold stream (7 °C) and two in a warm stream (23 °C) in Iceland.

Diapause, dormancy

In some cases (especially in brooks with very low discharge or low oxygen content in summer) the larvae are totally absent in summer. Possibly the eggs can go into dormancy, as has been stated in the genus *Orthocladius*. In other cases we found low numbers of larvae and pupae during summer and autumn. In early winter most larvae seemed to be in **diapause** in second or third instar. Madder et al. (1977) found large numbers of *E. claripennis* in cocoons in the Harris River (Canada) as early as September. The ventral side of the cocoons was completely attached to the shales on the bottom of a riffle. Elsewhere the larvae in winter are free living, without cocoons.

OVIPOSITION

Nolte (1993) found an egg mass of a species of gr. *claripennis* among moss at the water's edge. This egg mass contained 85 eggs, arranged in one line which was folded into loops.

FEEDING

See under the genus.

MICROHABITAT

The highest densities of larvae are stated on rough stones and on plants, in lower numbers on gravel and only very scarcely on sandy or silty bottoms (Pinder, 1980, 1987; own data). Millet et al. (1987) collected the species in a stream transect of a Spanish river most numerously at the edge of the vegetation in the middle of the stream. Nolte (1989) found the larvae numerously in moss cushions on stones, mainly near and above the water surface in the upper spray zone. In streams near Winterswijk (the Netherlands) Tolkamp (1980) stated the highest densities on coarse detritus and dead leaves, probably owing to the absence of plants and stones.

DENSITIES

On stones in riffles in southern France Maasri et al. (2008) collected more than 140 larvae/100 cm^2 stone area. In lowland streams with current velocities lower than 50 cm/s we usually found very low densities of larvae (maximum less than 10 larvae/100 cm^2 stone area). On stones in a polluted fast-flowing stream densities were higher, locally more than 20 larvae/100 cm^2 . On grasses the densities were sometimes nearly the same. Nolte (1989) stated more than 1000 larvae/100 cm^2 bottom area in moss cushions in an unpolluted small stream in Germany, but this density is much less per cm^2 leaf and stem area and it may apply to another species.

WATER TYPE

Current

In montane areas the larvae live mainly in the lower parts of the **streams** (Laville & Vinçon, 1991) and they are scarce at high current velocities (e.g. Michiels, 1999, 2004). The species can also be found scarcely in lowland streams with current velocities less than 50 cm/s. In such streams the larvae are most common in riffles or at sites where the current is a little faster. Moreover, the larvae live mainly where the current velocity is higher than the mean velocity of the stream. In springs where the current is very slow, but permanent, the larvae can live in very low densities (e.g. Lindegaard et al., 1975).

The occurrence of *E. claripennis* in **lakes** seems to be restricted to the boreal and subalpine zone. Brundin (1949) collected adults at several lakes in Sweden, locally in large numbers. Pankratova (1970: 27) found large numbers of larvae in a lake in Karelia. Reiff (1994) collected the exuviae of two specimens in a lake in Bavaria.

Dimensions

E. claripennis can live in very small brooks less than 0.5 m wide, but is more common in wider brooks and streams, also because of current velocity (see above). However, in larger streams the densities diminish gradually (e.g. Morris & Brooker, 1980; Pinder, 1980; see also Schmid, 1993). The larvae are rather scarce in the larger headwaters of the river Seine (Klink, 2010) and very scarce in the Rhine and Meuse (Klink & Moller Pillot, 1982; Smit, 1982; Becker, 1995).

In stagnant water the species is confined to lakes. There are no records from small water bodies such as pools or stagnant ditches.

Temperature

Rossaro (1991) collected this species (in 27 samples) at relatively low mean water temperature (10.6 °C). This does not apply to the Netherlands, where *E. claripennis* is almost the only species of the genus living in slow-flowing lowland brooks. Although the larvae are found here mainly in winter and spring, they can also be present in summer, when temperature goes up to more than 20 °C, if the oxygen content is high enough.

Hannesdóttir et al. (2012) found *E. claripennis* to be the dominant species in cold streams and in warm streams (up to 23 °C).

Permanence

There are very few records from temporary upper courses.

pH

Simpson (1983) collected larvae of the *claripennis* group in goodish numbers in brooks with a pH between 4.5 and 5 in New York state. The species *E. claripennis* was found in a very small number (2 exuviae) in a small stream at a pH of approx. 4 (Wilson, 1988). Orendt (1999) did not collect larvae of the *claripennis* group in brooks with pHs lower than 6. The absence in most acid upper courses in the Netherlands is also caused by slow current and other factors.

TROPHIC CONDITIONS AND SAPROBITY

From many **investigations outside the Netherlands** it appears that *E. claripennis* can live in oligotrophic to more or less polysaprobic conditions. Brundin (1949) caught the species at the oligotrophic lake Innaren in Sweden. On the other hand, the larvae appear to be most numerous in very polluted streams (Zavřel, 1939: 26; Bazerque et al., 1989; Casas & Vilchez-Quero, 1989; Maasri et al., 2008). We should mention here that the larvae are found hardly on the bottom of streams and live on plants and stones, where pollution has less impact than on the bottom (see Microhabitat). Davies & Hawkes (1981), who took samples from the stony stream bed of the river Cole in England, collected no more larvae in summer at stations where the (measured) oxygen content dropped below 2 mg/l. Maasri et al. (2008) suggest that the presence of many diatoms and other algae in enriched streams may be the most important factor allowing this species to reach high densities.

Without doubt more factors are important in polluted environments: e.g. in six rivers in England Wilson (1989) stated no higher numbersof pupal exuviae at places below sewage effluent discharges.

In the Netherlands Peters et al. (1988) called the species characteristic of very polluted lowland streams (*Chironomus* zone). However, in lowland streams the presence of the larvae is almost totally confined to winter and spring, when current velocity and oxygen content is not very low. From a large number of data the species appears to be confined to brooks and streams more or less rich in oxygen (see Moller Pillot & Buskens, 1990).

The numbers of larvae are often much higher in fast-flowing streams in southern Limburg than in lowland streams. Cuijpers & Damoiseaux (1981) found the larvae most numerously in the most polluted lower part of the rather fast-flowing river Geul (BOD5 4–23). We found larvae and pupae of this species more abundantly still in the Jeker near Maastricht, where this stream was severely polluted (BOD5 2–11, oxygen saturation 65–85%). See also under Densities.

The data on the influence of organic pollution and oxygen content are **summarised** in table 5 in Chapter 9.

Eukiefferiella clypeata (Kieffer, 1922)

SYSTEMATICS AND IDENTIFICATION

E. clypeata is related to the *claripennis* group. What Moller Pillot (1984) called *clypeata* agg. is only one species. Identification of adult males, exuviae and larvae (see under the genus) does not present much difficulty. However, worn central mental teeth of other species are a little wider and resemble the mentum of *clypeata*. Brennan & Mclachlan (1980) stated that the larvae are very distinctive because of a very dark head and purple fat bodies running the length of the larva.

DISTRIBUTION IN EUROPE AND THE NETHERLANDS

The species has been collected in many countries across Europe (Saether & Spies, 2010). In the Netherlands *E. clypeata* is rather common in streams in the central part of Limburg (own data); exuviae have been collected rarely in the Rhine (Klink & Moller Pillot, 1982). Other records have to be verified.

LIFE CYCLE

The life cycle of *E. clypeata* is not known exactly. Brennan & McLachlan (1980) suppose that in Northumberland (England) the species is bivoltine, emerging in spring and autumn. Exuviae and adults have been collected from April to September

(Lehmann, 1972; Klink & Moller Pillot, 1982; Wilson, 1988; own data). Pinder (1980) collected relatively many larvae in September, fewer in July and least of all in May. As stated for *E. ilkleyensis*, the life cycle might be different in different circumstances.

MICROHABITAT

Pinder (1980, 1987) found the larvae more on gravel than on plants and not on soft sediments. We collected them from stones. According to Potthast (1914: 293–294), the larvae live freely or in loose tubes, the pupae under a smooth shield-shaped cocoon, attached to the stone. Brennan & McLachlan (1980) observed that the larvae live under flat protective sheets of salivary secretion over cracks or depressions on the upper surfaces of stones.

DENSITIES

Pinder (1980) stated up to more than 2000 larvae/m^2 on gravel in September.

WATER TYPE

Current

Most authors call *E. clypeata* a species of fast-flowing streams (e.g. Pankratova, 1970; Kownacka & Kownacki, 1972; Gendron & Laville, 1997, 2000). This is consistent with the fact that the species is almost absent from the Netherlands and not rare in fast-flowing streams in adjacent parts of Germany. However, Zavřel (1939: 26) collected the larvae in a stream with a very slow current. According to Moog (1995), the larvae are sometimes found in the littoral zone of lakes.

Dimensions

The larvae are very scarce in large rivers (see Klink & Moller Pillot, 1982; Wilson & Wilson, 1984; Caspers, 1991; Schmid, 1993; Becker, 1995). In medium-sized rivers their presence seems to depend on the availability of suitable substrate (Morris & Brooker, 1980; Pinder, 1980; Klink, 2010). Most records are from streams from 5 to 25 m wide.

Temperature

Rossaro (1991) collected *E. clypeata* at higher mean water temperatures than all other species of the genus (16.5 °C). Brabec (2000) also found this species more numerously in the lower reaches of the river Srvatka in the Czech Republic, where the temperature was higher (annual mean 10.3 °). As mentioned elsewhere (section 2.10), these results are typical of mountainous landscapes; the distribution of this species is probably not heavily influenced by temperature.

OR

TROPHIC CONDITIONS AND SAPROBITY

Several authors stated the occurrence of this species in slightly to more or less heavily polluted streams (e.g. Laville & Viaud-Chauvet, 1985; Pinder & Farr, 1987). Zavřel (1939) even collected larvae in a stream heavily polluted by a leather factory. Conversely, the species can be common in very clean water (Morris & Brooker, 1980; own data). From these data it is not possible to quantify the relation between occurrence and saprobity, but it seems to lie between that of *E. brevicalcar* and *E. claripennis* (see table 5 in Chapter 9), i.e. the optimum is probably β- to α-mesosaprobic conditions. Moog (1995) gives oligosaprobic to β-mesosaprobic water as optimum quality.

TRACE METALS

Wilson (1988) collected the exuviae in goodish numbers in a stream heavily polluted with zinc (1–2 mg/l).

Eukiefferiella coerulescens Kieffer, 1926

IDENTIFICATION

The identification of adult males and exuviae do not present any problems. Brundin (1956) gives a short description and a figure of the hypopygium of the male.
The larva cannot be identified using Cranston (1982) or Moller Pillot (1984). Schmid (1993) gives a description and figures of the larva. It differs from the larva of *E. claripennis* because the setal mark lies proximal to the midpoint of the antenna and because the head is much smaller: in *claripennis* (and *E. lobifera*) > 300 μm long; in *coerulescens* (in our material) up to 230 μm. *E. coerulescens* cannot be distinguished from third instar larvae of *E. claripennis* because the setal mark of these larvae also lies more proximally.
The differences between *E. coerulescens* and *E. fuldensis* are small: see Schmid (1993).

DISTRIBUTION IN EUROPE AND THE NETHERLANDS

The species lives in nearly the whole of Europe (Saether & Spies. 2004). In the Netherlands the species has been collected in the Grensmaas (Meuse in southern Limburg) (Klink, 1985) and in some small brooks in southern Limburg (unpublished own data).

LIFE CYCLE

There are three generations a year (Lehmann, 1971; Reiff, 1994). The summer generation (in June) in the river Fulda was hardly developed and in the Bavarian lakes very well developed. In Austria Caspers (1983) collected adults from April to November and one male in December. We found the exuviae as early as March. Larvae of all instars can be found in winter.

MICROHABITAT

The larvae live on stones in fast running water (Reiss, 1968; Lehmann, 1971); however, they can also be found among mosses, on plants and on wood (Lehmann, 1971; own data).

WATER TYPE

Several authors stated the occurrence of *E. coerulescens* mainly in the upper courses of fast-flowing streams (Lehmann, 1971; Moog, 1995; Gendron & Laville, 2000; Punti et al., 2009). The species has been collected rarely in larger streams (Wilson & Wilson, 1984; Klink, 1984; Klink 2010), but seems to be not very scarce in upper courses of small streams in the hilly landscape of southern Limburg in the Netherlands and adjacent parts of Germany (own data). Moreover exuviae and adults have been collected from a number of lakes in England and Bavaria (Reiss, 1968; Reiff, 1994; Ruse, 2002).
The larvae seem not to be confined to stretches with a fast current, because Schmid (1993) collected them also in mosses in crenal areas. Neither can their distribution be termed cold-stenothermic (Caspers & Reiss, 1987; Gendron & Laville, 2000), because there are several records from lakes and less cold streams (see above). The species is absent from typical lowland streams.
As we have stressed in section 2.10 it is necessary to review all factors together, especially current, temperature, saprobity and oxygen content. A species can live in places where one factor seems to be unfavourable as long as one or more other factors are optimal.

TROPHIC CONDITIONS AND SAPROBITY

The larvae are collected mainly in clean upper courses of streams, even in xenosaprobic conditions (Schmid, 1993; Moog, 1995; Gendron & Laville, 2000). Ruse (2002) found the species in lakes with low conductivity. However, there are several records in more or less polluted water with high oxygen content (Izvekova et al., 1996; Maasri, 2008; own data).

Eukiefferiella cyanea Thienemann, 1936

E. cyanea is only known from mountain brooks and streams (Lehmann, 1972; Kownacka & Kownacki, 1972) and is not treated in this book.

Eukiefferiella devonica Lehmann, 1972

SYSTEMATICS AND IDENTIFICATION

E. devonica and *E. ilkleyensis* together form the *devonica* group (Lehmann, 1972: 397). Both species are very similar and it is not always possible to distinguish between these species as adult male, because the only distinguishing character, the antennal ratio, seems to show some overlap. The pupae can be identified using Lehmann (1972) and Langton (1991). However, as Langton wrote, the row of hooks at the posterior edge of sternite VII in *E. devonica* can be absent and it is not yet clear if the median antepronotal setae provide a reliable discriminating character.

According to Schmid (1993), the larvae of *E. devonica* can probably be identified by their shorter head length. Larvae with a slightly paler mentum and head capsule, in Cranston (1982: 74) named *Eukiefferiella* species B, probably belong to *E. devonica*.

DISTRIBUTION IN EUROPE

The species has been recorded in a number of countries across Europe (Saether & Spies, 2010). Near Aachen, not very far from the Dutch border, the species is rather common (own data). There are no records from the Netherlands.

LIFE CYCLE

Caspers (1983) stated emergence in Austria from (April) May until October (November), with the largest numbers in autumn. Ringe (1974) found emergence in central Germany during the same period, probably in three generations, with maximum numbers in July.

MICROHABITAT

The larvae have been found among mosses and on stones in fast-flowing streams (Lehmann, 1971; own data).

WATER TYPE

E. devonica is a common species in the upper courses of alpine streams and streams of the secondary mountain chain (Lehmann, 1971, 1972; Wilson & Wilson, 1984; Schmid, 1993; Michiels, 1999; Orendt, 2002a). According to Moog (1995), the larvae can be found more rarely in the lower courses of these streams (metapotamal). However, Caspers (1991) collected the species in the Middle Rhine.

The species was almost absent from the Rohrwiesenbach in central Germany, but very numerous in the colder and faster flowing Breitenbach (Ringe, 1974). Rossaro (1991) collected the exuviae in northern Italy only in very cold water (mean temperature 7.0 °C). Reiff (1994) found the exuviae in two lakes in Bavaria.

TROPHIC CONDITIONS AND SAPROBITY

According to Moog (1995), the species is almost confined to oligosaprobic and β-mesosaprobic water. Our own data suggest that *E. devonica* is a little more tolerant of organic pollution, but this is based on few exact data. The larvae need a high oxygen content.

Eukiefferiella dittmari Lehmann, 1972

SYSTEMATICS AND IDENTIFICATION

According to Lehmann (1972), *E. dittmari* cannot be classified easily within the existing groups of the genus. Halvorsen (pers. comm.) incorporates the species in the *devonica* group. Adult males can be identified using Lehmann (1972) and Langton & Pinder (2007); these keys use different characters to arrive at this species. Identification of exuviae is most reliable using Langton (1991). The larva is unknown.

DISTRIBUTION IN EUROPE AND THE NETHERLANDS

The species lives throughout Europe, but has been collected only in relatively few countries (Saether & Spies, 2010). In Finland it is one of the most common and widely distributed species (Paasivirta, 2012). In the Netherlands there is only one record: exuviae found in the river Roer in Limburg (unpublished own data).

WATER TYPE

E. dittmari has been collected in a number of rather fast-flowing streams in Germany (Lehmann, 1972; Orendt 2002a). In the lower course of the river Fulda the larvae lived among mosses.

Eukiefferiella fittkaui Lehmann, 1972

This species lives only in mountain springs and streams (Lehmann, 1972; Laville & Lavandier, 1977; Caspers, 1983; Braukmann, 1984: 211; Rossaro, 1991; Laville & Vinçon, 1991; Gendron & Laville, 2000). The larva has been described and figured by Schmid (1993). Pupae and larvae cannot be distinguished from those of *E. minor* (Langton, 1991; Schmid, 1993).

Eukiefferiella fuldensis Lehmann, 1972

SYSTEMATICS AND IDENTIFICATION

Schmid (1993: 102) writes that *E. fuldensis* is most probably synonymous with *E. alpestris* in Zavřel (1939), Pankratova (1970) and Moller Pillot (1984). Adult males can be identified using Lehmann (1972), and exuviae using Lehmann (1972) or Langton (1991). The larva has been keyed and described by Schmid (1993). The differences between the larvae of *E. fuldensis* and *E. coerulescens* are small.

DISTRIBUTION IN EUROPE

E. fuldensis has been recorded in a number of countries in western and central Europe. The distribution is still insufficiently known (Saether & Spies, 2010).

WATER TYPE

The species has been collected in a number of fast-flowing streams (Lehmann, 1972; Laville & Vinçon, 1991; Rossaro, 1991; Schmid, 1993; Orendt, 2002a), usually identified from exuviae. According to Laville & Vinçon (1991) the species does not occur in the potamal zone.

Eukiefferiella gracei Edwards, 1929

Eukiefferiella longicalcar Potthast, 1914: 290–291; Zavřel, 1939: 1–29; Pankratova, 1970: 28, 152–154, fig. 83
Eukiefferiella potthasti Lehmann, 1972: 376–378

SYSTEMATICS AND IDENTIFICATION

The nomenclature of *E. gracei* has been solved by Cranston (1974). Together with *E. minor*, *E. similis* and *E. fittkaui* this species forms the *gracei* group, which can be recognised as larvae very easily because of the wide central mental tooth and five lateral teeth. Also, the setae on the abdomen are more or less characteristic (Moller Pillot, 1984). For identification and description of the larva see Schmid (1993).
The hypopygium of the male is characteristic (Lehmann, 1972; Langton & Pinder, 2007). Identification of the pupa is less easy (see Lehmann, 1972; Langton, 1991).

DISTRIBUTION IN EUROPE AND THE NETHERLANDS

The species lives in almost the whole of Europe, and is possibly less common in northern countries (Saether & Spies. 2004; Paasivirta, 2012). In the Netherlands the species has been recorded in several brooks and streams in the outermost eastern part of the country (Twente) and in southern Limburg.

LIFE CYCLE

Drake (1985) stated emergence in southern England only in winter and spring (November to December and March to June). Most larvae are collected in May–June and in winter (Pinder, 1980; Grzybkowska & Witczak, 1990; own data). However, Krasheninnikov (2011) collected the adults in the Middle Urals in Russia from May until the end of October.

MICROHABITAT

The larvae inhabit mainly stones and gravel overgrown by mosses and filamentous algae, and are found scarcely on plants (Potthast, 1914; Pinder, 1980; Grzybkowska & Witczak, 1990; own data). They build loose tubes of the algal filaments.

DENSITIES

Pinder (1980) stated densities of 2000 larvae/m^2 .

WATER TYPE

E. gracei has been collected mainly in brooks and streams in secondary mountain chains, also in rather fast-flowing water (see Lehmann, 1972; Moog, 1995). Ertlová (1970) found it even in very fast-flowing water in de Danube. Gendron & Laville (1997, 2000) call it a plain species, living in the lower courses of streams in the Pyrenees. These authors and Grzybkowska & Witczak (1990) stated a sharp decrease in numbers after spates in summer, autumn and winter. We found larvae and exuviae in a small number of lowland brooks in western Germany and Twente (the Netherlands); the currents in the Dutch brooks were relatively slow. The species is absent

from Dutch lowland brooks elsewhere in the country, probably under the influence of other factors than current velocity. Nevertheless, their preference for faster currents is clear (see table 7 in Chapter 9).
Klink (2010) collected the exuviae in low numbers in four of the larger headwaters of the river Seine, but the species has been found rarely in large rivers (e.g. Caspers, 1991). In lakes the exuviae have been collected rarely (Langton, 1991; Reiff, 1994).

TROPHIC CONDITIONS AND SAPROBITY

The larvae are collected mainly in more or less pure upper courses of streams. We found them also in very eutrophic conditions. Izvekova et al. (1996) stated their presence also in slightly more polluted water between plants and algae, when the oxygen content was high enough.

Eukiefferiella ilkleyensis Edwards, 1929

Eukiefferiella quadridentata Chernovskij, 1949: 126 pro parte
Eukiefferiella lutethorax Ertlová, 1971: 139–142

SYSTEMATICS AND IDENTIFICATION

E. ilkleyensis belongs to the *devonica* group (Lehmann, 1972). The adult males of *E. ilkleyensis* and *E. devonica* can be distinguished only by their antennal ratio (Langton & Pinder, 2007). Lehmann (1972: 272) supposed that there is some overlap between these species. Identification of the exuviae can also present problems (Langton, 1991: note on p. 124). The same can be said about the larvae (Schmid, 1993: 90, 92).

DISTRIBUTION IN EUROPE AND THE NETHERLANDS

The species has been recorded in almost the whole of Europe (Saether & Spies, 2010). In the Netherlands *E. ilkleyensis* is common in southern Limburg and very rare in Noord-Brabant and the eastern provinces, but absent from other parts of the country (Limnodata.nl).

LIFE CYCLE

The life cycle of *E. ilkeyensis* seems to be different in different streams. Most authors found by far the largest numbers in late spring (Pinder, 1980; Tokeshi, 1986a), but Drake (1982) stated emergence in southern England only in February and March. Possibly the life cycle can be modified by the quality of food, as suggested by Storey (1987) and demonstrated for *Paratendipes albimanus* by Ward & Cummins (1979). Other environmental factors can also play a role in suppressing of the numbers in summer (Vallenduuk & Moller Pillot, 2007: 16–18).
As a rule there seem to be three or more generations a year, but the data also indicate the possibility of a dormancy period in summer and autumn (see under the genus). More probable is a diapause in autumn, as in many other Orthocladiinae. Tokeshi (1986) found only very young larvae in December and January, hardly or not growing until the end of February. In southern Limburg many larvae were already in third instar at the end of January (own data).

FEEDING

As treated in more detail in the introduction to the genus, the food of *E. ilkleyensis* consists mainly of diatoms and miscellaneous fine particulate organic matter. The larvae are free-ranging scrapers (Tokeshi & Pinder, 1985). Most probably the presence of a large population of larvae in spring is based on the large quantity of diatoms in this period (Storey, 1987). Williams (1981) stated that third and fourth instar larvae in spring fed to a greater extent on macrophyte tissue.

MICROHABITAT

Pinder (1980) found high numbers of larvae on plants, rather few on gravel and no larvae on soft sediments. We collected them on plants as well as on stones and gravel, and in one case among mosses under a weir.
Brennan & McLachlan (1980) found the larvae in large numbers in the leaf terminals of the moss *Eurhynchium riparoides*. Tokeshi & Pinder (1985) stated a significant trend of decreasing density from the apex towards the base of the stem of two *Potamogeton* species. According to Tokeshi (1986), the larvae do not construct tubes but forage freely on the leaf surfaces.

WATER TYPE

Current
Ertlová (1971) collected the larvae from artificial substrata in the river Danube at a current of 170–250 cm/s; Tokeshi & Pinder found them numerously in a small stream with a current velocity no more than 50 cm/s. Moog (1995) and Gendron & Laville (2000) regard *E. ilkleyensis* as a euryzonal species. However, in general the records apply to rather fast current velocities and in the Netherlands the species is almost absent from brooks and streams with velocities lower than 60 cm/s (see table 7).
On one occasion Reiff (1994) collected the exuviae in a lake in Bavaria.

Dimensions
The larvae live in water courses ranging in size from narrow brooks (1 or 2 m wide) to small rivers (more than 25 m wide) (Tokeshi & Pinder, 1985; Gendron & Laville, 2000; Orendt, 2002a; Klink, 2010; own data), but they are rare in large rivers (Ertlová, 1971; Klink & Moller Pillot, 1982; Caspers, 1991; Becker, 1995).

TROPHIC CONDITIONS AND SAPROBITY

The larvae seem to be absent from heavily polluted water (Moog, 1995; own data). However, we did not find the species in brooks without any enrichment, and most of the records in literature apply to more enriched watercourses, such as the river Frome (Pinder & Farr, 1987). Bazzanti & Bambacigno (1987) collected the species more numerously at the most polluted station in the upper course of the river Mignone (Italy). These data do not permit us to give a complete picture of the tolerance of the species.

Eukiefferiella lobifera Goetghebuer, 1934

nec *Eukiefferiella lobifera* Zavřel, 1939: 7–8, 16, fig. 3–4; Pankratova, 1970: 159; Moller Pillot, 1984: 78–79

IDENTIFICATION

The adult male is absent from the key by Langton & Pinder (2007), but can be identified using Lehmann (1972). Identification of the exuviae using Langton (1991) can present problems. The larvae are hardly different from those of *E. claripennis* and some other species (see Schmid, 1993: 91). According to Schmid, the median teeth of the mentum are commonly worn, which may be why Zavřel (1939) saw only one median tooth.

DISTRIBUTION IN EUROPE

E. lobifera has been found in many countries across Europe, but seems to be absent from Scandinavia and the British Isles (Saether & Spies, 2010). There are no reliable records from the Netherlands.

LIFE CYCLE

Reiff (1994) stated three generations a year in the Chiemsee in Bavaria: she collected very few exuviae at the end of April, very many from the end of June to the middle of July and again very few in October.

MICROHABITAT AND DENSITY

Very little is known about the habitat preference of the larvae. Fesl (2002) collected the larvae in large numbers on the bottom of the river Danube near Vienna (mean density 268 larvae/m^2).

WATER TYPE

The species has been recorded from many, mainly submontane, fast-flowing streams (Lehmann, 1972; Rossaro, 1991; Gendron & Laville, 2000; Orendt, 2002a). According to Syrovátka et al. (2009), the species is characteristic of runs and riffles. Fesl (2002) found the larvae numerously in the Danube near Vienna and Becker (1995) caught some males along the Rhine. In the Garonne the species is not very scarce (Garcia & Laville, 2001). Reiff (1994) collected the exuviae numerously in the Chiemsee in Bavaria.
According to Gendron & Laville (1997), *E. lobifera* is a foothill species, living in higher zones only when water temperature there is higher than normal.

Eukiefferiella minor Edwards, 1929

IDENTIFICATION

Exuviae and larvae cannot be distinguished from those of *E. fittkaui* (Langton, 1991; Schmid, 1993). Brundin (1956) gives a short description and a figure of the hypopygium of the male. See further under the genus.

DISTRIBUTION IN EUROPE

The species has been recorded in many countries across almost the whole of Europe (Saether & Spies, 2010). There are no records from the Netherlands.

LIFE CYCLE

Three generations a year are possible: in spring, early summer and late summer or autumn (Lehmann, 1972; Pinder, 1980). However, Laville & Lavandier (1977) stated emergence in the Pyrenees only in summer and autumn, and Fahy (1973) in Ireland from November until March. In the Danish spring Ravnkilde, pupae and adults were collected only in May (Lindegaard et al., 1975). In Iceland found Hannesdóttir et al. (2012) few fourth instar larvae in winter; they stated emergence almost all year round and overlapping generations in summer in warm streams (13° C. and more), but no winter or early spring emergence from cold streams.

MICROHABITAT

The larvae have often been found among submersed mosses (Lehmann, 1972; Lindegaard et al., 1975; Braukmann, 1984; Nolte, 1989). They also inhabit the stones among the moss cushions. Pinder (1980) stated their presence on gravel and on *Ranunculus*, but not on soft sediments.

WATER TYPE

Braukmann (1984) called *E. minor* a characteristic and numerous species in streams in mountains and hills (cf. Kownacka & Kownacki, 1972; Michiels, 2004). Laville & Lavandier (1977) found it mainly in the lower parts of mountain streams. The species was almost absent from the Rohrwiesenbach in central Germany, but very numerous in the colder and faster flowing Breitenbach (Ringe, 1974). In Scandinavia the larvae also live in brooks with slower currents (Brundin, 1956: 88).
In Iceland it is an abundant lake dweller (Lindegaard, 1980). Ruse (2002) collected the species rarely in lakes with low conductivity in England and Wales. Reiff (1994) probably collected two exuviae in a lake in Bavaria.

SAPROBITY

According to Moog (1995) *E. minor* lives only in xenosaprobic to β-mesosaprobic water.

Eukiefferiella pseudomontana Goetghebuer, 1935

IDENTIFICATION

The adult male has been described and keyed by Lehmann (1972). The species is absent from the key by Langton & Pinder (2007). The exuviae can be identified using Langton (1991). The larva is unknown.

DISTRIBUTION IN EUROPE

E. pseudomontana has been recorded in a number of countries in central and southern Europe and in Germany near Bonn, but seems to be absent from other regions (Caspers, 1980a; Serra-Tosio & Laville, 1991; Saether & Spies, 2010).

WATER TYPE

Langton (1991) called *E. pseudomontana* a montane species. Geijskes (1935) found the species only in hygropetric environment. However, Caspers (1980a) collected many adults in a small woodland brook near Bonn (Germany). Garcia & Laville (2001) collected only one specimen in the middle of the river Garonne in France.

Eukiefferiella similis Goetghebuer, 1939

SYSTEMATICS AND IDENTIFICATION

Within the genus, *E. similis* can be incorporated into the *gracei* group (Halvorsen, pers. comm.). The adult male is absent from the key by Langton & Pinder (2007), but it can be identified using Lehmann (1972). Exuviae can be identified using Langton (1991). The larvae can be distinguished from other species of the *gracei* group when the mental teeth have not been worn (see Pankratova, 1970; Schmid, 1993).

DISTRIBUTION IN EUROPE

The species has been recorded from many countries across Europe, but seems to be absent from Scandinavia and the British Isles (Saether & Spies, 2010). It has not been recorded in the Netherlands.

WATER TYPE

The species has been collected mainly in very fast-flowing streams (Braukmann, 1984: 211; Ertlová, 1970: 296). However, Michiels (2004) collected exuviae in the lower

course of the Elz stream in the Black Mountains in Germany.
According to Gendron & Laville (1997), *E. similis* is a foothill species, living in higher zones only when the water temperature there is higher than normal.

SAPROBITY

According to Moog (1995), *E. similis* lives only in xenosaprobic to β-mesosaprobic water.

Eukiefferiella tirolensis Goetghebuer, 1938

SYSTEMATICS AND IDENTIFICATION

The species is very similar to *E. brevicalcar*, but can be identified as adult, pupa and larva (see Lehmann, 1972; Langton & Pinder, 2007; Langton, 1991; Schmid, 1993).

DISTRIBUTION IN EUROPE

The species has been found only in relatively few European countries and seems to be absent from Scandinavia. It is not to be expected in the Netherlands.

WATER TYPE

E. tirolensis lives in fast-flowing streams in montane areas (Lehmann, 1972; Braukmann, 1984; Rossaro, 1991; Orendt, 2002a; Michiels, 2004). There are also some records from lakes in Switzerland and Bavaria (Lehmann, 1972; Reiff, 1994).

SAPROBITY

According to Moog (1995) *E. tirolensis* lives only in xenosaprobic to β-mesosaprobic water.

Eurycnemus crassipes (Meigen, 1810)

SYSTEMATICS AND IDENTIFICATION

Only a few species of the genus *Eurycnemus* have been named, of which only one in Europe. Some unnamed species are also known (e.g. Kobayashi, 1995). Orthocladiinae gen.? *shadini* (Pankratova, 1968) may be a second European species. This larva differs in its wider central mental tooth and probably bifid S I on the labrum.
The genus can be identified in all stages using the normal keys, but the larva is absent from Cranston (1982).

DISTRIBUTION IN EUROPE

The species has been collected throughout Europe (Ashe et al., 2000; Saether & Spies, 2010). In most countries it has been found rarely (e.g. in Germany: Orendt, 2002; Michiels, 2004; in France: Moubayed-Breil, 2008). In the Netherlands only two males have been caught, in the nineteenth century, one near Arnhem and one near Heelsum (see Ashe et al., 2000).

FEEDING AND MICROHABITAT

The fourth instar larva has been observed to parasitise on pupae of the trichopteran species *Hydropsyche siltalai* (Ashe et al., 2000). The Japanese species described by Kobayashi (1995) is a parasite or predator of trichopteran pupae or prepupae of the genus *Goera*.

WATER TYPE

Most records are from rather fast-flowing streams (Ashe et al., 2000; Orendt, 2002; Michiels, 2004), suggesting that *Hydropsyche angustipennis*, a common species in lowland streams, is not used as host.

Georthocladius luteicornis (Goetghebuer, 1941)

The larvae of this species live terrestrially in wet organic soils. We found them at many localities in grassland and woods, in the Netherlands and in Belarus. However, Schnell (1991) collected the species only in truly aquatic habitats (lakes and rivers) in Norway. As stated by this author and also by me, the number of squama setae can be more than 30 and cannot be used to distinguish the species from the Nearctic *G. fimbriosus*. It has to be noted that emergence in Norway has been stated in summer, whereas in the Netherlands it is confined to early spring, and in Belarus until the end of April.

Gymnometriocnemus Goetghebuer, 1932

This genus lives only in terrestrial habitats. Data of Ekrem et al. (2010) indicate that there are still several undescribed species. Up to now the larvae of some species cannot be distinguished from those of *Bryophaenocladius*. Figures of the larva are given by Cranston (1982) and Epler (2001). The most common species, *G. brumalis*, flies in autumn and winter in all types of woodland, even in small wood patches in cities. See also Delettre et al. (1992).

Halocladius Hirvenoja, 1973

SYSTEMATICS

The genus *Halocladius* is related to *Cricotopus* (Hirvenoja, 1973; Saether, 1977a). Adults, pupae and larvae resemble this genus. Two subgenera are recognised: *Psammocladius* with one species and *Halocladius* s.s. with five European species.

OR

IDENTIFICATION

The four species living in northwestern Europe can be identified as adult male using Langton & Pinder (2007), as exuviae using Langton (1991) and as larva using Cranston (1982) or Moller Pillot (1984). Descriptions of all stages can be found in Hirvenoja (1973). The larvae are most easily distinguished from *Cricotopus* because of the long curved end of the ventromental plates, extending far behind the bases of the setae submenti (see fig. 8).

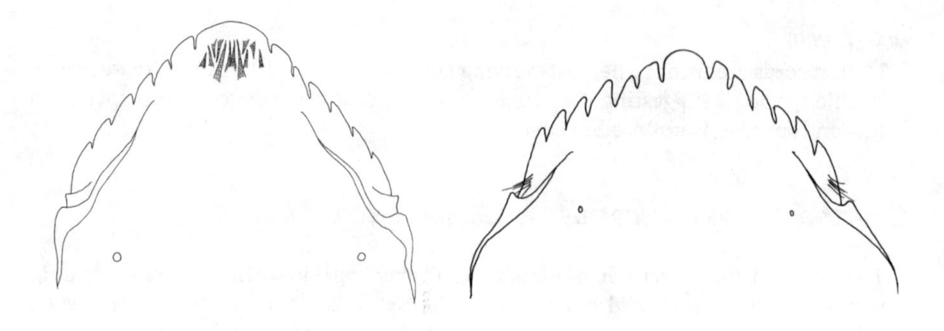

8 Mentum Halocladius varians (r) and Cricotopus bicinctus (l)

DISTRIBUTION

The whole genus lives mainly in coastal areas and can be found only rarely in inland saline waters. Four species occur around the North Sea and only these are treated below. Two other species live around the Mediterranean Sea (*H. mediterraneus* Hirvenoja, 1973 and *H. millenarius* Santos Abreu, 1918 = *H. stagnorum*).

Halocladius (Psammocladius) braunsi (Goetghebuer, 1942)

Trichocladius psammophilus Remmert, 1953: 236–237, fig. 21; 1955: 36

DISTRIBUTION IN EUROPE

H. braunsi has been found in western Europe in most countries along the North Sea and Atlantic Ocean, from Germany to Spain (Saether & Spies, 2010). In the Netherlands it has been collected only once (February 27, 1994; see Water type).

FEEDING

The larvae feed on algae growing on the sand grains (Remmert, 1955).

WATER TYPE

The larvae have been found in large numbers (up to 2 or 3 larvae/cm^2) in tidal flats in the German Shallows (Farbstreifenwatt) (Remmert, 1953, 1955). The only Dutch larva has been collected in the Moksloot, a brackish creek on the island of Texel (leg. A. Klink).

Halocladius fucicola (Edwards, 1926)

DISTRIBUTION IN EUROPE AND THE NETHERLANDS

H. fucicola has been found in a small number of countries along the coasts of western and southern Europe (Saeter & Spies, 2010). There is one record from the Netherlands in recent times, in the province of Friesland (Nijboer & Verdonschot, 2001); this record has to be verified. Palaeoecological remains of larvae of (probably) this species have been found near Vlaardingen (Zuid-Holland) (A. Klink, in prep., as *H. stagnorum*).

WATER TYPE

All records are from salt water at rocky coasts (Hirvenoja, 1973: 127). Cranston (1982) collected *H. fucicola* in mid littoral rock pools in Britain. The Dutch records (see above) are from different (not clearly described) habitats.

Halocladius variabilis (Staeger, 1839)

Trichocladius vitripennis Remmert, 1955: 36 (pro parte)

IDENTIFICATION

In older literature, *H. variabilis* and *H. varians* are not always correctly distinguished (see under the latter species). Therefore, ecological data in literature have to be used with caution.

DISTRIBUTION IN EUROPE AND THE NETHERLANDS

H. variabilis has been recorded in many countries along the Atlantic Ocean and other sea coasts around the whole of Europe (Saether & Spies, 2010). In the Netherlands the larvae have been collected on the Brouwersdam (province of Zeeland) and in the North Sea Canal and the ports of Amsterdam and Europort (Limnodata.nl; own data).

WATER TYPE

From the text in Hirvenoja (1973) it is not clear which records apply definitely to *H. variabilis* and in which water type the larvae lived. Cranston (1982) collected the larvae in mid littoral rock pools in Britain. Possibly they live only in rocky environments, because this species has been found in the Netherlands mainly in canals and ports and not in many other habitats where *H. varians* has been collected. This has to be investigated further.

Halocladius varians (Staeger, 1839)

? *Trichocladius halophilus* Potthast, 1914: 306–311
Trichocladius vitripennis Remmert, 1955: 36 (pro parte)
Cricotopus vitripennis Pankratova, 1970: 201–203, fig. 22 (pro parte)

IDENTIFICATION

Identification of the adult males and pupal exuviae is not always easy, because *H. varians* and *H. variabilis* closely resemble each other (Pinder, 1978; Langton, 1991; Langton & Pinder, 2007). The larvae can be identified more easily (Hirvenoja, 1973; Moller Pillot, 1984). See also under het genus. Some of the Dutch material has been identified as adults (Krebs, 1981, 1990), but mainly as larvae.

DISTRIBUTION IN EUROPE AND THE NETHERLANDS

H. varians has been collected in many countries along the coasts of the North Sea, the Atlantic Ocean and the Mediterranean Sea (Saether & Spies, 2010). Most probably this species can be found also in brackish or polluted water far from the coast, as in Westphalia (Potthast, 1914: 307). In the Netherlands the species is common on the Frisian Islands and in the delta region, but is rather scarce on the mainland of the provinces of Groningen, Friesland, Noord-Holland and Zuid-Holland, except for some ship canals. There are also some records from the lower courses of the river Rhine and small streams in western Noord-Brabant (Krebs, 1981, 1990; Limnodata.nl; own data).

LIFE CYCLE

The adults fly in the Camargue (southern France) all year round (Tourenq, 1975: 166) and in Norway from April to September (Hirvenoja, 1973).

SWARMING AND OVIPOSITION

Swarming and oviposition are described by A.Thienemann in Potthast (1914: 308–310); this description applies to *H. variabilis* and/or *varians*. He saw only females swarming low over the water. Males were seen much higher, near buildings. Egg masses were deposited on piles.

FEEDING

In the guts of the larvae we found mainly diatoms and/or detritus, probably depending on local availability.

MICROHABITAT

The larvae live in long tubes on stones and plants, among algae and on mineral and organic bottoms (own data; see also Tourenq, 1975, sub *H. stagnorum*). Remmert (1955) stated that the larvae cannot swim.

WATER TYPE

The larvae live in sea, rock pools, estuaries, canals, ditches and pools, rarely also in the lower courses of rivers (Potthast, 1914; Hirvenoja, 1973; Krebs, 1981, 1990; Limnodata.nl). In the sea most larvae live near the low water mark (Potthast, 1914; T. Kaiser, pers. comm.) and, according to Remmert (1955: 36), most commonly in the *Fucus* zone and not in the surf zone.

SAPROBITY AND OXYGEN

Besides brackish water, *H. varians* can also inhabit polluted fresh water (Hirvenoja, 1973), which explains why the larvae are found in the lower course of the river Rhine and in the Steenbergse Vliet in the Netherlands. In the Netherlands the species has often been collected in water with a BOD of more than 10 mg O_2/l and even higher than 20 mg O_2/l (Limnodata.nl). However, Tourenq (1975) stated that the larvae of the related *H. stagnorum* do not tolerate severe oxygen deficit. Steenbergen (1993) collected no larvae of *H. varians* in water with low oxygen content (lower than 40% saturation) and there are very few other records in the Netherlands with such low oxygen contents (Limnodata.nl).

SALINITY

The larvae withstand chloride concentrations higher than seawater and are rare at concentrations lower than 2000 mg Cl/l (Krebs & Moller Pillot, in prep.). However, they have been found in polluted fresh water too (see above). They tolerate wide fluctuations in chlorinity (de Kroon et al., 1985; Tourenq, 1975) and can be reared to emergence even in fresh water (Potthast, 1914; 308; cf. Tourenq, 1975: 166).

Heleniella Gowin, 1943

SYSTEMATICS AND IDENTIFICATION

Only two species of *Heleniella* can be expected in the western European lowland: *H. ornaticollis* and *H. serratosioi*. The exuviae of both species can be identified using Langton (1991). The adult males and females are described in Ringe (1976). The larvae can be identified only using Schmid (1993) or Bitušik (2000). Because in the past other keys have been used to identify larvae, older identifications of *H. ornaticollis* larvae have to be verified.

Heleniella ornaticollis (Edwards, 1929)

Heleniella thienemanni Pankratova, 1970: 273–274, fig. 174

DISTRIBUTION IN EUROPE AND THE NETHERLANDS

This species has been found throughout Europe, but has not yet been recorded in many countries (Saether & Spies, 2010). In the Netherlands larvae have been collected at eight localities in southern and northern Limburg, in Twente and on the Veluwe (Moller Pillot & Buskens, 1990; Nijboer & Verdonschot, unpublished). Three records have been verified at species level.

LIFE CYCLE

Kawecka & Kownacki (1974) collected the pupae in Poland in September; Michiels (1999) collected exuviae in Bavaria mainly in March and August. Schmid (1992) stated peak abundances in Austria in late spring and late summer. Both Schmid and Kawecka & Kownacki collected no fourth instar larvae in October (and rather many in winter), suggesting a diapause in autumn.

FEEDING

Kawecka & Kownacki (1974) found no diatoms in the guts of the larvae, but only detritus, mainly from rotting beech leaves, mineral matter and algal cells. Moog (1995) called all species of *Heleniella* detritus feeders and Ferrington (1987) stated that the American species *H. parva* consumed fine particles of detrital material (see Microhabitat).

MICROHABITAT

Schmid (1992) collected the larvae of all instars in the gravelly bottom of a stream in Austria, with a significant preference for depths between 20 and 30 cm in the substrate. We collected the larvae also in a sandy bottom and among mosses. In this context, it is interesting that Ferrington (1987) found many larvae of the American species *H. parva* among the root systems of water plants, where water movement was constant and supplied fine particulate organic matter.

WATER TYPE

The larvae are found mainly in fast-flowing spring brooks and upper courses of mountain streams (Lehmann, 1971; Laville & Vinçon, 1991; Gendron & Laville, 1997) and even more downstream in the same area (Schmid, 1993; Michiels, 1999, 2004). In the Netherlands the species has also been collected in helocrene springs at lower current velocities. According to Moog (1995), *H. ornaticollis* is more a rhithral than a crenal species.
The species seems to be truly cold-stenothermic (Gendron & Laville, 2000); Rossaro (1991) stated for the genus as a whole a mean water temperature of 10 °C.

pH

Syrovátka et al. (2012) stated a preference for basic conditions.

SAPROBITY AND OXYGEN

According to Moog (1995), *H. ornaticollis* is a typical inhabitant of oligosaprobic water. Its preference for deeper layers of the substrate and its occurrence in helocrenes suggest that a good supply of oxygen within the substrate in combination with a low temperature is most important.
Wilson (1988) collected the exuviae in a small brook polluted with zinc.

Heleniella serratosioi Ringe, 1976

DISTRIBUTION IN EUROPE

H. serratosioi has been recorded in relatively few countries, mainly in central and southern Europe, but also in Finland. It seems to be absent from the British Isles (Saether & Spies, 2010; Paasivirta, 2012). Its occurrence in the Netherlands is not impossible. There are no records from northern France (Serra-Tosio & Laville, 1991) or from Belgium, but many adults have been collected in a woodland brook near Bonn (Germany) (Caspers, 1980a) and we collected the exuviae in the Eifel.

LIFE CYCLE

Ringe (1976) stated emergence only in spring, but Caspers (1980a) collected adults at a woodland brook near Bonn from April until October, with the highest numbers in spring and late summer, suggesting two generations a year.

WATER TYPE

Ringe (1976) found this species mainly in a spring marsh, and Schmid (1993) and Stur et al. (2005) in crenal regions of mountain streams. Other authors stated its occurrence in rhithral brooks and streams (Caspers, 1980a; Bitušik, 2000; Orendt, 2002).

Heterotanytarsus apicalis (Kieffer, 1921)

DISTRIBUTION IN EUROPE AND THE NETHERLANDS

The species is distributed throughout Europe, but seems to be scarce in many countries (Serra-Tosio & Laville, 1991; Bitušik, 2000; Saether & Spies, 2010). It is more common in Scandinavia (e.g. Brundin, 1949; Paasivirta, 2012). In the Netherlands it has been collected at many localities on the Veluwe and more rarely near Nijmegen, in Twente and Limburg (Verdonschot & Schot, 1987; Limnodata.nl).

LIFE CYCLE

In Scandinavia *H. apicalis* emerges from April to October and has two and probably sometimes more generations (Brundin, 1949; Särkkä, 1983).

FEEDING AND TUBE BUILDING

The larvae build transportable cases, which is an exception in Orthocladiinae. They feed mainly on detritus (Moog, 1995).

MICROHABITAT

Verdonschot & Lengkeek (2006) found the larvae mainly on detritus bottoms and less often between dead leaves. Janzen (2003) collected them almost only in organic lowland streams. Brundin (1949) also collected some specimens from stony bottoms and from *Isoetes* vegetations in Swedish lakes.

DENSITY

Särkkä (1983) collected up to 400 larvae/m^2 in a Finnish lake. Such densities have also been found in lowland brooks, but usually the larvae are very scarce there (Janzen, 2003: 9; own data).

WATER TYPE

Flowing water

In central Europe and in the Pyrenees the species has been found in springs and (forest) brooks with slow or rather fast currents (Lehmann, 1971; Gendron & Laville, 1997; Orendt, 1999; Michiels, 2004; Stur et al., 2005). Janzen (2003) collected them in West Germany mainly in organic lowland brooks and rarely in foothill brooks. In the Netherlands most records are from springs and small brooks, usually with rather slow current and often in woodland (Limnodata.nl). See table 7 on p. 287.

Stagnant water

In Scandinavia and the British Isles the species lives mainly in lakes (Brundin, 1949; Paasivirta, 1974; Saether, 1979; Cranston, 1982; Särkkä, 1983; Palomäki, 1989; Ruse, 2002). There and in central Europe it is also found in bog pools (Thienemann, 1944) and montane pools (Janecek, 2007). The larvae are most common in the littoral zone of lakes, but they also live in the profundal zone (Brundin, 1949; Paasivirta, 1974; Brodin & Gransberg, 1993).

In the Netherlands there are no recent records from moorland pools and only one from a sand pit (Buskens, unpublished), but Klink (1986) found the larvae in mesotrophic moorland pools during palaeolimnological investigations.

pH

H. apicalis was more common in a Scottish lake after acidification during the twentieth century (pH dropping below 5) (Brodin & Gransberg, 1993). However, Raddum et al. (1984) stated an obvious increase after liming of a lake in Scandinavia (pH increasing from 4.8 to more than 6.7). Eriksson & Johnson (2000) found no difference in a lake in Sweden after the pH decreased from 5.9 to 4.7. Many other lakes inhabited by this species are slightly acid (see e.g. Macan, 1949; Särkkä, 1983; Palomäki, 1989). The species lives in oligotrophic oligohumic lakes as well as in mesohumic and polyhumic lakes (Saether, 1979).

Janecek (2007: 29) called *H. apicalis* a typical inhabitant of brooks rich in humic acid and/or poor in nutrients and also of acid helocrenes and limnocrenes. In the acid brooks studied by Orendt (1999) exuviae and larvae were found at pH 3.8 and higher. Janzen (2003) collected the larvae in several acid lowland brooks in Germany, but only numerously in a circumneutral brook. Most Dutch brooks inhabited by this species lie on the Veluwe and are not or only slightly acid (Limnodata.nl), but we also found the larvae in the Netherlands and Belgium at a pH of about 5. On the Veluwe the species lives in brooks fed by deep to undeep ground water, but it is absent when this water has taken up much calcium, this in contrast to the flatworm *Polycelis felina* (Klink, 2012).

In Dutch Pleistocene regions *H. apicalis* is absent from most of the acid upper courses and from acid moorland pools. This may be because these water bodies are extremely poor in calcium and trace nutrients, or because they have high ammonium contents, as discussed under *Heterotrissocladius marcidus* and in section 2.12. Everywhere the species is absent from calcareous streams.

In summary, we may conclude that *H. apicalis* tolerates very acid conditions, but that it can be common in circumneutral water. The absence in nutrient poor acid water bodies, rich in ammonium, is not caused by direct influence of the pH on the larvae.

SAPROBITY AND OXYGEN

In Scandinavia *H. apicalis* is a characteristic inhabitant of oligotrophic oligohumic lakes and occurs also in mesohumic and polyhumic lakes (Brundin, 1949; Saether, 1979; Särkkä, 1983). Ruse (2002) found it in English lakes with low conductivity. As mentioned under Water type, the species also inhabits unpolluted upper courses of brooks that are poor in nutrients and/or rich in humic substances, often in woodland (see also Fahy & Murray, 1972). It has been found in brooks with a BOD up to 3 mg O_2/l (Limnodata.nl) and lives in acid brooks, often in bottoms where decomposition may be dominant, but very slow. Several data indicate that a somewhat lowered oxygen content is no problem in winter, under ice in lakes or in acid woodland brooks. The figures given in table 5 (p. 283) have to be understood in this context.

Heterotanytarsus brundini Fittkau, 1956

This species is only known from northern Scandinavia and is not treated in this book.

Heterotrissocladius Spärck, 1923

SYSTEMATICS AND IDENTIFICATION

Most species of *Heterotrissocladius* live in alpine, boreal and arctic lakes. In the Netherlands and adjacent countries only *H. marcidus* is more or less common and one or two other species can be expected. Identification of exuviae of the European species (except for *H. scutellatus*) is possible using Langton (1991). Adult males, pupae and larvae have been keyed and described by Saether (1975), but *H. brundini* has been described later by Saether & Schnell (1988). A comparison between the larvae of *H. marcidus* and *H.* cf. *scutellatus* with good figures can be found in Schmid (1993). A pupa (probably belonging to this genus) collected by H. Cuppen in a moorland pool on the Veluwe could not be identified.

Heterotrissocladius brundini Saether & Schnell, 1988

IDENTIFICATION

See under the genus. The *H. maeaeri* type 2 in Brooks et al. (2007) could be *H. brundini*.

DISTRIBUTION IN EUROPE

The species has been collected in Scandinavia (also in southern Norway) and in Scotland.

WATER TYPE

H. brundini lives in oligotrophic acid lakes.

Heterotrissocladius maeaeri Brundin, 1949

IDENTIFICATION

See under the genus. The *H. maeaeri* type 2 in Brooks et al. (2007) could be *H. brundini*.

DISTRIBUTION IN EUROPE

The species has been collected in Scandinavia (also in southern Sweden and southern Finland).

WATER TYPE

The larvae live in very oligotrophic lakes (Saether, 1979).

Heterotrissocladius marcidus (Walker, 1856)

DISTRIBUTION IN EUROPE AND THE NETHERLANDS

H. marcidus is the only species of the genus that has been found throughout Europe (Saether & Spies, 2010). In the Netherlands it is distributed throughout the Pleistocene part of the country; it is very common on the Veluwe, but almost absent from large parts of Noord-Brabant, Salland and Drenthe (Limnodata.nl; Klink, unpublished). We also collected the larvae scarcely in southern Limburg.

LIFE CYCLE

In southern Sweden, Germany and Austria the species usually has two generations a year, emerging in April–June and September–October, or even until December (Brundin, 1949: 706; Ringe, 1974; Caspers, 1983). According to Cranston (1982), three or four generations could be possible. In a montane lake in the Pyrenees the larvae born in spring hibernate in fourth instar and the larvae born in late summer hibernate in third instar (Laville & Giani, 1974).

EGGS

Nolte (1993) described and figured an egg mass attached to wood at the water's edge. It contained 231 eggs.

FEEDING

Moore (1979, 1979a) stated that the related American species *H. changi* fed mainly on diatoms and other benthic algae, but did not ingest *Scenedesmus* and other phytoplankton species. According to Moog (1995), *H. marcidus* is a detritus feeder. In the guts of larvae of *H. marcidus* we found mainly detritus and plant remains and hardly any diatoms.

MICROHABITAT

The larvae live mainly on organic bottoms, especially in mud, and only scarcely in sand (Pinder, 1980; Wiley, 1981; Palomäki, 1989; Verdonschot & Lengkeek, 2006). Janzen (2003) collected the species almost only in organic lowland streams. Schmid (1993) stated that the larvae in brooks in Austria showed a preference for deeper sediment layers. According to Brundin (1949), the larvae also live on substrates with vegetation. They build more or less vertical tube-like burrows into the sediment.

WATER TYPE

The pH (or a pH related factor) appears to play an important role; see below.

Flowing water

In foothill and mountain streams *H. marcidus* lives mainly in the spring zone and upper courses (Lehmann, 1971; Gendorn & Laville, 1997). However, it can be found there also in somewhat lower stretches, in lentic curves, where detritus has been deposited (e.g. Braukmann, 1984: 212, 455). This may be the reason why the species

has often been mentioned as an inhabitant of fast-flowing streams (Bitušik, 2000; Orendt, 2002a; Michiels, 2004). It is usually absent from larger rivers, possibly in connection with pH and/or detritus type.
In lowland streams the larvae seem to prefer organic bottoms (Janzen, 2003). They are found mainly in springs and small brooks (often in woodland), but also regularly in larger brooks and small streams (Limnodata.nl; own data).

Stagnant water
According to Brundin (1949), *H. marcidus* prefers the littoral zone of lakes and is rare in the profundal zone (cf. Reiff, 1994). Brundin cited many records from other authors in montane lakes and pools in different parts of Europe and also mentioned the absence of the species in lakes in northern Germany. There are many records from lakes by later authors (e.g. Särkkä, 1983; Ruse, 2002). In the Netherlands the species has been found only once in stagnant water (Apeldoorn Canal, leg. H. Cuppen). See under pH and Saprobity.

Temperature
According to Saether (1975), in most of the European localities where the species is found the water temperature never exceeds 18 °C throughout the year. Although the species has been found in Germany and the Netherlands also at higher temperatures, lower temperatures could be optimum (cf. Reiff, 1994).

pH

Brodin & Gransberg (1993) and Eriksson & Johnson (2000) stated an obvious increase in lakes after pH had decreased below 5. Syrovátka (2012) stated a preference for acid water in springs in the Czech Republic. Mossberg & Nyberg (1980) found *H. marcidus* in forest lakes at pHs lower than 5; Janzen (2003) and Hawczak et al. (2009) collected the species in acid brooks in Germany (pH 4.2–4.4) and England (pH 5–6) respectively. In contrast, almost all records from the Netherlands are from brooks with a pH around 7 (mainly on the Veluwe, see Limnodata.nl) and we even sometimes found larvae in calcareous springs in southern Limburg. Elsewhere, some specimens have been collected at pH 8 or higher (e.g. Pinder, 1980; Reiff, 1994; own data). The species is conspicuously absent from acidified moorland pools and strongly acidified upper courses of streams in the Netherlands, although Klink (1986: 17) found remains of the species in sediments of Dutch moorland pools during palaeolimnological investigations. As discussed in section 2.12 an important aspect could be that most acid upper courses on Pleistocene sandy soils have become extremely poor in calcium and trace elements. The high ammonium contents and strong accumulation of organic material in moorland pools may also play a role. Therefore the figures given in table 6 in Chapter 9 must be regarded as more or less typical of the Netherlands.
The relation with pH could be indirect, possibly because this detritus feeder is dependent on particular decomposition products, for example from woodland; pH is certainly not the key factor for this species.

TROPHIC CONDITIONS AND SAPROBITY

In the littoral zone of oligohumic lakes *H. marcidus* is characteristic of oligotrophic water (Saether, 1979). However, the species has often been collected in mesotrophic lakes (Palomäki, 1989; Orendt, 1993) and it occurs scarcely in eutrophic lakes (e.g. Reiff, 1994). Ruse (2002) found the species in a number of lakes at low conductivity. In streams it is also characteristic of unpolluted water (e.g. Moog, 1995; Michiels, 2004), but it is not totally intolerant of organic pollution (e.g. Wilson & Ruse, 2005). In Dutch brooks and streams it is found up to BOD 3.5 (Limnodata.nl). It is rather

common in humic water and tolerates some decrease in oxygen content (probably under ice as well as in summer nights).
A low water temperature could be optimum (see Water type). In some regions the larvae are found mainly in more or less acid water. As discussed under pH, all these factors together could be important because they play a role in the process of decomposition.

Heterotrissocladius scutellatus (Goetghebuer, 1942)

DISTRIBUTION IN EUROPE

H. scutellatus has been collected in the alpine region and as far north as Luxemburg (Saether & Spies, 2010).

EGGS

Nolte (1993) described the egg masses of (most probably) this species. They were attached to wood at the water's edge and contained 190–256 eggs per mass.

WATER TYPE

The larvae live in the littoral and profundal zones of lakes (Saether, 1979) and possibly also in streams (Nolte, 1993; Orendt, 2002a).

TROPHIC CONDITIONS AND SAPROBITY

Saether (1979) called *H. scutellatus* a characteristic inhabitant of the profundal zone of oligotrophic alpine lakes. According to Moog (1995), it is an indicator of oligosaprobic to β-mesosaprobic conditions in streams.

Hydrobaenus Fries 1830

Trissocladius Thienemann, 1944: 587, 634 (pro parte); Brundin, 1956: 18, 73 (pro parte); Pankratova, 1970: 134–145 (pro parte)

SYSTEMATICS

Before the revision of the genus *Hydrobaenus* and related genera by Saether (1976) there was much confusion about the relationships and the generic names. Many authors used the name *Trissocladius* for all or most species of this whole group of genera. According to Saether (1976: 24) in Europe the group consists of three genera: *Hydrobaenus*, *Trissocladius* and *Zalutschia*. Saether divided the genus *Hydrobaenus* into four species groups.

IDENTIFICATION

Most species can be identified as male, female, pupa and larva using Saether (1976). However, the characters of the larvae are not always reliable because the variation in mentum and the length of setae is much greater than mentioned in the key. Moreover, not all species have been described. For instance, in Belarus we found at least two undescribed species. Pupae and exuviae can be identified using Langton (1991). Adults are not keyed in Langton & Pinder (2007), because no species has been found in the British Isles.

DISTRIBUTION IN EUROPE

The genus *Hydrobaenus* is widely distributed in the northern part of the Holarctic.

Further south the number of species is low and only very few species (e.g. *H. distylus*) have reached the Mediterranean Sea. The whole genus is absent from Britain and Ireland (Saether, 1976; Saether & Spies, 2010).

LIFE CYCLE

Possibly all species of this genus aestivate in second (rarely in third) instar in a cocoon, as has been stated for *H. lugubris* and *H. pilipes* (see under these species), *H. kondoi* (Kondo, 1996) in Japan and an unknown American species (as *H. pilipes*, Grodhaus, 1980). Larvae in cocoons more or less survive desiccation, but the larvae do not depend on desiccation for further development, as has been stated for *H. lugubris* (Steinhart, 1998) and can be concluded from the occurrence in permanent water of e.g. *H. distylus* and *H. pilipes*. Strictly speaking, the species have no diapause, but rather, more accurately, a parapause or dormancy, because it is not induced by photoperiod. There is one generation a year, emerging in spring, or sometimes two (emerging in winter and spring, e.g. in *H. kondoi*).

Some species do not swarm and copulate on the water surface or on banks (e.g. *H. lugubris*, *H. pilipes*), but other species do swarm (e.g. *H. kondoi*).

FEEDING

The larvae can feed on detritus, as has been stated for *H. kondoi* (Kondo, 1996) and is very probable for *H. pilipes*; Steinhart (2000) fed the larvae of *H. lugubris* with epiphytic algae.

WATER TYPE

Many species of the *Hydrobaenus* group seem to live only in temporary water. The aestivation in second instar will be an adaptation to this environment. However, not all species need a period of drought (e.g. *H. distylus* and *H. lugubris*).

Hydrobaenus conformis (Holmgren, 1869)

H. conformis is a boreal species with circumpolar distribution. It has been collected in deep lakes in (also southern) Scandinavia (Saether, 1976) and also once in the southern part of France (Moubayed-Breil, 2007).

Hydrobaenus distylus (Potthast, 1914)

Trissocladius distylus Lehmann, 1971: 490

DISTRIBUTION IN EUROPE AND THE NETHERLANDS

More than other species of the genus, *H. distylus* is distributed over a large part of Europe, southwards to Spain, Italy, Hungary and the Near East. It is absent from Britain and Ireland (Saether & Spies, 2010). In the Netherlands there are only two records of exuviae: in the provinces of Limburg (Susteren) and Gelderland (Zelhem).

LIFE CYCLE

Emergence has been stated only in spring, from March to early May (Lehmann, 1971; Saether, 1976; own data). The species most probably has a period of dormancy in second instar, as in other species of the genus. However, it seems not to be restricted to temporary brooks and streams (see below).

MICROHABITAT

We collected the larvae and pupae from stones and plants, but there are too few data to give an indication of any preference.

WATER TYPE

Most data apply to slow-flowing brooks and streams, but Serra-Tosio & Laville (1991) reported occurrence of the species in the Alps above 1000 m and it has also been recorded in lakes and ponds (Lehmann, 1971; Saether, 1976). We collected larvae and pupae in permanent brooks near Coesfeld (Westphalia, Germany). Apparently desiccation in summer is not necessary for further development (see the Life cycle section in the introduction to the genus).

SAPROBITY

We found the species in moderately polluted brooks as well as in brooks without any anthropogenic pollution.

Hydrobaenus lugubris Fries, 1830

DISTRIBUTION IN EUROPE AND THE NETHERLANDS

The species lives in the northern and central part of Europe, southwards to Hungary, Austria and northern France. It is absent from Britain and Ireland (Saether & Spies, 2010; Serra-Tosio & Laville, 1991). In the Netherlands, where natural floodplains are absent, there are relict populations near the rivers Rhine, IJssel and Scheldt in the provinces of Gelderland and Zeeland.

LIFE CYCLE

Steinhart (1998) stated emergence in the floodplain of the Oder from the end of February until early May. In the Netherlands and in Belarus we found pupae and adults mainly in April. Only one generation a year has been stated, but Steinhart (1998: 86; 2000: 422) could not exclude that a second generation is possible from eggs laid in late winter. Later in spring she observed that the larvae go into dormancy and form a cocoon, as a rule in second instar (rarely in third instar). Cocoon formation is not induced by a photoperiod, temperature or wetness. The end of the dormancy in winter is induced by low temperatures. Desiccation is not necessary for further development. See further under the genus.

OR

MATING AND OVIPOSITION

The adults do not swarm; copulation takes place on still water surfaces (Saether, 1976; Steinhart, 1998: 53). Steinhart observed that the females preferred water for egg deposition, but they also deposited eggs in moist substrate. Twelve egg masses contained 80 to 221 eggs.

MICROHABITAT

Steinhart (1998) gives hardly any information about microhabitat, but probably most larvae lived on grasses and other plants in the floodplain. Ertlová (1970) collected the larvae (as *H. distylus*) among mosses on stones in the river Danube.

DENSITIES

Steinhart (1998: 39) stated densities of up to 20,592 larvae/m^2 in the floodplain of the river Oder. Only small populations have been found in the Netherlands.

WATER TYPE

H. lugubris seems to be specially adapted to the floodplains of large rivers (Lechthaler, 1993; Steinhart, 1998; 2000; own data). The larvae are found in early spring in the inundated floodplain and can live there as long as the bottom is not completely dried up. In summer and autumn the young larvae survive in cocoons in temporary conditions as well as in permanent water (see Life cycle). We also collected the species in a temporary brooklet in woodland near the river Pripyat in Belarus. Schnabel & Dettinger-Klemm (2001) stated its occurrence in temporary pools in and near the floodplain of the river Lahn in Germany. In the Netherlands all verified records (about 15) apply to ditches (1 to 3 m wide) and small brooklets near to large rivers and estuaries.

DISPERSAL

Steinhart (1998) observed that copulation took place without swarming, shortly after emergence, and that the females were ready for oviposition just two minutes later. The wings are relatively short and the flight capacity is low. This may be the reason why the species is confined to historical life regions.

Hydrobaenus pilipes (Malloch, 1915)

Trissocladius grandis Meuche, 1938: 481; Brundin, 1949: 733–734; Mozley, 1970: 433–451; Lehmann, 1971: 490
nec *Hydrobaenus pilipes* Grodhaus, 1980: 314–322

SYSTEMATICS AND IDENTIFICATION

Saether (1976) stated much variation within this species, especially in the male hypopygium and in the total sizes of the adults. Possibly more than one species is involved. Identification of the larvae using Saether (1976) or Moller Pillot (1984) is not always reliable, because the length of the mental teeth is more variable.
The observations by Grodhaus (1980) in California apply to another (related) species (Grodhaus, pers. comm.). His results are treated under Life cycle in the introduction to this genus.

DISTRIBUTION IN EUROPE AND THE NETHERLANDS

H. pilipes has been recorded in many countries in northern and central Europe. It seems to be absent from Britain and Ireland and from most countries near the Mediterranean Sea, but there are records from southern France, European Turkey and southern Russia (Moubayed-Breil, 2008; Saether & Spies, 2010). In the Netherlands larvae have been identified from about 15 localities, mainly in the southeastern and eastern part of the country and the Veluwe; only three of them have been confirmed by identification of pupae or exuviae (Limnodata.nl; own data).

LIFE CYCLE

Emergence has been stated in March and April (Brundin, 1949; Schleuter, 1985; own data). Mating takes place on a substratum. The larvae aestivate in second instar in a cocoon. There is one generation a year (Saether, 1976).

MICROHABITAT

The larvae live in lakes on plants and stones (Meuche, 1938). We collected them in small brooks on wood and on sandy bottoms.

WATER TYPE

H. pilipes has been collected in lakes in North America, Germany and Sweden (Meuche, 1938; Brundin, 1949; Saether, 1976). Schleuter (1985) stated emergence from a woodland pool in Germany. It also lives in streams (Lehmann, 1971; Saether, 1976), but it seems to be absent from fast-flowing brooks and streams. In the Netherlands all records apply to small (rarely larger) non-acid, often temporary brooks in woodland or agricultural environments. In Belarus we collected large numbers of larvae, pupae and adults in a wooded part of the floodplain of the river Pripyat, in small brooklets, pools and a dead river branch.

pH

All records from the Netherlands are from non-acid upper courses, but Schleuter (1985: 135) caught many adults emerging from a pond with pH 5.0. Because many Dutch upper courses are acid and the species has never been found there, *H. pilipes* seems to be almost absent from acid water. However, also nutrient availability can play a role here (see section 2.12).

TROPHIC CONDITIONS AND SAPROBITY

The species has been collected in oligotrophic to polytrophic lakes (Brundin, 1949; Mozley, 1970). In the Netherlands it appears to be absent from brooks with substantial organic pollution.

Hydrobaenus rufus (Kieffer, 1922)

H. rufus is known only from the Czech Republic and France, and possibly from the Near East (Saether & Spies, 2010). According to Saether (1976: 113) it could be only a form of *H. lugubris*. Seven exuviae collected by us in the floodplain of the river Pripyat in Belarus seemed to belong to this species, but some of them were transitional to *H. lugubris* according to characters given by Saether (1976: 76) and Langton (1991: 98–99). Because Saether based the separate status of *H. rufus* mainly on the larva, no definitive answer can yet be given as to whether the two names are synonyms.

OR

Hydrosmittia Ferrington & Saether, 2011

Five European species of the former genus *Pseudosmittia* s. ampl. belong to the new genus *Hydrosmittia*: *H. brevicornis*, *H. montana*, *H. oxoniana*, *H. ruttneri* and *H. virgo*. The larvae of all these species are terrestrial or semiterrestrial, living mainly in pioneer situations or on mosses at the edge of rain puddles, pools and lakes, and more scarcely along streams, just above the water level (Strenzke, 1950; own data). The larvae and exuviae can sometimes be found below and on the water surface respectively (Langton, 1991; Reiff, 1994).

The genus is not treated in this book.

Limnophyes Eaton, 1875

The larvae of the genus *Limnophyes* live mainly terrestrially and therefore only the most common species can be treated here briefly. Their ecology will probably be treated in future in a separate article.

SYSTEMATICS AND IDENTIFICATION

The genus is rich in species in Europe. Most males and females can be identified using the review by Saether (1990). The key to males by Langton & Pinder (2007) is also very useful. Identification of exuviae is more difficult; Langton (1991) gives a key to most species. Some species can be identified as larvae and we hope to publish a key in future. The differences given by Pankratova (1970: 243) are not reliable.

LIFE CYCLE AND SWARMING

Probably most or all species of *Limnophyes* can emerge during the whole year and have several generations a year (Lloyd, 1941; Pinder, 1974; own data). However, development is slow in winter and emergence of most species has not been stated in this season. Parthenogenetic forms exist in a number of species (Saether, 1990; Dettinger-Klemm, 2003). Swarming seems to be necessary for sexual forms (Gibson, 1945: 269–270). Swarming has been seen up to 1 m above the ground or the water surface in *L. minimus* and *L. asquamatus*, but up to 2 m high near shrubs and trees in *L. natalensis* (e.g. Gibson, ibid.; own data).

FEEDING

The larvae feed on decaying organic matter, bacteria and sometimes diatoms (Lloyd, 1943; Crisp & Lloyd, 1954; Moller Pillot, 1984; Langton, 1991: 144).

MICROHABITAT

The larvae of most species live mainly terrestrially in the upper layers of wet bottoms. In water bodies they are often found on plants or wood near the water surface or along the water's edge. As far as is known the larvae are always free living, without any tube. After inundation some larvae can be found floating on the water surface and other larvae can remain in the water column for weeks (Moller Pillot, 2005: 117, 119). However, when rearing the larvae Thienemann (1926) saw them often creep out of the water.

WATER TYPE

Limnophyes larvae can be found in aquatic and terrestrial habitats, but most species are more or less semiterrestrial and live mainly in wet soil, flooded grassland, temporary waters, springs, banks of streams and hygropetric environments. Few species seem to be truly aquatic (*L. pumilio*, ? *L. spinigus*). Nevertheless, larvae are often found in (usually small) stagnant and flowing water bodies and possibly most or all species are able to develop there. It is not known if such larvae spend their whole life there and if survival is lower than in terrestrial habitats (see Microhabitat).

pH

Limnophyes has been found in waters ranging from very acid to very alkaline (Leuven et al., 1987: pH 3.46; Steenbergen, 1993: pH > 8.5). Even at extreme pHs the larvae can be numerous. However, in acid bottoms and pools the genus can be locally dominant (e.g. Wilson, 1988; Moller Pillot, 2003), whereas in very alkaline water the larvae are relatively scarce (Steenbergen, 1993).

SALINITY

Steenbergen (1993) collected the larvae scarcely in brackish water and very rarely in water containing more than 1000 mg Cl/l. In the Baltic Sea they tolerate much higher salinities; Paasivirta (2000) found the genus there widely distributed and sometimes very abundant.

DISPERSAL

Many *Limnophyes* species live in very dynamic environments and can be called circulating species, in which successive generations move from one place to another because the preferred wet bottom is usually not permanently available at one site (Delettre, 1986; Moller Pillot, 2003).

Limnophyes asquamatus Andersen, 1937

SYSTEMATICS AND IDENTIFICATION

Dettinger-Klemm (2003: 84–107) described and keyed the different ecotypes and the differences from other species in the larval and pupal stage. He found one sexual and two parthenogenetic ecotypes and at least one closely related undescribed species.

DISTRIBUTION IN EUROPE AND THE NETHERLANDS

L. asquamatus has been recorded in almost the whole of Europe, but could be absent from the Mediterranean area (Saether & Spies, 2010). In the Netherlands it is a common species in the whole country.

ECOLOGY

The larvae live terrestrially in wet soils or are aquatic. Dettinger-Klemm (2003: 93–95) stated that the aquatic form is parthenogenetic and often bears no lanceolate prescutellars. The larvae are often very abundant in wet soil and along the edges of temporary water bodies, but can also be found below the water level in pools, peat cuttings, trembling bogs and upper courses of brooks, rarely more downstream (own data). They are usually common in acid environments, but also in more alkaline poor grassland, especially in pioneer vegetation. They are able to survive a period of drought in relatively dry mud without a true diapause (Dettinger-Klemm & Bohle, 1996; Dettinger-Klemm, 2003). However, there is no survival in soils with very low water content (Dettinger-Klemm, 2001a; own data).

Limnophyes minimus (Meigen, 1818)

DISTRIBUTION IN EUROPE AND THE NETHERLANDS

L. minimus has an almost worldwide distribution and is widely distributed in Europe (Saether & Spies, 2010); the species is common in all parts of the Netherlands.

ECOLOGY

This is one of the most common chironomid species, living almost everywhere in wet soils and at the margins of brooks and pools (Saether, 1990; own data). The larvae can also be collected in pools and small streams, but seem to be very scarce in larger water bodies. Orendt (1993) and Reiff (1994) collected no exuviae of this species in Bavarian lakes, and the exuviae were also rare in Bavarian streams classificated by Orendt (2002a). Becker (1995) reared only two males from stones along the river Rhine. The species is rather common in springs (e.g. Stur et al., 2005). Lloyd (1941, 1943) found the larvae in large numbers in sewage filter beds. Without doubt the larvae do not as a rule depend on the oxygen content of a water body, because they rarely stray far from the water surface. There are also records from waterworks (own data).
Because the larvae cannot be identified, it is not known whether a significant number of the many records from plants and bottoms in ditches, small lakes and moorland pools (e.g. in Leuven et al., 1987; Steenbergen, 1993; Limnodata.nl) apply to this species or not.

Limnophyes natalensis (Kieffer, 1914)

L. natalensis has been collected in a small number of countries scattered over Europe; most probably the species is common everywhere, but has been insufficiently investigated (Saether & Spies, 2010). In the Netherlands it is common throughout the country.
The species lives mainly terrestrially, but we found it regularly in inundated grassland, marshes and trembling bogs. Stur et al. (2005) collected it from a number of spring habitats in Luxembourg. I also reared many specimens from the gutter of my house.

Limnophyes pentaplastus (Kieffer, 1921)

Limnophyes prolongatus Pinder, 1978: 88, fig. 129A

IDENTIFICATION

Identification of the larvae is not yet possible. Larvae identified as *L. prolongatus* using Pankratova (1970) could also belong to other species. However, in contrast to most other species the colour of thorax and abdomen is violet; in nature it is rather dark violet, but gradually loses its colour in alcohol. The value of other characters (antenna, labrum) has to be verified.

DISTRIBUTION IN EUROPE AND THE NETHERLANDS

L. pentaplastus has been collected in many countries throughout Europe (Saether & Spies, 2010). In the Netherlands it is a common species in southern Limburg and Winterswijk and it has been found scarcely in other Pleistocene regions.

MICROHABITAT

The larvae live mainly at the water's edge along springs and small streams, mainly above the water level. In a Danish spring Lindegaard et al. (1975) collected most larvae in the madicolous zone in the moss carpet. They can also be found on branches and other wood in streams (own data). When rearing the larvae, Thienemann (1926) saw that they often creep out of the water.

WATER TYPE

L. pentaplastus is a common inhabitant of small upland streams, springs and hygropetric mosses (Thienemann, 1926; Caspers, 1980; Lindegaard, 1995; Stur et al., 2005; own data). It can be found scarcely in large rivers (Caspers, 1991) and was even found in the Baltic Sea (Paasivirta, 2000). It seems to be very scarce in slow-flowing lowland streams in the Netherlands. The species is also scarcely collected in woodland pools (Schleuter, 1985; Langton, 1991; Dettinger-Klemm, 2003) and in lakes (Strenzke, 1950; Otto, 1991).

SAPROBITY

The species is a typical inhabitant of springs and small streams and has been found usually in unpolluted environments. However, the larvae live within decaying material; Crisp & Lloyd (1954) reared them even numerously from a patch of woodland mud. Most probably the larvae are very tolerant of high decomposition levels if oxygen is available and anthropogenic poisonous products are absent (cf. *L. minimus*).

Limnophyes pumilio (Holmgren, 1869)

Limnophyes globifer Pinder, 1978: 88, fig. 43D (pro parte); Cranston, 1979: 19–25 (pro parte)

DISTRIBUTION IN EUROPE AND THE NETHERLANDS

L. pumilio has been collected in many European countries, but there are no records from the Mediterranean area (Saether & Spies, 2010). In the Netherlands there are only five records of adult males and some doubtful records of larvae from locations scattered across the country (own data).

MICROHABITAT AND WATER TYPE

Cranston (1979) collected a larva at a depth of 1.5 m in an English lake. Reiss (1968) calls the species characteristic of stones in the surf zone of lakes. We reared the adults from mosses at the banks of a canal in Noord-Brabant at two localities. Most records are from lakes and canals, even in cities (Cranston, 1979; Koskenniemi & Paasivirta, 1987; Langton, 1991; Reiff, 1994; own data). Probably it lives also along river banks (Becker, 1995: 106). The species seems to be more aquatic than most other species of the genus. We never reared it from moist bottoms further from the water.

Limnophyes punctipennis (Goetghebuer, 1919)

DISTRIBUTION IN EUROPE AND THE NETHERLANDS

L. punctipennis is known only from western and central Europe (Saeter & Spies, 2010). In the Netherlands it is a common species throughout the country (own data).

LIFE HISTORY

The species has several generations a year (Steinhart, 1998). Only parthenogenetic multiplication has been stated and probably males do not exist (Saether, 1990; Steinhart. 1998). The eggs are deposited on moist substrata or in water (Steinhart, 1998: 51).

ECOLOGY

Steinhart (1998) collected this species only in temporarily flooded parts of the floodplain of the river Oder and not in permanent water bodies. However, in the laboratory the larvae developed in water without problems and in the aquaria two or three generations were obtained. In nature we reared the species mainly from moist to wet bottoms, but also numerously from wood in a small stream. Reiff (1994) collected the exuviae in several lakes in Bavaria, but rarely in large numbers.

Mesocricotopus thienemanni (Goetghebuer, 1940)

Acricotopus thienemanni Brundin, 1949: 695, 826
nec *Trichocladius thienemanni* Thienemann, 1944: 606, 650
Limnophyes karelicus Pankratova, 1970: 249, fig. 157

M. thienemanni is the only species of this genus. The larvae live in oligotrophic lakes in Russia and Scandinavia as far as southern Sweden and southern Finland (Brundin, 1949; Saether, 1979; Paasivirta, 2012), and, according to Pankratova (1970), also in rivers. The species is not treated in this book.

Mesosmittia flexuella (Edwards, 1929)

Most probably the larvae of *M. flexuella* live mainly in terrestrial habitats; we collected them (in the Netherlands) only from grassland. However, there are several records from adult males reared from submerged aquatic habitats (e.g. Cranston, 1982: 92). We observed several small groups of this species swarming and copulating above our trays of water when we collected larvae and exuviae in nature. Elsewhere we also caught adult males along ditches in grassland. The species is not treated here.

Metriocnemus van der Wulp, 1874

Most species of *Metriocnemus* live mainly in terrestrial habitats. Because this book only covers aquatic species, species which are not found in aquatic habitats, or only very rarely, are mentioned briefly. Much attention has been paid to problems with identification and nomenclature.

SYSTEMATICS AND IDENTIFICATION

Saether (1989, 1995) has revised the genus and described a significant number of the Palaearctic species. However, many species (e.g. *terrester* and *inopinatus*) were not studied and many identification problems have been left. At this moment the genus can be divided into two subgenera: *Metriocnemus* and *Inermipupa*, the latter being monospecific (Langton & Cobo, 1997). In this book no further division into groups are used.
A significant number of the adult males can be identified using Saether (1995) or Langton & Pinder (2007). Most descriptions of pupae can be found in Langton & Visser (2003). Very incomplete keys to larvae are given by Golubeva (1980) and Moller Pillot (1984). Special problems are treated below.

ECOLOGY

When larvae live in aquatic environments they never build a tube. Nolte (1991) stated that the larvae in moss cushions on stones in a German brook were restricted to mosses with permanently emersed areas. As far as is known, prepupae or pupae creep out of the water for pupation. The pupae lie freely or in a gelatinous cocoon. The males usually swarm like other chironomids, but some species (*hygropetricus* type, *tristellus*) can mate without swarming.

Metriocnemus albolineatus (Meigen, 1818)

Metriocnemus atratulus atratulus Strenzke, 1950: 236, 238

IDENTIFICATION

According to Thienemann (1937a, 1944) and Strenzke (1950: 238) the larvae of this species are violet (apparently without white rings). A further difference from *M. eurynotus* could be that the shape of the labral setae S I and S II are different. However, Thienemann (1944: 642, note 28) doubted whether the difference in the labral setae was reliable. As to the colour or the size of the larvae, we hesitate to rely on these characters alone. The procercus has a dorsoanal spur, as in *eurynotus* (Thienemann, 1944; 641). *M.* cf. *albolineatus* of Schmid (1993) is another species because of the very short first antennal segment and aberrant labral setae (see Saether, 1989).
Male adults and exuviae are easily distinguished from those of *M. eurynotus*, although otherwise identification in this genus is difficult (see Saether, 1995; Langton, 1991).

DISTRIBUTION IN EUROPE

M. albolineatus has been collected in many countries throughout Europe (Saether & Spies, 2010). There are at least five records of adult males from the Netherlands (own data); probably the species can be found in the whole country.

ECOLOGY

As far as is known the species lives mainly in terrestrial habitats. Stur et al. (2005) stated emergence from two springs in Luxembourg. We reared a male from wood in a small lowland stream.

Metriocnemus atriclava Kieffer, 1921

NOMENCLATURE

According to Pagast et al. (1941), *M. terrester* is a different species from *M. atriclava*, because:

1. of the absence of a lobe on the gonocoxite in *atriclava*;
2. the pupa of *atriclava* has two distal setae on the anal lobe, whereas that of *terrester* three (Pagast et al. p. 211; Thienemann, 1937a: 182; Langton, 1991: 150).

These authors did not see any *atriclava* specimens and based their knowledge on a bad figure by Goetghebuer. The gonocoxite lobe is present in *atriclava* too (see Saether, 1989: fig. 16; 1995: 62). Probably the type material of *atriclava* has been lost (Saether, 1995: 62) and no new pupal material with two setae on the anal lobe is available.
At this moment *terrester* is absent from all recent keys to adults and *atriclava* from all keys to larvae and pupae. It is questionable whether there are indeed two species.

ECOLOGY

Because the adults cannot be distinguished from those of *M. terrester* and pupae or exuviae have not been found, nothing can be said about their ecology. See under *M. terrester*.

Metriocnemus beringiensis (Cranston & Oliver, 1988)

SYSTEMATICS AND IDENTIFICATION

Saether (1995: 59) wrote that *M. beringiensis* may be no more than a small form of *M. fuscipes*. However, Ekrem et al. (2010) found that the name of the latter species seems to cover four species and that some synonyms might have to be revalidated.
According to Saether (1995: 40), *M. beringiensis* can be separated from *M. fuscipes* by the absence of the anal point (or at most 15 μm long), its wing length of 1.5–1.7 mm, and squama with 18–26 setae (in *fuscipes*: anal point 23–53 μm, wing length 1.8–2.5 mm and squama with 28–40 setae). However, Dutch specimens without anal point sometimes have a wing length of 1.8–1.9 mm; one specimen with anal point 25 μm had only 18 squama setae, and once a specimen with an anal point of 22 μm was caught in a population of only *beringiensis*. This does not mean that *beringiensis* is not a separate species, but it appears to be hardly possible to identify all specimens on the basis of the given characters. See further under Ecology below.

DISTRIBUTION IN EUROPE AND THE NETHERLANDS

M. beringiensis has been collected in Canada and Norway (Saether, 1995). In the Netherlands the species appears to be fairly common in the fen peat regions in Overijssel and Holland/Utrecht (leg. H. Siepel); we caught one male in the dune region.

ECOLOGY

Most populations in the Netherlands have been found in the fen peat regions in nutrient poor marshy vegetation rich in mosses (especially *Sphagnum*). It has never been found in the most common habitat of *fuscipes*, which is drier and often rich in nutrients (see there). From this it seems plausible that *beringiensis* is a separate species or ecotype.

Metriocnemus (Inermipupa) carmencitabertarum Langton & Cobo, 1997

Orthocladiinae gen.? sp.? Pe 3 Langton, 1991: 82

SYSTEMATICS AND MORPHOLOGY

Langton and Cobo (1997) erected a new subgenus for this species mainly on the basis of the armament of the pupal tergites (see also Langton & Visser, 2003). The adult male is absent from Langton & Pinder (2007). The larva has been described only briefly, but is conspicuous because of the very long first antennal segment (126–140 µm long, 22–35 µm wide; AR 4) and the brown pattern on the head. The back of the procercus is brown sclerotised without dorsoanal spur.

DISTRIBUTION

The species has been collected in Spain, Portugal, Poland, Estonia and the Azores. In the Netherlands larvae have been reared to adults from a bucket in Appingedam (Kuper & Moller Pillot, 2012) and there are two other records from different parts of the country.

ECOLOGY

In Spain the larvae have been found living in rain-filled pools in rocks (Langton & Cobo, 1997); the Dutch material from Appingedam was collected in a bucket placed under a roof gutter (Kuper & Moller Pillot, 2012). The last authors give a description of the biology.

Metriocnemus caudigus Saether, 1995

This species is known from Norway, Finland and Germany (Saether & Spies, 2010) and from springs in Luxembourg (Stur et al., 2005). The pupae and larvae have not been described.

Metriocnemus cavicola Kieffer, 1921

Metriocnemus martinii Thienemann, 1944: 557, 643; Kitching, 1972: 53–62

DISTRIBUTION IN EUROPE AND THE NETHERLANDS

M. cavicola has been found in few countries (Saether & Spies, 2010), possibly because of the fact that tree holes are not investigated everywhere. In the Netherlands it has been collected at only nine localities (Moller Pillot & Buskens, 1990: fig. 128), but most probably it lives throughout the country.

ECOLOGY

The larvae live exclusively in water-filled tree holes, mainly in beeches, but sometimes also in other tree species. The biology of the species has been studied by Kitching (1972).

Metriocnemus corticalis Strenzke, 1950

Metriocnemus atratulus corticalis Strenzke, 1950: 235–238, fig. 25b

SYSTEMATICS

Saether (1989) stated that as male imago *corticalis* is separable from other species, such as *M. albolineatus* and *eurynotus*. The pupa has been described by Saether (1989); see also Langton & Visser (2003). The larva is reddish brown, but is so far not separable from the *hirtellus* type of *eurynotus* (see there).

DISTRIBUTION IN EUROPE

The species is only known from Germany (Strenzke, 1950: 238) and France (Moubayed-Breil, 2008: 21).

ECOLOGY

As far as is known *M. corticalis* is a terrestrial species (Strenzke, 1950: 238). It is not treated further in this book.

Metriocnemus eurynotus (Holmgren, 1883)

Metriocnemus obscuripes Saether, 1989: 410–417, figs 7, 8

SYSTEMATICS AND NOMENCLATURE

Pinder (1978: 89) distinguished *M. hygropetricus* from *M. hirticollis* on the basis of the colour of the halteres. According to Saether (1989), *Metriocemus hirticollis* as used by Edwards (1929) and Pinder (loc cit.) is not *M. hirticollis* Staeger. *M. hirticollis* sensu Pinder has to be placed together with *hygropetricus* under *M. eurynotus* (as *obscuripes*). However, the larvae of the former species have a colour, biology and ecology very different from those of the *hygropetricus* type (see below). Without doubt these two types represent different species. Ekrem et al. (2010) stated, using DNA barcoding, that *M. eurynotus* is an aggregate of at least four species, affirming, as previously suggested by Saether (1989: 416), that possibly more than one species is involved. It is not known if the colour of the halteres of the adult male is useful as a character to distinguish the different species.

For the moment only **two ecotypes can be described** in this book, each of which can represent more than one species. Because the name *hirticollis*, as used in most of the literature, can cause confusion (see above), we have chosen for the name *hirtellus*. *M. hirtellus* Goetghebuer (belonging to *eurynotus* s. ampl.) has green larvae and a procercus with dorsoanal spur as mentioned by Thienemann (1937a: 180, 181). Within the *hirtellus* type this author describes differences in the colour of the larvae (p. 181) and the posterior transverse row of teeth on the tergites of the pupa (p. 182). Our division is **based on the colour of the larvae**. Most probably more than two species are involved (see above). We are not certain that our *hirtellus* type is *hirticollis* sensu Pinder (*Metriocnemus* sp. in Langton & Pinder, 2007: 116) and that our *hygropetricus* type is identical with *Metriocnemus eurynotus* sensu Langton & Pinder (2007: 116).

We must note here that it is not known with any certainty whether the **larvae** of both types can be distinguished from those of *M. albolineatus* (see there); those of the *hirtellus* type cannot be distinguished from those of *M. corticalis*. Most probably *M. albolineatus* and *M. corticalis* inhabit more terrestrial biotopes, but identification of larvae alone to species level is as yet not very reliable. It is possible that yet other species resemble *eurynotus* in the larval stage.

The two ecotypes are treated separately here.

Metriocnemus hirtellus type (within *M. eurynotus*, see above)

Metriocnemus hirticollis agg. Moller Pillot, 1984: 98–100, fig. VI.27; Steenbergen, 1993: 530 (pro parte)
nec *Metriocnemus hirticollis* Saether, 1989: 402–404
Metriocnemus atratulus Pankratova, 1970: 254 (pro parte)

IDENTIFICATION

See the introduction under *M. eurynotus*. The larvae of this type have an abdomen without alternating dark violet and white rings; the colour can be greenish, brownish, rarely pale violet or white; the head is rarely dark brown, usually more yellowish. The males are not easily distinguished from those of the *hygropetricus* type, but the females are yellowish brown in colour, while those of the *hygropetricus* type are black (Tomlinson, 1946: 15).

DISTRIBUTION IN EUROPE AND THE NETHERLANDS

According to Thienemann (1937a) this ecotype has been collected in northern and western Europe. In the Netherlands larvae of *M. hirticollis* agg. are common in the whole country (Moller Pillot & Buskens, 1990; Limnodata.nl); most of them will belong to the ecotype under consideration.

LIFE CYCLE

According to Lloyd (1937, 1943), adults can be collected during every month of the year. In the sewage filter beds studied by Lloyd, adults were most numerous during April to June. The
adults need to swarm in the air in order to mate (see also Tomlinson, 1946: 15). Lloyd attributed the scarcity of the species in filter beds in summer and autumn to the activities of *M. hygropetricus* larvae (see there). In more natural habitats we also found pupae and adults mainly in spring and rarely in winter.

FEEDING

According to Lloyd (1943), the larvae in filter beds fed on algae, bacteria, fungi and Protozoa. In the guts of larvae living in natural habitats we found mainly detritus with sand grains and sometimes fungi.

MICROHABITAT

The larvae often live on plants in aquatic habitats, mainly near the water surface, e.g. on floating leaves of *Glyceria*, very rarely on duckweed. They can be found also at the water's edge and sometimes on the bottom of the water body. Further they live in hygropetric conditions in mosses, in springs, in sewage filter beds, etc., and in terrestrial habitats such as wet grassland (Cranston, 1982; own data). According to Lloyd (1943), the larvae of this ecotype live deeper in the sewage filter beds than those of the *hygropetricus* type, although the eggs are laid by the female at the water surface. During rearing in Petri dishes, the prepupae crept upwards and pupated above the water surface in a gelatinous cocoon (see also Tomlinson, 1946: 15), this in contrast to *M. tristellus*.

WATER TYPE

The *hirtellus* ecotype is most common in semiaquatic habitats, such as helocrenes, edges of water bodies and wet grasslands. The larvae are often found in ditches, pools and in the vegetation zone of lakes; sometimes in slow-flowing brooks and streams (Cranston, 1982; Limnodata.nl; own data). Steenbergen (1993) recorded the aggregate in 223 aquatic localities in the Dutch province of Noord-Holland, in ditches, brooks, canals and lakes. The larvae are very rare in fast-flowing water.

pH

The *hirtellus* type has been found only very rarely in acid water with a pH lower than 5.5 (Leuven et al., 1987; van Kleef, 2010) and sometimes at pHs about 6 (own data). Steenbergen (1993) and Limnodata.nl give a mean pH of 8.

TROPHIC CONDITIONS AND SAPROBITY

The larvae usually live in relatively clear water with low orthophosphate content (Steenbergen, 1993; Limnodata.nl). However, they can be found also in sewage filter beds, polysaprobic ditches and mud flats, where much more food is available (Lloyd, 1943; Tomlinson, 1946; Crisp & Lloyd, 1954; own data). The absence of oxygen in such cases is no problem, because the larvae live mainly near the water surface (Steenbergen, 1993; Limnodata.nl; own data).

SALINITY

Steenbergen (1993) found the larvae regularly in water containing more than 300 mg Cl/l and very rarely at concentrations higher than 1000 mg Cl/l.

Metriocnemus hygropetricus type (within *M. eurynotus*, see above)

Metriocnemus hygropetricus Thienemann, 1944: 567, 643; Pinder, 1978: 89, fig. 130D
Metriocnemus violaceus Thienemann, 1944: 567, 643
Metriocnemus hygropetricus agg. Moller Pillot, 1984: 98, 100 (pro parte)
Metriocnemus longitarsus Lloyd, 1937: 1–16; Tomlinson, 1946: 15

IDENTIFICATION

See under *M. eurynotus* and *hirtellus* type above. According to Thienemann (1944: 567) the pupae of *hygropetricus* and *violaceus* are different. Oliver & Sinclair (1989) described the male, female, pupa and larva. The larvae are characterised by a dark brown head and an abdomen with conspicuous dark violet and white rings. Oliver & Sinclair stated that newly mounted larvae have two median teeth, which after some time are often worn to a single rounded tooth. The difference from the larvae of *M. albolineatus* is possibly slight (see under this species).

DISTRIBUTION IN EUROPE AND THE NETHERLANDS

It is not clear which records of *eurynotus* apply to this ecotype, but *hygropetricus* has certainly been collected throughout Europe (Strenzke, 1950) and in Canada. In the Netherlands many records from southern Limburg, Twente, the Veluwe and the dune region of Noord-Holland belong to this ecotype (Moller Pillot & Buskens,

1990; Steenbergen, 1993). Some larvae of this type from terrestrial habitats in Noord-Brabant could belong to *M. albolineatus*, although a male caught on 8 March 1997 with a sweep net near Riethoven belonged to *M. hygropetricus* sensu Pinder (1978).

LIFE CYCLE

Lloyd (1937, 1943) stated that *M. longitarsus* emerged from filter beds throughout the year, most abundantly from December to March and again in June. In brooks, emergence has been stated relatively more in summer (Lehmann, 1971; Ringe, 1974; Pinder, 1974). The species is capable of mating in a confined space, in contrast to the *hirtellus* type (Tomlinson, 1946: 15), but normally swarm at a rather high level, e.g. around the top of a building at 4 m (Gibson, 1945: 264).

FEEDING

In filter beds the larvae feed mainly on algae, but also on bacteria, fungi and Protozoa and on eggs and small larvae of chironomids and psychodids, the last in contrast to the larvae of the *hirtellus* type (Lloyd, 1943; Tomlinson, 1946). Oliver & Sinclair (1989) studied the species in natural hygropetric environment and saw that the gut contents were mainly green, probably derived from algal chlorophyll.

MICROHABITAT

The larvae live in different types of madicolous (e.g. hygropetric) zones, in springs, among mosses and on stones in small brooks, in city fountains and sewage filter beds, possibly also in more terrestrial environments (see distribution). According to Lloyd (1937, 1943), they live more in the upper layers of the filter beds than the *hirtellus* type. Nolte (1989) found them only in the spray zone of moss cushions in a spring brook. Crisp & Lloyd (1954) stated that they are not tolerant of submergence, but Lindegaard et al. (1975) found the larvae in the interior areas as well as in the borders of the moss carpet in the Ravnkilde spring in Denmark. Cranston (1982) found this type also on plants in watercress beds in springs. They pupate in a gelatinous cocoon as the *hirtellus* type (Tomlinson, 1946).

WATER TYPE

The larvae inhabit mainly springs, small brooks and several types of madicolous habitats (see Microhabitat) (Thienemann, 1926, 1944; Lehmann, 1971; Oliver & Sinclair, 1989). In contrast to the *hirtellus* type, the larvae are rarely reported from ditches, pools or lakes (Limnodata.nl) and often live in fast-flowing water (Lehmann, 1971). According to Lloyd (1943), they thrive at much lower temperatures than the *hirtellus* type.

SAPROBITY

The larvae are found mainly in very clean springs and spring brooks, but thrive also in sewage filter beds (see above). As stated for the *hirtellus* type, the absence of oxygen in such cases is no problem, because the larvae live mainly near the water surface.

Metriocnemus fuscipes (Meigen, 1818)

SYSTEMATICS AND MORPHOLOGY

Saether (1989) called *M. fuscipes* a well defined species. However, Ekrem et al. (2010) found that DNA barcoding of their *fuscipes* material indicated the existence of four species. Also, *M. beringiensis* might not be just a small form of *fuscipes* (see there). Because the larvae of *beringiensis* are unknown, larvae identified as *fuscipes* may belong to this species.

As a rule, the abdomen of young larvae is white with very dark violet rings (as in *hygropetricus*). Older larvae are usually yellow.

DISTRIBUTION IN EUROPE AND THE NETHERLANDS

M. fuscipes has been collected throughout Europe (Saether & Spies, 2010). In the Netherlands the species lives in all parts of the country and is probably common everywhere (see Moller Pillot & Buskens, 1990).

ECOLOGY

M. fuscipes is rarely found in water bodies. The larvae are common in moist grassland and feed mainly or exclusively on mosses. They are often found in helocrene springs (e.g. Thienemann, 1926), but live in such cases mainly in the madicolous zone just above the water surface (Lindegaard et al., 1975). The species is rare in very acid environments (own data). More information is given by Strenzke (1950) and Moller Pillot (1984).

Metriocnemus inopinatus Strenzke, 1950

IDENTIFICATION

The adult male is absent from all recent keys. The descriptions of the male and the female by Strenzke (1950a) are very complete and accompanied by two figures. The pupae are characteristic because of the long anal macrosetae (Strenzke, 1950a; Langton, 1991: 148). The larvae can be identified easily using Moller Pillot (1984: 98); see also Strenzke (1950a: 166).

DISTRIBUTION IN EUROPE AND THE NETHERLANDS

The species has been collected in very few countries, probably because of the absence of the adult male from the usual keys. It has been found in all regions of the Netherlands and seems to be very common (Moller Pillot & Buskens, 1990; Moller Pillot, 2005; Limnodata.nl). At seven Dutch localities identification was based on adults or exuviae; in all other cases only larvae had been caught.

OR

LIFE CYCLE

As in some other species of the genus, most larvae are collected from December to April and a small number in summer. Possibly there is a diapause in autumn (own data).

ECOLOGY

We collected most larvae in grassland, sometimes along ditches or pools. Most probably they live there always near the surface. When the grassland becomes inundated the larvae creep around actively like the other species of the genus. We found never larvae in the leaf axils of *Scirpus sylvaticus*, although Strenzke (1950a: 137) found them there numerously. In some cases the larvae lived in horse dung. Larvae and exuviae have also been collected scarcely in springs and streams (Limnodata.nl; own data).

Metriocnemus pankratovae Golubeva, 1980

This species is only known from a spring in Russia (Yaroslav province). The adult male, pupa and larva have been described by Golubeva (1980). The male has a very large gonocoxite lobe. The larva resembles *M. scirpi*, but has a very small spur on the procercus.

Metriocnemus picipes (Meigen, 1818)

IDENTIFICATION

Saether (1995) has described and keyed the male and female (the females were tentatively associated). The pupa has been keyed in Langton & Visser (2003); see also Langton & Moller Pillot (1997). The larva cannot be distinguished from that of *M. terrester* (see there).

ECOLOGY

The larvae live mainly in grassland, sometimes in moist to wet bottoms near the margins of pools and ditches, rarely in marshland (Langton & Moller Pillot, 1997; own data). Specimens found in aquatic environments had probably been transported there by flooding (e.g. Otto, 1991: 42).

Metriocnemus scirpi (Kieffer, 1899)

NOMENCLATURE AND IDENTIFICATION

Because the original description was not available, Strenzke (1950a) made a new description of the male female, pupa and larva. The name *scirpi* is now considered to be a nomen dubium and a new revision is desirable before further decisions are made about the nomenclature.
The larva is very characteristic and has been keyed by Moller Pillot (1984: 98, 102). It resembles the larva of *M. pankratovae* (see there).

ECOLOGY

Strenzke (1950a) found the larvae often in the leaf axils of *Scirpus sylvaticus*. In the Netherlands there are three records of the larvae: two from a small stream and one from a spring (Cuppen & van Nieuwenhuijzen, 2005).

Metriocnemus terrester Pagast, Thienemann & Krüger, 1941

NOMENCLATURE AND IDENTIFICATION

As has been treated under *M. atriclava*, the adult male of *terrester* cannot be distinguished from that species. It is questionable whether these two species are different. The adult male, pupa and larva of *M. terrester* have been described by Pagast et al. (1941). The pupa can be identified using Langton (1991). The larva has been keyed and described by Moller Pillot (1984). However, as the larva of *terrester* still cannot be distinguished from that of *picipes* (Langton & Moller Pillot, 1997), larvae identified as *M. terrester* could belong to this species. The pupae of these two species can be distinguished easily (Langton & Visser, 2003).

ECOLOGY

The larvae and pupae, sometimes associated with adults, have often been collected in terrestrial habitats. In most cases the larvae could belong to *picipes*, and the adults could be *atriclava* if this is a separate species. Many records apply to edges of water bodies, but probably all aquatic finds had been transported there by flooding (e.g. Nolte, 1989: 253).

Metriocnemus tristellus Edwards, 1929

IDENTIFICATION

The adult male and pupa are keyed in almost all the usual keys. The larva is described only briefly in Moller Pillot & Buskens (1990: 87). The procercus is not higher than wide and has a brown sclerotisation at the back, without dorsoanal spur. The head is brown, the abdomen violet.

DISTRIBUTION IN EUROPE AND THE NETHERLANDS

M. tristellus probably lives throughout Europe, but has been collected only in a few countries, most in Western Europe (Saether & Spies, 2010). In the Netherlands it lives in all regions. In the provinces of Noord-Holland and Noord-Brabant, where wet grasslands have been studied more intensively, it appeared to be present almost in every suitable locality, often very abundantly (Moller Pillot, 2005: fig. 4).

LIFE CYCLE

In the Netherlands the larvae and pupae are found from September to early May. It is not known if there is a diapause in early summer. Moller Pillot (2003) called *M. tristellus* a sheltering species, surviving dry periods in low numbers at the edges of ditches when the normal habitat (wet grassland) is too dry. It is also possible to call it a habitat-shifting species.
Mating is possible also without swarming (own observation).

FEEDING

In the laboratory the larvae were seen feeding on decaying plant material, but also scraping a leave surface.

HABITAT

M. tristellus lives in semiaquatic habitats, but has been collected scarcely in ditches and pools (Limnodata.nl; own data). It is extremely common on very wet (inundated) grassland, but has been found also at the edges of water, in trenches and wheel tracks (own data). Steinhart (1998) did not find the species on the floodplain of the river Oder; we caught the larvae and pupae on floodplains only when the water layer was no deeper than a few centimetres and remained for a long time. As a rule, the larvae and pupae (which creep around very actively) are caught with a hand net used for sampling aquatic invertebrates or in pitfall traps for carabid beetles.

Metriocnemus ursinus (Holmgren, 1869)

M. ursinus lives in and around springs in arctic and alpine regions (Thienemann, 1944; Serra-Tosio & Laville, 1991; Saether & Spies, 2010). In Finland the distribution is confined to the northern provinces (Paasivirta, 2012). Pinder (1974) caught the species in a sticky trap in a small stream in England (this record has to be verified). All stages are described in Saether (1989); the male and female have been keyed by Saether (1995). The larva resembles *M. tristellus*.

Molleriella calcarella Saether & Ekrem, 1999

This species has been collected only in the Netherlands. It is probably related to *Paraphaenocladius*; only the male and female have been described (Saether & Ekrem,

1999). In the Netherlands it seems to be not very rare. The larvae appear to live semi-terrestrially in wet woodland, a biotope that has been investigated rarely. There are four records from locations scattered across the country: Bergen op Zoom, Weerribben, Nieuwkoop and Tilburg.

Nanocladius Kieffer, 1913

Microcricotopus Fittkau & Lehmann, 1970: 391–402

SYSTEMATICS AND IDENTIFICATION

Saether (1977) revised the Nearctic and Ethiopian species of the genus. All European species belong to the subgenus *Nanocladius*, which is divided into three groups. Because the name *N. bicolor* has been replaced by *N. dichromus* (Spies & Saether, 2004: 24) the group and species names are now:
gr. *dichromus*: *N. dichromus*, *N. distinctus*;
gr. *parvulus*: *N. parvulus*, *N. rectinervis*;
gr. *balticus*: *N. balticus*.
There is no complete key to all European male adults and larvae (see under the species). The exuviae of all European species can be identified using Langton & Visser (2003).

FEEDING

The guts of the larvae have been found to contain mainly detritus (own data). However, the mouth parts, especially the mandibles, indicate a predatory life. Possibly the larvae suck out chironomids and Oligochaeta including the contents of their guts. As far as is known there are no observations about the feeding habits.

WATER TYPE

The larvae live in brooks, streams and rivers and in (not very small) stagnant water bodies. In lakes they can be found in the littoral and upper profundal zone.

TROPHIC CONDITIONS AND SAPROBITY

Cranston et al. (1983) suppose that larvae of the genus live only in oligotrophic to mesotrophic lakes, sporadically also in moderately eutrophic lakes. However, in the Netherlands the genus appears to be not rare in eutrophic stagnant water bodies, although it is more common in less eutrophic water. Orendt (1993) found *N. dichromus* more in eutrophic lakes. In streams the larvae are often numerous in polluted water, especially at fast currents. Obviously the most important condition is the oxygen content.

Nanocladius balticus (Palmén, 1959)

Microcricotopus balticus Palmén, 1959: 61–65, fig. 1–4

IDENTIFICATION

The adult male is keyed in Langton & Pinder (2007), the exuviae in Langton (1991). The larva is absent from the key by Moller Pillot (1984), but is keyed, described and figured in Cranston et al. (1983) and Klink & Moller Pillot (2003).

DISTRIBUTION IN EUROPE AND THE NETHERLANDS

N. balticus has been recorded in a number of European countries, but might be absent from parts of southern and eastern Europe (Saether & Spies, 2010). There is one possible record from Russia (Zinchenko, 2002: 58). In the Netherlands there are several records from the southern and eastern provinces (Moller Pillot & Buskens, 1990).

LIFE CYCLE

Reiff (1994) collected the exuviae from the end of April until early October. Probably the life cycle will be more or less the same as in *N. dichromus* and *N. rectinervis*, i.e. about three generations a year (see her figures 129–130, p. 252).

FEEDING

Moog (1995) called *N. balticus* a detritus feeder. However, the mouthparts suggest a predatory life. Possibly the larvae feed mainly on Oligochaeta, which are difficult to recognise in the guts.

MICROHABITAT

Palmén (1959) supposed that the larvae live on plants like other species of the genus, possibly also on stones or on the weed *Fucus vesiculosus*. Verneaux & Aleya (1998) collected the larvae in low numbers from artificial substrate.

WATER TYPE

N. balticus lives in the littoral zone of lakes (Saether, 1979; Orendt, 1993; Reiff, 1994), up to 7 m depth (Palmén, 1959). It has rarely been found in rivers (Langton, 1991; Klink, 2010). In the Netherlands the species has been collected in sand pits (Buskens & Verwijmeren, 1989), a fen peat lake (Limnodata.nl), two moorland pools (van Kleef, unpublished) and in a small stream (own data).

pH

In the lakes in Bavaria where the larvae have been found the pH is about 8 or higher (Orendt, 1993; Reiff, 1994). Buskens (unpublished) collected the larvae and exuviae only at a pH of 6.7 or higher. In Dutch moorland pools some larvae and exuviae have been found at pH 6.4 and 5.5 (van Kleef, unpublished).

TROPHIC CONDITIONS AND SAPROBITY

The larvae live in mesotrophic to eutrophic lakes (Fittkau & Lehmann, 1970: 397; Saether, 1979; Janecek, 2000: 455). In lakes in Bavaria the optimum appears to lie in more mesotrophic conditions (Orendt, 1993; Reiff, 1994: 294). Also Buskens (unpublished, cf. Buskens & Verwijmeren, 1989) collected no *N. balticus* in severely eutrophicated sand pits in the Netherlands. The less eutrophicated sand pits always have a very high oxygen content and the absence of *N. balticus* in most other stagnant water bodies in the Netherlands leads to the supposition that the oxygen content is particularly important, as in other *Nanocladius* species.

SALINITY

Palmén (1959) and Paasivirta (2000) collected the species fairly often in brackish water in Finland.

Nanocladius dichromus (Kieffer, 1906)

Nanocladius bicolor Saether, 1979: 69; Moller Pillot & Buskens, 1990: 13, 28, 66 (pro parte)
Nanocladius bicolor agg. Moller Pillot, 1984: 107 (pro parte)

SYSTEMATICS AND IDENTIFICATION

The group and species names are explained under the genus. Only since 1990 has it become clear that both *N. distinctus* and *N. dichromus* are European species. Therefore most data in the European literature under the name *bicolor* or *dichromus* can apply to both species.

DISTRIBUTION IN EUROPE AND THE NETHERLANDS

N. dichromus is known from nearly the whole of Europe (Saether & Spies, 2010). In the Netherlands the species is rather common everywhere, but might be absent from the province of Zeeland.

LIFE CYCLE

From experiments Mackey (1977) calculated a larval development time of 5.4 days at 15 °C, which was substantially shorter than in larger Orthocladiinae species. Nevertheless, in nature three generations a year seems to be the rule (see e.g. Reiff, 1994). Pupae and adults were collected from April to October, more rarely in March and November (Reiss, 1968; Lehmann, 1971; Wilson, 1977; Becker, 1995; Reiff, 1994; own data), and in Swedish lakes from May to September (Brundin, 1949). Probably the larvae have a diapause in early winter in second and third instar (own data). Incidental emergence appears to be possible in winter (December, January; see Becker, 1995: 107).

MICROHABITAT

In lakes and streams the larvae live on plants and stones, but also on open bottoms, probably depending mainly on the availability of food (Brundin, 1949; Lehmann, 1971; Srokosz, 1980; own data). Smit (1982) collected the larvae in more or less equal (low) numbers from the upper and under sides of stones in the river Meuse.

WATER TYPE

N. dichromus lives in stagnant and running water (Fittkau & Reiss, 1978). As can be seen in table 7 (p. 287), it becomes scarcer when **current velocity** increases, and the species is absent from most fast-flowing streams (Laville & Viaud-Chauvet, 1985; Laville & Vinçon, 1991; Orendt, 2002a; Michiels, 2004; own data). However, Lehmann (1971) collected the species in the river Fulda more numerously than *N. rectinervis*. In Dutch lowland streams *N. dichromus* is a common species in middle and lower courses, but is scarce in narrow upper courses. *N. dichromus* is rather common in large rivers (Móra, 2008), but some of the records mentioned in the literature apply to *N. distinctus* (see under this species).

In **stagnant water** the larvae are usually not numerous; Steenbergen (1993) collected them rarely in ditches less than 4 m wide and mainly in lakes more than 100 m wide. The larvae appeared to be rare in water bodies with more than 25% submersed or floating vegetation. *N.* gr. *dichromus* is also scarce in sand pits and storage reservoirs (Mundie, 1957; Mol et al., 1982; Kouwets & Davids, 1984; Buskens & Verwijmeren, 1989; Kuijpers et al., 1992; Schmale, 1999; Ruse, 2002a).

Temperature

In Italian streams *N. dichromus* has been collected at a mean temperature of 22.2 °C. (Rossaro, 1991). This is very high for a rheophilous species. See also under *N. rectinervis*.

pH

The species is almost absent in acid water. Buskens (unpublished) collected no *N.* gr. *dichromus* in sand pits with a pH lower than 6.1. Koskenniemi & Paasivirta (1987) found this group in a lake in southwestern Finland at a pH of 5.5–6 and Leuven et al. (1987) collected it once in a moorland pool with a pH of 5.2. Steenbergen (1993) stated a mean pH of 8.1 for records from the province of Noord-Holland. The species has also been found in the very alkaline Neusiedler See in Austria (Wolfram, 1996).

TROPHIC CONDITIONS AND SAPROBITY

Srokosz (1980; see Kownacki, 1989) collected substantially more *N.* gr. *dichromus* in the river Nida in Poland following pollution (BOD5 5.8); the larvae were still moderately abundant at a site with 45% oxygen saturation (during the day). In English and French streams the exuviae of this group are more numerously collected at polluted sites (Wilson, 1987; Bazerque et al., 1989). In Dutch eutrophic lowland streams *N. dichromus* is moderately abundant; in faster flowing streams the species seems to be more abundant following rather severe pollution.

In **lakes** *N. dichromus* was thought to be characteristic of more or less mesotrophic conditions (Saether, 1979). Steenbergen (1993) collected the larvae significantly more in water with low orthophosphate content (less than 0.15 mg P/l). However, Orendt (1993) called gr. *dichromus* an indicator for eutrophic lakes in Bavaria. Ruse (2002) found the species more in lakes with higher conductivity.

The preference for larger stagnant water bodies (see Water type) and increasing preference for polluted sites in streams as currents increase indicates that **oxygen supply** is an important factor (see table 5).

TRACE METALS

Arslan & Emiroglu (unpubl.) stated that *N. dichromus* was more tolerant to trace metals than other species of chironomids (see also par. 2.18).

SALINITY

Steenbergen (1993) collected the larvae in the Dutch province of Noord-Holland rarely in water with more than 300 mg Cl/l. In the Baltic Sea the species is rather common and tolerates more than 3000 mg Cl/l (Paasivirta, 2000).

Nanocladius distinctus (Malloch, 1915)

SYSTEMATICS AND IDENTIFICATION

The species belongs to the *dichromus* group and closely resembles *N. dichromus*. The adult male has been described and keyed by Saether (1977). The exuviae can be identified only using Langton & Visser (2003). The differences between the larvae of *N. distinctus* and *N. dichromus* have not been sufficiently studied. In the key by Moller Pillot (1984) both species key out as *N. bicolor* agg., which should be changed to gr. *dichromus*.

DISTRIBUTION IN EUROPE AND THE NETHERLANDS

The distribution of *N. distinctus* in Europe is still insufficiently known, because identified European material dates back no earlier than 1992 (Langton, pers. comm.). The species lives in the Nearctic and eastern Palaearctic and has been recorded in a few West European countries and in the rivers Tisza and Danube (Saether & Spies, 2010, Janecek, 2010; Klink, unpublished).

In the Netherlands there are records from the large rivers, the dune region and from the Wilhelmina Canal in Tilburg (see below).

WATER TYPE

Janecek (2010) caught the adults along the river Danube and Klink found the exuviae also in the river Tisza in Hungary. In the Netherlands *N. distinctus* has been collected in the rivers Rhine, Meuse and Grensmaas (Meuse in southern Limburg) (A. Klink, unpublished), in the Wilhelmina Canal (own data) and in a sand pit in the dune region (Schmale, 1999). In some cases the species lived together with *N. dichromus*. In small streams only *N. dichromus* has been collected. However, Driver (1977) collected the species in five small prairie ponds in Canada.

TROPHIC CONDITIONS

According to Saether (1979), the species is confined to oligotrophic and mesotrophic lakes, in contrast to *N. dichromus*. Our data indicate more tolerance to eutrophication.

Nanocladius parvulus (Kieffer, 1909)

SYSTEMATICS AND IDENTIFICATION

N. parvulus is related to *N. rectinervis* (see under the genus). The adult male has been described and keyed by Fittkau & Lehmann (1970). Exuviae can be identified using Langton (1991). The larva is not known with certainty, but possibly has an AR of 2.35 (see Saether, 1977).

DISTRIBUTION IN EUROPE

The species has been collected in few countries, mainly in central and northern Europe, but also in Spain (Saether & Spies, 2010). It is probably absent from the Netherlands.

LIFE CYCLE

Ringe (1974) stated emergence of *N. parvulus* in central Germany from May to October, with rather high numbers only in August, suggesting only one generation.

WATER TYPE

Lehmann (1971) collected *N. parvulus* in the rhithral and potamal zone of the river Fulda. Other authors stated its presence also in fast-flowing streams (e.g. Ringe, 1974; Michiels, 2004) or rivers (Caspers, 1980, 1991). However, the species seems to be scarce and has not been found by many other authors, including Laville & Vinçon (1991), Becker (1994) and Orendt (2002a).

Nanocladius rectinervis (Kieffer, 1911)

IDENTIFICATION

Because adult males of the related *N. parvulus* cannot be identified using Langton & Pinder (2007), the keys in Fittkau & Lehmann (1970) or Saether (1977) are recommended for identifying adult males. The larva has been described in detail by Lindegaard-Petersen (1972). The larva of *N. parvulus* is still unknown. See further under the genus.

DISTRIBUTION IN EUROPE AND THE NETHERLANDS

N. rectinervis most probably lives almost everywhere in Europe, although it has not been recorded in a number of countries (Saether & Spies, 2010). In the Netherlands there are many records from the southern and eastern parts of the country.

LIFE CYCLE

N. rectinervis emerges from March until November (Lehmann, 1971; Wilson, 1977; Drake, 1985; Lindegaard & Mortensen, 1988; Becker, 1995). Most probably there are three generations a year. Because of the very short theoretical development time (see under *N. dichromus*) more generations are not impossible.
The larvae may have a diapause in early winter in second and third instar (Lindegaard & Mortensen, 1988; own data; cf. Prat et al., 1983; Ladle et al., 1984). We collected many larvae in fourth instar as early as the end of February.

MICROHABITAT

The larvae are usually collected on stones and plants (Lehmann, 1971; Drake, 1982; Becker, 1995; Verdonschot & Lengkeek, 2006). We found the larvae in many brooks and streams on plants or stones and only rarely on the open bottom. Lindegaard-Petersen (1972) observed that the larvae build short sand tubes and pupation occurs in semiovoid gelatinous houses covered with sand and detritus.

DENSITIES

The densities of the larvae are always low. Lindegaard & Mortensen (1988) estimated the density of *N. rectinervis* in a 3 m wide upper course to be 209 larvae/m^2 . We found a lower maximum on stones in the middle courses of streams. Becker (1995) reared only very low numbers from stones in the littoral zone of the river Rhine.

WATER TYPE

Current
The species is very scarce in brooks and streams with a current velocity less than 40 cm/s (own data), but seems to be common at faster currents (e.g. Laville & Vinçon, 1991; Orendt, 2002a; Michiels, 2004). Both *N. dichromus* and *N. rectinervis* can be found in lowland streams in fish ladders (own data). See table 7.
In many European regions the species seems to be absent from **lakes** (Brundin, 1949; Cranston, 1982; Orendt, 1993; Reiff, 1994). However, Saether (1979) reported the presence of the species in oligotrophic lakes, most probably in Scandinavia. Otto (1991) collected a small number of specimens in a eutrophic lake in northern Germany.

Dimensions
N. rectinervis settles scarcely in narrow upper courses of streams, but mainly in brooks and streams between 3 and 20 m wide (Lehmann, 1971; Lindegaard & Mortensen, 1988; own data). In large rivers the larvae are widespread but the numbers are rather low (Klink & Moller Pillot, 1982; Becker, 1995; Klink, 2010).

Temperature
According to Gendron & Laville (1997), *N. rectinervis* is a foothill species and lives in higher zones only when the water temperature there is higher than normal.
Rossaro (1991) collected the pupae and exuviae in Italian streams at a mean water temperature of 12.7 °C, which is rather low in comparison with *N. dichromus* (mean 22.2 °C). Because *N. rectinervis* lives there mainly in upper courses it is difficult to decide whether temperature plays a role (as suggested by Rossaro), or only current velocity, or other factors. In the Netherlands and Germany the species is rather common at summer temperatures of 18–20 °C (own data).
Klink (1989) found no remains of the species in the river Rhine before the 20th century; the lower temperatures in previous centuries cannot be the reason.

TROPHIC CONDITIONS AND SAPROBITY

Bazerque et al. (1989) called the species moderately pollution resistant and found the exuviae especially following pollution in de rivers Selle and Somme in northern France. Wilson (1989: 374) also stated increasing numbers of the exuviae following discharge of sewage effluent in English rivers. In the Netherlands and adjacent parts of Germany we found the species often in unpolluted or slightly polluted slow-flowing brooks and streams. However, in streams with faster current the larvae were equally or more numerous in rather heavily polluted water.

In lakes *N. rectinervis* is an indicator of oligotrophic conditions (Saether, 1979). However, Otto (1991) collected a small number of specimens in a eutrophic lake in northern Germany (see Water type). All the data together indicate a high tolerance to organic pollution, but a rather high oxygen demand (cf. *N. dichromus*).

ORTHOCLADIINAE GEN.? L. *ACUTICAUDA* CHERNOVSKIJ, 1949: 23, 95, FIG. 73

Orthocladiine aus Fluszsand Pagast, 1936: 275–278, figs 8–12

SYSTEMATICS AND IDENTIFICATION

The larva has been described and figured by Pagast (1936), Chernovskij (1949) and Pankratova (1970). Pagast also gives a short description of the pupa. In 2012 adult males were reared by A. Przhiboro; a publication is in preparation.

DISTRIBUTION IN EUROPE

The species has been found in Latvia (Pagast 1936), Russia (Przhiboro, pers. comm.) and in the Wümme, a lowland stream in western Germany (leg. M. Siebert, U. Haesloop).

ECOLOGY

Pagast (1936) and Przhiboro (pers. comm.) collected the larvae in lowland brooks in sandy bottoms without fine detritus and sometimes with traces of coarse detritus (never abundant). One of Przhiboro's sites had a stony-gravel-sandy bottom. Current velocity at the time of sampling was between 15 and 50 cm/s. Also Siebert (pers. comm.) and Haesloop (pers. comm.) found the larvae in sandy bottoms in a lowland brook.

Most of these brooks were not more than 5 m wide, but the stony stream investigated by Przhiboro was about 10 m wide and up to 70 cm deep.

Orthocladius v.d. Wulp, 1874

SYSTEMATICS

Some decennias ago the genus was divided into three or four subgenera. Currently six subgenera are generally accepted, delimited by Saether (2005):

Orthocladius
Euorthocladius
Eudactylocladius
Mesorthocladius
Pogonocladius
Symposiocladius.

The last three subgenera are represented in the western European lowland by only a few species:

Mesorthocladius:	*O. (M.) frigidus*
Pogonocladius	*O. (P.) consobrinus* (monotypic subgenus)
Symposiocladius	*O. (S.) holsatus, lignicola.*

Because not everyone is acquainted with this classification, we treat all species of the genus in alphabetic order and mention the subgenus in the heading. In the text we also use the above-mentioned classification when this is not given in the articles cited.

IDENTIFICATION

Identification of the European species has been a problem for a long time. In particular, some adult males can be hardly identified and the keys are difficult to use and not always reliable. The appearance of characters such as the shape of the hypopygium and virga depends on their position on the slide. The revision of the subgenus *Euorthocladius* by Soponis (1990) is mainly based on pupal exuviae, but the adult males and larvae are described as well.
The last review of the subgenus *Orthocladius* (Rossaro et al., 2003) is not generally accepted (Spies & Saether, 2004: 24–26) and the names may possible be changed again in future. Depending on the keys, the names used in the literature apply to different species. In many cases exuviae can be identified better than adults, but it is necessary to use multiple keys, for example Langton (1991) or Langton & Cranston (1991) together with Rossaro et al. (2003). As yet there is no single reliable key for most larvae.
We consistently use the latest nomenclature (including Rossaro et al., 2003), even where authors use a different name. In some cases we could verify identifications, e.g. many slides of Becker (1995).

LIFE CYCLE

The life cycles of the species of *Orthocladius* exhibit obvious differences between species and also within species in different circumstances (fig. 9). Many species can be univoltine, emerging in late winter and spring, or bivoltine, with a large or small second generation in autumn. Williams & Hynes (1974) suggested a univoltine life cycle with a 'diapausing' egg stage lasting from July to November. Such a one-year life cycle, with an egg dormancy period from April to February, has been stated for *Orthocladius rhyacobius* (as Pe 9) by Tokeshi (1986). However, elsewhere the same species also emerged from October to December (usually in smaller numbers) (Lehmann, 1971; Becker, 1995; own data).
We stated the same difference between two populations of *Orthocladius thienemanni* in two small streams in Germany. It is possible that the eggs or some of the eggs do not hatch in summer under the influence of temperature. Rossaro et al. (2003: 232) mentioned that *O. rubicundus* emerges in spring, but 'also in summer if the water temperature is under 15 °C'. Possibly water temperature is not the only factor influencing such differences. It is not clear if the fact that the second generation is usually small is the result of prolonged dormancy of some of the eggs or caused by factors such as predation, food shortage (Tokeshi, 1986a) and larval mortality.
Other variations are also possible. For instance, Schmid (1992) found *O. rivulorum* larvae during the whole year and hardly any indication of a diapause.

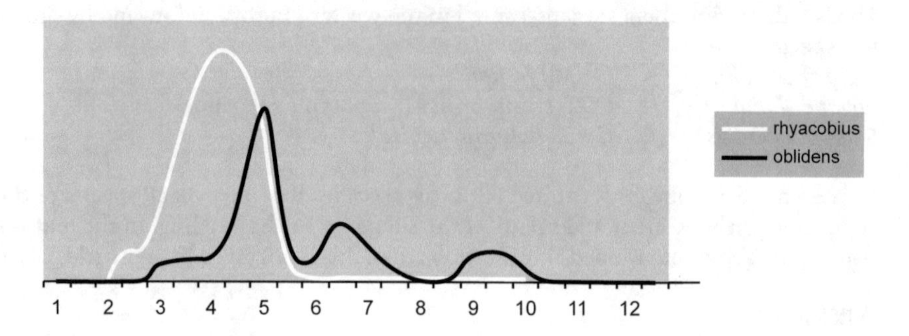

9 *Schematic representation of the emergences of Orthocladius rhyacobius and oblidens in rather fast-flowing upper courses of the Berkel near Coesfeld, Westphalia*

EGGS

Nolte (1993) described the egg mass and the eggs of a species of subgenus *Orthocladius.* She found 170–190 eggs per mass, but mentioned, on the base of the literature (p. 42), 600 eggs in a long tube in *O. obumbratus.*

FEEDING

Tokeshi (1986a) stated that *O. rhyacobius* and *O. oblidens* fed mainly on diatoms. Their density was significantly correlated with diatom abundance. Their annual life cycle, with a short larval period in spring, coincided with maximum abundance of diatoms. Both species exhibited only a small overlap in time because *O. oblidens* larvae appeared in April, when most *O. rhyacobius* already had emerged. Thienemann (1954) reported that *O. rivulorum* feed only on diatoms and filamentous algae. A. Klink (unpublished) found almost only diatoms (*Melosira varians*) in the guts of *oblidens* larvae (in April and August). However, Zinchenko et al. (1986) found 82% detritus in the guts of this species and the larvae seemed to prefer detritus to filamentous algae. She also found some animal remains in the guts. We have also stated that the guts of many larvae of different species of *Orthocladius* from lowland streams had very few diatoms in their guts and much detritus. For instance, one *O. oblidens* larva had consumed mainly filamentous algae, while others had consumed mainly detritus or diatoms. *O. thienemanni* larvae in a stream in Germany had mainly diatoms in the guts, and in a small stream near Aalten (province of Gelderland) almost only detritus.

Our tentative **conclusion** is that at least some species are more or less opportunists or grow better on diatoms (or algae in general), but that they can live and emerge also when feeding on detritus. It is possible that certain species can maintain themselves only when enough diatoms are available. Although Moog (1995) suggested that the larvae of subgenus *Orthocladius* feed mainly on detritus, there is most probably no obvious difference between the subgenera *Euorthocladius* and *Orthocladius.*

MICROHABITAT

Most *Orthocladius* larvae are usually found on stones and plants and few on sandy sediments (Pinder, 1980; Tokeshi, 1986, Becker, 1995; own data). However, such differences depend on the availability of food and oxygen and on current velocity, and can therefore vary between locations and seasons (see Pinder, 1980). Williams & Hynes (1974) found many larvae also in the sediment: young larvae only in the upper layers, older larvae up to 70 cm deep. In general, different species in the same water body appear to inhabit more or less the same microhabitat and seem to avoid concurrence more through differences in life cycle, preferred current, temperature regime, etc. Tokeshi & Townsend (1987) stated that the coexistence of different species seems to be facilitated mainly by random patch formation and the absence of strong interspecific competition.

WATER TYPE

Punti et al. (2009) investigated almost pristine mountain streams in Spain and called *Orthocladius* a tolerant and opportunistic genus, not related to a well-defined range of environmental variables. Notwithstanding this, many species appear to exhibit very different distribution patterns. Larvae of the genus *Orthocladius* inhabit mainly flowing water and many species are even confined to fast-flowing streams. Table 2 gives an impression of the differences between some species of the genus. A typical exception is *Orthocladius holsatus*, which is an inhabitant of more or less stagnant water. Some more species can be present in lakes with a stable oxygen regime in the littoral zone (Reiff, 1994). It is not clear if trophic conditions or oxygen cause the absence of the genus from most lakes in the Netherlands. The genus is very scarce in pools and ditches. *Orthocladius fuscimanus* in particular can be found in hygropetric environments, even along the edges of stagnant water bodies.

Table 2. The occurrence of some Orthocladius species in small streams and ditches with different current velocities

	current (cm/s)				
	< 10	10–25	25–50	50–75	75–100
Orthocladius excavatus	1	1	3	3	2
Orthocladius holsatus	7	2	1	0	0
Orthocladius oblidens	+	1	3	3	3
Orthocladius rhyacobius	0	0	1	5	4
Orthocladius rivulorum	0	0	+	3	7
Orthocladius rubicundus	0	0	+	3	7
Orthocladius ruffoi	0	0	0	1	9
Orthocladius thienemanni	0	+	1	5	4

pH

The whole genus *Orthocladius* appears to be rare in acid environments. Raddum & Saether (1981) did not collect the genus in Norwegian lakes with pHs between 4 and 5 and only scarcely in two lakes with pHs of 5.56 and 6.25. Leuven et al. (1987) and van Kleef (unpublished) found no larvae of the genus in acid moorland pools and Orendt (1999) did not find the larvae in acidified brooks in Germany. In an upper course in the Netherlands with a pH of about 6 the larvae were very rare (Moller Pillot, 2003). However, Laville & Vinçon (1991) collected *Orthocladius* in Pyrenean streams with low alkalinity (from 0.3 meq/l).

TROPHIC CONDITIONS AND SAPROBITY

Without doubt the trophic state of a water body is an important factor for the presence of *Orthocladius*. In this paragraph we discuss the complications arising from interaction with other factors.

Lakes

Maasri et al. (2008) stated that the numbers of larvae of *Orthocladius (Orthocladius)* in riffles in a Mediterranean stream in France decreased when eutrophication increased and were almost absent at the most eutrophicated site, although the numbers of diatoms increased with eutrophication. The water temperature did not differ significantly between the reaches and the oxygen content in the most eutrophicated reach was still above 5 mg/l. The authors suggest that the relative densities of different groups of algae can influence the presence of chironomid species.

Langdon et al. (2006) reported the presence of *Orthocladius* larvae mainly in less eutrophic lakes, but Reiff (1994) found differences between species in Bavarian lakes: *O. rubicundus* more in eutrophic lakes and *O. oblidens* mainly in mesotrophic lakes. The whole genus seems to be absent from the very eutrophic lakes in the Netherlands and even from mesotrophic fen peat lakes, but it is rather common in the mesotrophic dune lakes (Schmale, 1999). The larvae are rather common in moderately eutrophic Danish lakes in the stony littoral zone (Brodersen et al., 1998), so the substrate could play an important role here. Clearly, the presence or absence of *Orthocladius* is not just a question of trophic conditions alone.

Streams

For saprobity, Moog (1995) stated that β-mesosaprobic water in streams provides optimum conditions for almost all species of the genus, but that at least two species of the subgenus *Orthocladius* (*O. oblidens* and *O. rubicundus*) could be encountered sometimes in polysaprobic water (cf. Bazzanti & Bambacigno, 1987; Bitušík, 2000: 93; Wilson & Ruse, 2005). In lowland streams all *Orthocladius* species appear to be less tolerant to organic pollution and the larvae are usually absent from α-mesosaprobic streams. This was stated also for rather fast-flowing streams in southern Limburg and northern France (Cuijpers & Damoiseaux, 1981; Bazerque et al., 1989; own data). Our data suggest that *Orthocladius* species cannot go through their whole life cycle in polluted streams, but that older larvae from elsewhere can survive on plants in hypertrophic water above heavily polluted bottoms. Milošević (pers. comm.) also found the larvae in organically polluted downstream reaches in Serbia, but only in March.

Table 3. The occurrence of some Orthocladius species in streams with different saprobity and oxygen contents (outside mountain regions)

	saprobity							oxygen			
	ol	ol/B	B	B/A	A	A/p	p	stab	unst	low	rott
Orthocladius frigidus	3	4	3	0	0	0	0	10	0	0	0
Orthocladius oblidens	1	2	5	2	(+)	0	0	8	2	0	0
Orthocladius rhyacobius	1	3	4	2	(+)	0	0	9	1	0	0
Orthocladius rivulorum	1	4	4	1	0	0	0	10?	?	0	0
Orthocladius rubicundus	1	3	4	2	+	0	0	9	1	0	0
Orthocladius thienemanni	1	3	4	2	(+)	0	0	9	1	0	0

SALINITY

There are hardly any records of the genus from brackish water in the Netherlands. Tempelman (unpublished) found *O. excavatus* regularly in slightly brackish ditches on the islands in the province of southern Holland. Ruse (2002) collected the exuviae in England almost exclusively in lakes with low conductivity.
In the Baltic region the genus is not rare in brackish environments. Besides lake dwelling species, such as *O. consobrinus*, species of flowing water can also be found in the Baltic Sea (Paasivirta, 2000).

Orthocladius spec. Aalten

Orthocladius ?? rivulorum Moller Pillot, 1984: 115; Moller Pillot & Buskens, 1990: 28, 67

IDENTIFICATION

The species is only known as larva (description in Moller Pillot, 1984). It resembles the American species *O. roussellae* (Soponis, 1990: 34) because of the 7 or 8 pairs of lateral mental teeth. However, the premandible has only one apical tooth. A further character is the rather long l_4 (150 µm or more).

DISTRIBUTION IN THE NETHERLANDS

The species has been collected at seven localities in the eastern and southeastern part of the Netherlands (Moller Pillot & Buskens, 1990: map 142).

WATER TYPE

All records are from small to very small upper courses, usually with very slow current. Most of these small streams ran through woodland, and in most cases *Hydrobaenus* larvae were also present.

Orthocladius (Euorthocladius) ashei (Soponis, 1990)

Orthocladius rivicola Langton, 1984: 142, fig. 49c; Klink & Moller Pillot, 1982: 52

IDENTIFICATION

O. ashei was recognised by Lehmann (1971) as a type of *O. rivicola* (see also Rossaro, 1982: fig. 31: *rivicola* β), but it was only described in 1990. The adult male has not been recognised before. Becker (1995) most probably overlooked it. Langton (1984) keyed the exuviae of the species as *O. rivicola*; most pupae and exuviae identified as *rivicola* between 1984 and 1991 belong to *ashei*. Larvae cannot be identified to the species level (Soponis, 1990: 13, 19; Schmid, 1993: 133, 144).

DISTRIBUTION IN EUROPE AND THE NETHERLANDS

The species has been recorded in all parts of Europe (Saether & Spies, 2010). In the Netherlands there are only records from the large rivers (coll. Klink; reference coll. Het Waterlaboratorium).

LIFE CYCLE

Michiels (1999) collected the exuviae in the river Salzach in Bavaria from late winter to the autumn, most numerously in summer and autumn. See also the introduction to the genus *Orthocladius*.

MICROHABITAT

Schmid (1993) stated the presence of larvae on stones and on gravel with mosses (and sometimes plants) and suggested that the species might be associated with epilithic and epiphytic algae.

WATER TYPE

Based on data from Schmid (1993) and Gendron & Laville (2000), *O. ashei* is a montane euryzonal species in the Alps and Pyrenees. Langton (1991) reported its occurrence in streams in hilly regions. Garcia & Laville (2001) collected the exuviae in rather large numbers in the lower part of the river Garonne in France. It also appears to be common in the Dutch stretch of the river Rhine (Klink & Moller Pillot, 1982, as *O. rivicola*).
There are no records from lakes or other stagnant water bodies.

Orthocladius (Euorthocladius) calvus Pinder, 1985

IDENTIFICATION

The adult male of *O. calvus* is very similar to that of *O. thienemanni* (Soponis, 1990: 19–20; Langton & Pinder, 2007: 124). The pupa can be rather easily identified using Langton (1991). The larva differs from *O. thienemanni* in having a distally bifid premandible (Pinder, 1985: 240).

DISTRIBUTION IN EUROPE

O. calvus has been collected only in the most westerly part of Europe, from Germany and the British Isles to Spain (Saether & Spies, 2010). There are no records from the Netherlands.

WATER TYPE

O. calvus has been collected only in running water. The early colonisation in an experimental recirculating stream in England was striking: after only 16 days a maximum population density was attained. Shortly thereafter the densities declined, while the larvae of other chironomid species increased (Ladle et al., 1985). Langton & Casas (1997) stated that the species settled very quickly in a Mediterranean temporary mountain stream after a long period of drought.

Orthocladius (Orthocladius) carlatus (Roback, 1957)

O. carlatus has been found in the river Garonne (Garcia & Laville, 2001) and possibly on Corsica. Elsewhere the species is only known from the Nearctic (Saether & Spies, 2010). The species is not treated here.

Orthocladius (Pogonocladius) consobrinus (Holmgren, 1869)

Orthocladius crassicornis Meuche, 1938: 482; Brundin, 1949: 712 (fide Brundin, 1956: 99–100; Pinder & Cranston, 1976)
Pogonocladius consobrinus Moller Pillot, 1984: 132, fig. VI.39

SYSTEMATICS AND IDENTIFICATION

O. consobrinus was previously placed in a separate genus, but the justification even for a separate subgenus is now under discussion (Rossaro et al., 2003: 227). Identification of the adult male is not very easy (see Pinder & Cranston, 1976; Rossaro et al., 2003: 227). Pinder & Cranston have figured the wing in comparison with that of *O. glabripennis*. Identification of pupae and larvae do not present any problems.

DISTRIBUTION IN EUROPE AND THE NETHERLANDS

The species has been found in many European countries, but seems to be absent from parts of the Mediterranean area (Saether & Spies, 2010). In the Netherlands the species has been collected in all provinces except Friesland, Zeeland and Limburg (Limnodata.nl; own data).

LIFE CYCLE

Brundin (1949) and Shilova (1976) stated in Sweden and Russia only one generation a year, emerging in early summer. Elsewhere two or three generations have been found, emerging from May to November (Reiss, 1968; Reiff, 1994).

FEEDING

Most probably the larvae feed on diatoms and filamentous algae.

MICROHABITAT

The larvae live mainly on plants and stones (Meuche, 1938; Shilova, 1976; Mol et al., 1982). Brundin (1949) found them also in the silty bottom of Swedish lakes and Lindegaard & Jónasson (1979) collected them from a bottom consisting of diatoms and *Cladophora* in a eutrophic lake in Iceland.

DENSITY

Lindegaard & Jónasson (1979) found densities from 26 to 3700 larvae/m^2 on the bottom in the profundal zone of Lake Myvatn on Iceland. It is not known in which degree larval densities are dependent on the presence of diatoms (see Microhabitat).

WATER TYPE

O. consobrinus lives in the littoral and sublittoral zone of lakes (Shilova, 1976; Saether, 1979; Reiff, 1994), and in Iceland also in the profundal zone (Lindegaard & Jónasson, 1979). It is also sometimes recorded from streams or rivers (e.g. Shilova, 1976: 97; Singh & Harrison, 1984; Scheibe, 2002; Orendt, 2002a). In the Netherlands it is a scarce species in fen peat lakes and large undeep remnants of the former lake IJssel (Randmeren), sometimes collected also in sand pits and other lakes (Limnodata.nl; own data). Around these lakes the larvae have sometimes been found in smaller peat cuttings and ditches, and there is one record from a moorland pool (van Kleef, unpublished).

TROPHIC CONDITIONS

Saether (1979) called the species characteristic of very oligotrophic lakes. Elsewhere in Europe it lives also in less oligotrophic water (Moller Pillot, 1984). Reiff (1994) collected the exuviae in Bavaria only in mesotrophic and eutrophic lakes, sometimes numerously. In the Netherlands it inhabits mesotrophic to very eutrophic lakes, although it is restricted in its distribution (see Water type). The reason for this difference is unknown.

SALINITY

In the Baltic region *O. consobrinus* can be found in brackish water. Tõlp (1971) collected the species in bays in Estonia in water with more than 3000 mg Cl/l (salinity > 6‰). Paasivirta (2000) found it at several locations in the Baltic Sea.

Orthocladius (Orthocladius) dentifer Brundin, 1947

Orthocladius oblidens Langton, 1984: 154 (misidentification)
Orthocladius Pe 4 Langton, 1991: 186, fig. 77e

IDENTIFICATION

The adult male is characterised by a tooth-like process on the gonostylus. According to Brundin (1947: fig. 42) and Rossaro et al. (2003: p. 217, fig. 3), this tooth is rounded, but in Langton & Pinder (2007: fig. 64A, 172D) it is more triangular. The pupa can be identified using Langton (1991). The larva is still undescribed, but can be recognised from the protruding median mental tooth, the backwards placed setae submenti and the relatively long l_4 setae. These characters are based on a pupa with attached larval skin and larvae associated with exuviae from another place (leg. H. Cuppen). The central mental tooth is different from that of Soponis (1977). Such differences in the central tooth have been stated also in *O. fuscimanus* and *O. frigidus* by Michailova (1982, 1985) and Soponis (1987).

DISTRIBUTION IN EUROPE AND THE NETHERLANDS

O. dentifer has been recorded in a number of countries scattered across Europe, except for the southeast (Saether & Spies, 2010). In the Netherlands the species has been collected in three small brooks on the Veluwe (leg. H. Cuppen, A. Klink) and in some lowland brooks in Noord-Brabant (own data) and eastern Groningen (A. Klink, unpublished).

FEEDING

A. Klink (pers. comm.) found the gut of a larva filled with filamentous algae.

WATER TYPE

Brundin (1949) collected *O. dentifer* in large numbers emerging from the oligotrophic lake Innaren in Sweden. Reiff found the exuviae of one specimen in the mesotrophic lake Brunnsee in Bavaria. Rossaro et al. (2003) wrote that the species is restricted to lakes. However, Scheibe (2002) found *O. dentifer* in the rhithral zone of a brook in the Taunus mountains in Germany and Orendt (2002) collected the exuviae in the lower course of the river Inn in Bavaria. Most records from the Netherlands are from small brooks.

Orthocladius (Orthocladius) excavatus Brundin, 1947

Orthocladius ticinoi Rossaro & Prato, 1991: 65–66, fig. 16–18
nec *Orthocladius excavatus* Lehmann, 1971: 486, figs 12, 13; ? Laville & Vinçon, 1991: 1777; ? Schmid, 1993: 139
Orthocladius glabripennis Langton. 1991: 188, fig. 78 e, f (pro parte)

NOMENCLATURE AND IDENTIFICATION

The name *excavatus* is still under discussion (Spies & Saether, 2004). Here we follow Rossaro et al. (2003), based on their study of the lectotype. The species differs from *O. rhyacobius* in having a well developed virga in the adult male and a very small and colourless apical taeniae in the pupa. For descriptions of adult male and pupa see Rossaro & Prato (1991) and Rossaro et al. (2003). However, the characters mentioned here based on the last two publications are not always reliable and it is not possible to distinguish between the exuviae of *excavatus* and *glabripennis* because the taeniae can be invisible or absent (own data). In fact, many publications use the name *O. glabripennis*. The species is absent from the key to adult males by Langton & Pinder (2007) and from the keys to pupal exuviae by Langton (1991) and Langton & Cranston (1991). The larva resembles that of *O. rhyacobius* and some other species of the genus; a provisional key is in preparation.

DISTRIBUTION IN EUROPE AND THE NETHERLANDS

Because of identification problems, Saether & Spies (2004) call most records of *O. excavatus* doubtful. Most probably the species lives in large parts of Europe, considering the many records from Italy, Germany, the Netherlands and Sweden.
In the Netherlands *O. excavatus* has been collected in brooks in the eastern part of the country and in southern Limburg, and also in the large rivers, in dune lakes and in a number of stagnant or slow-flowing ditches on the islands in the provinces of Zuid-Holland and Noord-Brabant (identification of exuviae verified). The species seems to be absent from large parts of the Pleistocene area.

LIFE CYCLE

Rossaro et al. (2003) saw adults with exuviae from Italy collected from January to April and in November, and from Sweden from March to early June and in October. Our data correspond with theirs (in spring most numerous in April; rare in autumn). Most probably the life cycle is the same as in *O. rhyacobius* (see there). See also the Life cycle section in the introduction to the genus.

FEEDING

A larva from a ditch in Zuid-Holland had many diatoms in the gut, together with much clay and detritus. See further under the genus.

WATER TYPE

Rossaro et al. (2003) mentioned records from Swedish lakes (cf. Brundin, 1956: 104) and rivers and lakes in Italy. The species is rather common in the river Rhine (Becker, 1995: 112, adult males, as *O. obumbratus*; Het Waterlaboratorium, reference coll. exuviae). Most probably some of the records of *O. glabripennis* exuviae from fast-flowing streams belong to *excavatus* (see under Identification). We collected the exuviae in the Netherlands and adjacent parts of Germany in a large number of brooks with slow to rather fast currents and in dune lakes. Most records from dune lakes under different names may belong to this species. The temperature of these lakes can rise to more than 20 °C; in the ditches the temperature can be still higher.
In contrast to almost all other species of the genus, *O. excavatus* has been collected in a large number of ditches in the western part of the Netherlands, mainly on clayish bottoms. Langton (1991) stated the presence of this (?) species also in drains and ditches (as *O. glabripennis*).
The figures on current preference in table 2 (p. 155) and table 7 (p. 287) are provisional.

SALINITY

In the western part of the Netherlands larvae and pupae have been collected in slightly brackish ditches (Grontmij | Aqua Sense, unpublished).

Orthocladius (Mesorthocladius) frigidus (Zetterstedt, 1838)

SYSTEMATICS AND IDENTIFICATION

In the past the species was thought to belong to the subgenus *Euorthocladius* (e.g. in

Pinder, 1978) or *Orthocladius* (e.g. Soponis, 1987; Langton & Pinder, 2007). Saether (2005) erected the new subgenus *Mesorthocladius* and provided diagnoses for all stages and both sexes. Soponis (1987) gives descriptions of male, female, pupa and larva of the species. Head length and head width of all instars are given by Schmid (1993); the head of young larvae is dark brown.
Identification of the adult male and the pupa does not present any problems. Using the key to larvae of the genus in Moller Pillot (1984), the larvae of *O. frigidus* can be confused with some species of *Euorthocladius*. The most reliable difference between these species is the seta interna of the mandible, which is absent in *O. frigidus* and present in all species of this subgenus. Contrary to Schmid (1993: 132), one species in the subgenus *Eudactylocladius* also appears to have no seta interna (Cranston, 1982: 100). Michailova (1985) and Soponis (1987) stated variation in the width of the central mental tooth in the larva of *O. frigidus*. Schmid (1993) found a population with a larger size than other populations.

DISTRIBUTION IN EUROPE AND THE NETHERLANDS

O. frigidus has been found in all parts of Europe (Saether & Spies, 2010), most commonly in mountain regions and only scarcely in most lowland regions (e.g. Bazerque et al., 1989). In the Netherlands the species has been collected on the Veluwe (leg. A.Klink, see Moller Pillot & Buskens, 1990). The exuviae of one specimen have been collected in the river Rhine (Klink & Moller Pillot, 1982), but the origin of these is unknown. All other records have to be verified.

LIFE CYCLE

Emergence has been reported in all months of the year, with the highest numbers in spring or early summer (from April to July) (Lehmann, 1971; Ringe, 1974; Caspers, 1983; Soponis, 1987; Schmid, 1992). Most data suggest at least a small second generation in the autumn. Schmid found young larvae during the whole year, but the fact that the autumn generation is very small suggests that a quiescence in the egg stage is possible, as in *O. rhyacobius*.

MICROHABITAT

The larvae live in mud tubes, younger larvae also freely, on stony substrate and among mosses (Lehmann, 1971; Cranston, 1982). Schmid (1992) found many larvae in the sediment in an alpine gravel brook: the highest densities in the first 10 cm of the sediment, but second instar larvae were most numerous between 10 and 30 cm deep.

WATER TYPE

The larvae live in springs and fast-flowing reaches of streams (Lehmann, 1971; Laville & Vinçon, 1991; Schmid, 1993). However, it is more a lotic than a crenophilic species and is scarcely mentioned from true springs (Lindegaard, 1995). In Italy the larvae are confined to streams and lakes with a maximum temperature of 10 °C (Rossaro, 1991; Rossaro et al., 2003). In streams in Germany a much higher temperature appears to be tolerated (Orendt, 2002a), although they are found there mainly in summer cold streams. Ringe (1974: 255) attribute the limited abundance in the Rohrwiesenbach to a combination of slow current and high temperature. Except for spring brooks, it is rarely found in lowland streams (e.g. Lindegaard & Mortensen, 1988).

TROPHIC CONDITIONS AND SAPROBITY

O. frigidus has been reported from oligotrophic lakes in Iceland (e.g. Lindegaard, 1980), Slovakia (Bitušik, 2000) and Italia (Marziali et al., 2008). In streams the species gives a strong indication of oligosaprobic to β-mesosaprobic conditions (Moog,

1995; Wilson & Ruse, 2005). In the river Rhine and in the Selle in France *O. frigidus* is confined to the oligosaprobic upper courses (Wilson & Wilson, 1984; Bazerque et al., 1989), but in other fast-flowing streams in Germany it is still present after some organic pollution (Orendt, 2002a; Michiels, 2004). It is rather difficult to separate the influence of oxygen content and temperature from that of organic pollution in a narrower sense (see also under Water type).

Orthocladius (Eudactylocladius) fuscimanus (Kieffer, 1908)

Orthocladius bipunctellus Lehmann, 1971: 484; Michailova, 1982: 82–91; Ringe, 1974: 238
Orthocladius hygropetricus Cranston, 1982: 100; 1984: 873–895

SYSTEMATICS AND IDENTIFICATION

Cranston (1984) studied the subgenus *Eudactylocladius* in Europe and redescribed the adult male and immature stages of *O. fuscimanus*. Older keys to the species of this subgenus were not reliable. The immature stages of some species are still unknown. Larvae of *fuscimanus*, *mixtus* and *olivaceus* can be identified using the key by Cranston, 1984: 891. However, as Michailova (1982) pointed out, the width of the middle tooth of the mentum in *O. fuscimanus* is very variable. For measurable features of male, female, pupa and larva see Cranston, 1999: 280–282.

DISTRIBUTION IN EUROPE AND THE NETHERLANDS

O. fuscimanus has been recorded from all parts of Europe, including several islands in the Mediterranean Sea and the Atlantic Ocean (Saether & Spies, 2010). In the Netherlands the species appears to be present almost everywhere in the eastern and southern part of the country and in the large rivers (mainly unpublished data). There are hardly any records from the western provinces.

LIFE CYCLE

From the limited number of data it can be concluded that in western and central Europe *O. fuscimanus* emerges from early March to October, probably most abundantly in early spring and/or in summer (Lehmann, 1971; Ringe, 1974; own data). These data suggest that the life cycle can be aberrant from that of other species of this genus (see the introduction to the genus).

FEEDING

In the guts of *O. fuscimanus* from different localities we found almost only detritus. The fact that the species has been collected from trickle bed sewage works (Cranston, 1984) suggests feeding on bacteria. See, however, the introduction to this genus. Moog (1995) suggests that *O. fuscimanus* is an active filterer. This seems to be very improbable, because the species often lives at the water's edge and even in stagnant water bodies.

MICROHABITAT

O. fuscimanus has been found in all types of hygropetric environments: in thin water film flowing over rocks, on wood and stones on the banks of canals and streams, in sewage filter beds, in fountains in cities, etc. (Cranston, 1984, 1999; Michailova, 1982; Hamerlik & Brodersen, 2010; Tempelman, unpublished data; own data). The larvae live more rarely on stones under the water surface in streams (Lehmann, 1971). A. Kuijpers (pers. comm.) found one larva at great depth in a water reservoir.

When the larvae are reared in vessels, the pupae climb up the glass wall before emergence (own data).

WATER TYPE

O. fuscimanus is most often reported from springs and brooks, relatively less from rivers (especially on dams, waterfalls and wood and stones along the banks). The larvae are much more common in fast-flowing streams with many stones than in lowland streams. They are locally common on the banks of canals and reservoirs (see Microhabitat) (Cranston 1984, 1999; own data).

SAPROBITY

The larvae can live in sewage filter beds and can therefore tolerate very polluted water, as long as they can creep around at the surface and therefore have no shortage of oxygen.

SALINITY

Cranston (1999) stated that the species tolerates waters of high conductivity, such as bodies contaminated by marine salt spray or in percolating filters of sewage works. Exact data are not available.

Orthocladius (Eudactylocladius) gelidorum (Kieffer, 1923)

O. gelidorum is an arctic species and is not treated here. For identification and descriptions see Cranston (1999).

Orthocladius (Eudactylocladius) gelidus Kieffer, 1922

IDENTIFICATION

Identification of the male does not present any problems with any key. Cranston (1999) described the male and female and listed many measurable features. The pupa can be identified using Cranston (1999); ? *O. gelidus* in Langton (1991: 182) probably applies to *O. gelidorum*. The larva cannot be distinguished from that of *O. fuscimanus* (Cranston, 1999: 282,289).

DISTRIBUTION IN EUROPE AND THE NETHERLANDS

O. gelidus has been recorded only from northern Europe, the British Isles and Italy (Saether & Spies, 2010). In the Netherlands A. Klink (unpublished) caught a male and a pupa at a waterfall on the Veluwe.

Orthocladius (Orthocladius) glabripennis Goetghebuer, 1921

IDENTIFICATION

As in other species of the genus, identification of adult males is not easy. The exuviae resemble those of *O. excavatus*, which is absent from the key by Langton & Cranston (1991). Owing to this, records of *O. glabripennis* in many publications (e.g. Wilson, 1989; Reiff, 1994; Orendt, 2002a) may apply to *O. excavatus*. According to Rossaro et al. (2003), the latter species can be recognised by the presence of small colourless taeniae; however, these are not always clearly visible. In all populations of *excavatus* we also found some exuviae which certainly had no taeniae. Given the ongoing discussion about *O. excavatus* (see there) and the fact that most investigations after 1991 are based on identification of exuviae, the ecology of *O. glabripennis* cannot be treated now.

DISTRIBUTION IN EUROPE AND THE NETHERLANDS

O. glabripennis has been recorded in almost the whole of Europe except for Russia (Saether & Spies, 2010). There are no reliable records from the Netherlands.

LIFE CYCLE

The life cycle of this species is most probably the same as that of other species of the genus: most emergences in spring and a small second (or third) generation in autumn (e.g. Mundie: 1957).

Orthocladius (Symposiocladius) holsatus Goetghebuer, 1937

Cricotopus holsatus Moller Pillot, 1984; 65, fig. VI.12.ak, al

SYSTEMATICS AND IDENTIFICATION

The systematic position of *O. holsatus* has been a problem for a long time. For students of larvae it seems to be a *Cricotopus* because of the setal tufts on the abdominal segments, but other systematic workers considered it to be a normal *Orthocladius*. At the present time there is no doubt it has to be placed in the subgenus *Symposiocladius*. The adult male and female, pupa and larva have been redescribed by Dettinger-Klemm (2001), and Saether (2003) reviewed the whole subgenus, with keys to adult males, females, pupae and larvae.

DISTRIBUTION IN EUROPE AND THE NETHERLANDS

O. holsatus has been recorded from most parts of Europe (Saether & Spies, 2010). However, there are no records from the Mediterranean area except for southern France (Moubayed-Breil, 2008). In the Netherlands it is a common species in most parts of the country, except for the province of Zeeland and parts of Noord-Holland (Moller Pillot & Buskens, 1990; Limnodata.nl).

FEEDING

The guts of larvae studied by Dettinger-Klemm (2001) were filled mostly with pennate diatoms and to a lesser extent with filamentous algae, presumably *Nostoc*.

WATER TYPE

Most records in the Netherlands are from stagnant water, such as small lakes and ditches, but the larvae are rather common also in lowland brooks, usually with a very slow current (Limnodata.nl; own data). In other countries the species seems to be less common and is collected mainly in lakes (Dettinger-Klemm, 2001; Moubayed-Breil, 2008).

The species is not known from temporary water.

pH

There are a few Dutch records from water bodies with a pH of about 6. In most cases the pH was 7 or higher.

TROPHIC CONDITIONS AND SAPROBITY

Almost all Dutch records apply to more or less eutrophic water with a rather low orthophosphate content (usually < 0.05 mg P/l). The BOD is rarely more than 5 mg/l and the oxygen content drops during the day, usually not lower than 4 mg/l (Limnodata.nl).

SALINITY

There are no records from brackish water.

Orthocladius (Symposiocladius) lignicola Kieffer, 1914

Orthocladiinae gen.? l. *acutilabis* Pankratova, 1970: 323, fig. 214

IDENTIFICATION

The adult male resembles species of the subgenus *Orthocladius* (see Cranston, 1982a; Langton & Pinder, 2007). Pupae and exuviae can be identified using Langton (1984, 1991). A description of the adult male and pupa can be found in Potthast (1914) and Cranston (1982a). The larva is very unique because of its distinctive mentum. In almost all larvae lateral setae l_4 are developed as a setal tuft. Specimens with single l_4 are mentioned by Saether (2003) and have been found also in the Netherlands (leg. O. Duijts). Saether (2003) reviewed the whole subgenus *Symposiocladius*.

DISTRIBUTION IN EUROPE AND THE NETHERLANDS

The species has been recorded in many countries scattered over Europe. In the Netherlands the larvae have been collected in a number of small brooks in southern Limburg, in the eastern part of the country (Twente, Nijmegen, Arnhem) and rarely in the large rivers (mainly unpublished data, see also Limnodata.nl). Some centuries ago *O. lignicola* was a common inhabitant of the large rivers, because in those times a large quantity of wood was present in the rivers (e.g. Klink, 1989).

LIFE CYCLE

In Oregon streams *O. lignicola* is bivoltine, flying in spring and autumn (Anderson, 1989). Ringe (1974: 238, 271, 300) investigated its emergence in two brooks in Germany. In both brooks he found two or three generations, emerging from April to October and a few specimens up to December. The pupal exuviae remain in the wood when emergence takes place from a tube above the water line (see Cranston, 1982a).

FEEDING

Wood fibres are the most important particles in the gut contents (Warmke & Hering, 2000). Without doubt the larvae are xylophagous, but it is not clear whether fungi or bacteria are the source of nutrition or if there is a symbiotic gut fauna responsible for the degradation of cellulose (Cranston, 1982a: 428). Anderson (1989) stated that the larvae were much more abundant in alder than in conifer wood. Warmke & Hering (2000) also found a preference for deciduous wood, but the larvae appeared to be more abundant in beech than in alder.

MICROHABITAT

The larvae live mainly in the outer layer of firm wood (Anderson, 1989). Early larval instars often stay in voluminous vessels in the wood (Warmke & Hering, 2000). They cannot be reared when removed from the wood. See also under Feeding.

WATER TYPE

O. lignicola lives in flowing water, in narrow upper courses as well in large rivers, even

in channels with a very slow current (Cranston, 1982a; Anderson, 1989; Klink, 1989). Most records are from fast-flowing brooks or from brooks with a steep gradient, but current velocity can be rather low and there are also records from lowland brooks (own data). Crisp & Lloyd (1954: 298) found the larvae even in decayed wood in a mud flat with seepage. Recently the species has been found mainly in small upper courses (Caspers, 1980a; Anderson, 1989; Warmke & Hering, 2000; Dutch data), probably because wood is quickly removed from lower courses. Woodland brooks in deciduous woodland are most frequently inhabited; probably they have to be more or less permanent and not acid.
The exuviae have sometimes been collected in lakes (Langton, 1991: 182).

SAPROBITY

There is no information about wood in polluted water. It is possible that the larvae die or disappear when the aeration within the wood is reduced because of silt deposition. However, Crisp & Lloyd (1954) found the larvae in a mud flat and A. Klink collected a larva from a willow branch in the moderately polluted river Waal. Such examples are rare and the species is possibly not very tolerant of pollution, as Wilson & Ruse (2005) suggest. See further under Water type.

Orthocladius (Euorthocladius) luteipes Goetghebuer, 1938

Orthocladius luteipes Langton, 1984: 144, fig. 49b (pro parte)

IDENTIFICATION

O. luteipes males are difficult to identify without associated exuviae (Soponis, 1990: 25); the species is still absent from the key by Langton & Pinder (2007). The exuviae can be identified using Soponis (1990) and Langton (1991). In Langton (1984) the name *luteipes* applies to *luteipes* and *rivicola* together. Identification of the larva to species level is not reliable (Soponis, 1990).

DISTRIBUTION IN EUROPE AND THE NETHERLANDS

O. luteipes is known from several countries in southern and central Europe (Saether & Spies, 2010). It is probably absent from the Netherlands. Records from the western European lowland have to be verified.

WATER TYPE

Schmid (1993) collected *O. luteipes* in the crenal and upper rhithral area of gravel brooks in Austria. Michiels (1999, 2004) found this species also in the lower parts of two fast-flowing streams in Germany, together with *O. rivicola*. According to Moog (1995), the species is absent from the lower parts of rivers, even from the epipotamal zone.

SAPROBITY

Moog (1995) called *O. luteipes* a typical species of xenosaprobic to β-mesosaprobic water, indicating hardly or no pollution. The two streams in Germany investigated by Michiels (1999, 2004) were slightly to moderately polluted.

Orthocladius (Orthocladius) maius Goetghebuer, 1942

DISTRIBUTION IN EUROPE

O. maius is known from a limited number of countries in central, western and southern Europe (Saether & Spies, 2010). There are no records from the Netherlands.

WATER TYPE

According to Langton (1991) and Rossaro et al. (2003), *O. maius* lives in lakes and streams. Klink (2010) collected pupae and exuviae from several upper courses of the Seine in France, sometimes in large numbers.

Orthocladius (Eudactylocladius) mixtus (Holmgren, 1869)

O. mixtus is an arctic species; other records are based on misidentification (Cranston, 1984: 890). The species is not treated here.

Orthocladius (Orthocladius) oblidens (Walker, 1856)

Orthocladius sp. A Pinder, 1978: 72, fig. 36E; Tokeshi, 1986: 433 et seq.

IDENTIFICATION

As adult male *O. oblidens* can be confused with *O. rubicundus*, see under this species. Identification of pupae presents no problems. The larvae have a very wide central mental tooth (> 33 µm) and the setae submenti are not so far lateral as the base of the submentum (as in figs 91 and 92 in Schmid, 1993), but this delivers no difference from some other species (own data).
The material published by Becker (1995) under *O. oblidens* appears at least partly to belong to this species, but most probably some could be *O. rubicundus*.

DISTRIBUTION IN EUROPE AND THE NETHERLANDS

O. oblidens lives in all parts of Europe (Saether & Spies, 2010). In the Netherlands the species is rather common throughout the Pleistocene part of the country and in southern Limburg; in many brooks and streams it is the only species of the genus. The species also lives in the large rivers. Elsewhere in the country the species is very rare. Most data are based on identification of exuviae.

LIFE CYCLE

Tokeshi (1986, 1986a) found one generation a year in a small lowland stream in England. The species (as sp. A) had a long quiescence in the egg stage from summer to early spring and a short larval period in spring during the period of maximum diatom abundance. The larvae developed mainly after the disappearance of *O. rhycobius* in the same stream. In small streams in southern Limburg and western Germany we found more overlap in time between these two species. In spring the adults emerged from March until June in one or two generations. In contrast to *O. rhyacobius*, almost everywhere some specimens emerged also in summer (see fig. 9 on p. 154) and there was a small second or third generation in autumn. (Compare the Life cycle section in the introduction to the genus *Orthocladius*). Comparable results have been obtained by other workers (Drake, 1982; Gendron & Laville, 1992; Reiff, 1994; Klink, 2010). In Dutch slow-flowing lowland brooks emergence seems to be scarce in autumn and we found one case of emergence in December.

FEEDING

Tokeshi (1986a) found 90% diatoms in the guts of *O. oblidens*. He suggested that the negative overlap in time with *O. rhyacobius* served for avoiding concurrence with this species (see Life cycle). In the guts of *oblidens* we sometimes also found many diatoms, in other streams mainly filamentous algae or detritus. The species seems to be a very opportunistic feeder. See further the introduction to the genus.

MICROHABITAT

The larvae live mainly on stones and plants in lakes, streams and rivers (Tokeshi, 1986; Becker, 1995; own data). In a Dutch lowland brook we collected 62 mostly fully grown larvae from 0.4 m^2 of a sandy bottom with filamentous algae (compare the Microhabitat section in the introduction to this genus).

DENSITIES

Lindegaard & Jónasson (1979) stated densities up to 185 larvae/m^2 in lake Myvatn in Iceland. Tokeshi (1986) found up to 100 larvae per apical 10 cm section of *Myriophyllum* plants in a lowland brook in England.

WATER TYPE

O. oblidens is a common species in running water, both in narrow brooks and in large rivers (e.g. Tokeshi, 1986; Becker, 1994; Rossaro et al., 2003; own data). In contrast to most other species of the genus it lives in slow-flowing as well as fast-flowing streams (see table 7 in Chapter 9). In mountain streams it is found more downstream than other species of the genus (Gendron & Laville, 1997). It has been recorded also from many lakes in Europe, for example from England, Sweden, Iceland, Bavaria and Italy (Lindegaard & Jónasson, 1979; Reiff, 1994; Ruse, 2002; Rossaro et al., 2003; Marziali et al., 2008). In the Netherlands it is almost entirely confined to flowing water, but Couwets & Davids (1984) caught more than 10 adult males near lake Maarsseveen.

Temperature

In Rossaro (1991) *O. oblidens* seems to be a cold water species, found at a mean temperature of 11 °C. However, this was a mean of only four records in Italy. Reiff (1994) collected the exuviae in Bavarian lakes numerously in June at temperatures of about 20 °C. In Iceland Hannesdóttir (unpublished) did not find *O.* cf. *oblidens* in the coldest streams (7 °C), but found it dominating in the warmer streams.

Permanence

O. oblidens seems to be rare in temporary water, in contrast to *O. rhyacobius*. Because the life cycle of these two species is not very different (a possible quiescence in the egg stage in summer and autumn) the life cycle of *O. oblidens* cannot be the reason why it has been rarely collected in temporary water. In the Netherlands a lower pH in many temporary upper courses can cause the absence of the species. Anyhow the figures for permanence given in table 7 have to be used with caution.

pH

See the introduction to the genus and table 6 on p. 284.

TROPHIC CONDITIONS AND SAPROBITY

In Bavarian lakes *O. oblidens* displayed a greater preference for mesotrophic water than other species of the genus (Orendt, 1993; Reiff, 1994). This seems to be the case only in lakes. Our data from Dutch lowland brooks and rivers suggest that the species there is more specific for eutrophic and β-mesosaprobic conditions and endures

a lower oxygen content than other species of the genus, although the differences in tolerance between the species are small. Probably there is a relation with the temperature preference of the species (see the investigations of Hannesdóttir et al. mentioned above).
For explanation of the figures of table 3 on p. 156 see the text in the introduction to the genus. We do not follow Orendt (2002: 117), who writes that the species is very tolerant of organic pollution.

Orthocladius (Eudactylocladius) olivaceus Kieffer, 1911

Orthocladius mixtus Cranston, 1982: 100, fig. 38f (not Holmgren)

IDENTIFICATION

Adult males and pupae can be identified using Cranston (1984) or any key published since (see under *O. fuscimanus*). The larva has been keyed in Cranston (1982) as *O. mixtus*. The mola of the mandible is rounded (see Cranston, 1984: 891); the width of the central mental tooth is most probably not a reliable feature.

DISTRIBUTION IN EUROPE AND THE NETHERLANDS

The species is known from a limited number of countries scattered across Europe (Saether & Spies, 2010). In the Netherlands pupae have been collected in two fast-flowing brooks in southern Limburg (H. Cuppen, unpublished; B. van Maanen & M. Korsten, unpublished).

LIFE CYCLE

The life cycle of *O. olivaceus* seems to be comparable with that of *O. fuscimanus*. Ringe (1974) stated emergence in central Germany from April to October, with the largest numbers in summer.

MICROHABITAT

Lehmann (1971) collected the species from stones (cf. Thienemann, 1954: 349). H. Cuppen (unpublished) found a pupa on a gravel bottom in southern Limburg.

WATER TYPE

O. olivaceus has been found only in fast-flowing brooks and streams (Thienemann, 1954: 349; Lehmann, 1971; Orendt, 2002a; Dutch data). Ringe (1974) stated emergence mainly from the fast-flowing Breitenbach and hardly from the slow-flowing Rohrwiesenbach in Germany.

The species also lives in lakes in Scotland and other countries in northern Europe and America (Langton, 1991; Cranston, 1999).

Orthocladius (Orthocladius) pedestris Kieffer, 1909

Orthocladius rhyacobius Langton, 1984: 158, fig. 55b nec Kieffer

IDENTIFICATION

Identification of the adult male is not easy and it is recommended to use Rossaro et al. (2003) and Langton & Pinder (2007) together. The pupa is very characteristic and has been described by Potthast (1914). It can be identified using all keys from 1991 and later. However, the species is very variable in the pupal stage (Langton, pers. comm.). The larva has been described briefly by Potthast (1914), but no distinguishing character has yet been found to separate it from all other species. Most probably the narrow central mental tooth (about twice as wide as the first lateral tooth) is an important character.

DISTRIBUTION IN EUROPE AND THE NETHERLANDS

O. pedestris has been recorded across almost the whole of Europe except for Russia (Saether & Spies, 2010). In the Netherlands some larvae, pupae and exuviae have been collected in the river Waal by A. Klink (Klink & Moller Pillot, 1996). All other records have to be verified.

WATER TYPE

According to Rossaro et al. (2003), *O. pedestris* is restricted to small running waters. However, the species has also often been collected in rivers: in the river Garonne (Garcia & Laville, 2001), in English rivers (Langton, 1984, 1991) and even (rarely) in the river Waal in the Netherlands (Klink & Moller Pillot, 1996). In brooks and small rivers the species is rather rare in central Europe (Orendt, 2002a; Michiels, 2004), but common in England (Langton, pers. comm.). Probably the larvae almost always live where the current is rather fast. We once found the exuviae in the upper course of the Berkel, a small lowland stream in Westphalia, but this specimen probably came from a riffle with fast current.
O. pedestris is very rare in lakes. Langton (1984, 1991) reported the species from mountain lakes, but it was absent in the Bavarian lakes (Orendt, 1993; Reiff, 1994).

Orthocladius (Orthocladius) rhyacobius Kieffer, 1911

Orthocladius excavatus Lehmann, 1971: 486, fig. 12, 13; Ringe, 1974; Becker, 1994: 110 nec Brundin
Orthocladius Pe 9 Langton, 1984: 155; Wilson & Wilson, 1984: 131; Tokeshi, 1986: 433 et seq.
Orthocladius obumbratus Langton & Cranston, 1991: 190; Reiff, 1994: 118, figs 139,140

NOMENCLATURE AND IDENTIFICATION

At the moment we follow Rossaro et al. (2003), but the name is still under discussion. According to Langton & Cranston (1991), *O. obumbratus* is the same species and this name has priority. A definitive decision has not yet been made.
Identification of the adult male is often difficult. The characters of the hypopygium are not very reliable, because the shape depends on the preparation. A virga can be difficult to see (or even absent) in species that should have it. Pupae and exuviae are very easy to identify. Only identifications before 1991 using Langton (1984) apply to *O. pedestris*. The larva cannot be distinguished from some other species. A key is in preparation.

DISTRIBUTION IN EUROPE AND THE NETHERLANDS

Most probably the species lives almost in the whole of Europe, although the records of some countries are considered as doubtful because of identification problems (Saether & Spies, 2010). In the Netherlands there are up to now only records from Limburg, the eastern part of Noord-Brabant and the large rivers. Most probably it lives also in the most eastern part of the country.

LIFE CYCLE

Tokeshi (1986) stated that in England the species has a long dormancy period in the egg phase and hatches from mid February to mid March. The larvae grew rapidly and emerged in April. In southern Limburg and in Germany we found third instar larvae from the end of January and emergence from early March until the end of May, possibly in two successive generations (cf. Berg & Hellenthal, 1992). Most authors collected few exuviae in autumn (Lehmann, 1971; Ringe, 1974; Becker, 1995) (as *O. excavatus*); Wilson & Wilson (1984) and Reiff (1994) collected exuviae also in July/August. See the Life cycle section in the introduction to this genus and under *O. oblidens*.

FEEDING

Tokeshi (1986a) stated that *O. rhyacobius* feeds mainly on diatoms. We also found many diatoms in the guts, together with detritus and filamentous algae. However, see the introduction to the genus.

MICROHABITAT

The larvae are found in large numbers on stones and plants (Potthast, 1914, sub *O. rhyacophilus* and *rhyacobius*; Lehmann, 1971; Tokeshi, 1986; Becker, 1995). In some cases we also collected large numbers from sandy/silty bottoms.

WATER TYPE

Current

The species has been found mainly in running water (Rossaro et al., 2003). It prefers faster currents than *O. oblidens* and is absent from most slow-flowing lowland brooks (Lehmann, 1971; Bitušik, 2000; Michiels, 2004; own data). See table 2 on p. 155. However, Tokeshi (1986) reported the species numerously from a lowland stream with flow velocities between 0.3 and 0.6 m/s. Reiff (1994) collected the species in several lakes in Bavaria, but it is less common in lakes than *O. oblidens*.

Dimensions

The larvae can be common in brooks, streams and rivers (Lehmann, 1971; Becker, 1995; Orendt, 2002a; Rossaro et al., 2003). They can be found abundantly even in narrow brooks no more than 1 m wide (own data).

Permanence

The species can live in brooks without any surface water in summer, probably because of the summer diapause; see Life cycle.

TROPHIC CONDITIONS AND SAPROBITY

Reiff (1994) collected the exuviae of *O. rhyacobius* mainly in more or less mesotrophic lakes in Bavaria, but sometimes also in eutrophic lakes. The larvae need well oxygenated water and the species is usually absent from polluted streams. However, at least the older larvae are able to survive on plants and stones in polluted environments (own data). In comparison with *O. oblidens*, they appear to be more common in oligosaprobic brooks and streams, but this is possibly more a question of temperature and oxygen content. For explanation of the figures in table 3 on p. 156, see the introduction to the genus.

Orthocladius (Euorthocladius) rivicola Kieffer, 1911

Orthocladius rivicola Wilson & Wilson, 1984: 130 (pro parte); Becker, 1995: 109 (pro parte)

nec *Orthocladius rivicola* Langton, 1984: 142, fig. 49c
Orthocladius luteipes Langton, 1984, 144, fig. 49b (pro parte); Klink & Moller Pillot, 1982: 52 (pro parte)

IDENTIFICATION

Before 1990 the differences between *O. rivicola*, *luteipes* and *ashei* were hardly known (see under last two species). Therefore data from older literature can apply to *O. luteipes* or *O. ashei*. Adult males without exuviae can hardly be identified to species level (Soponis, 1990: 25). Identification of exuviae using Soponis (1990) or Langton (1991) is possible.
The larvae of these three species are also very similar (see Soponis, 1990). The AR of the larvae is lower than in *O. thienemanni*, according to Soponis (1990) 1.38–1.80. Rossaro (1982) gives 1.2–1.3, but this is possibly a mistake.

DISTRIBUTION IN EUROPE AND THE NETHERLANDS

O. rivicola has been recorded from many countries scattered across Europe (Saether & Spies, 2010). In the Netherlands the exuviae have been collected only in the large rivers (Het Waterlaboratorium, reference coll.), but larvae possibly belonging to this species are known from some lowland streams.

LIFE CYCLE

In many regions *O. rivicola* has two generations: one in spring, from (January) March to May and one in autumn (October) (Lehmann, 1971; Soponis, 1990). The autumn generation can be absent, possibly mainly in lowland regions (Ringe, 1974; Soponis, 1990; Becker, 1995). At higher altitudes and in northern regions the emergence periods can be different (Laville & Lavandier, 1977; Caspers, 1983; Soponis, 1990). Michiels (1999) found three generations in the Untere Salzach in Bavaria, with peaks in February, May and October. A summer generation in July/August has also been stated in the river Rhine by Wilson & Wilson (1984).

MICROHABITAT

Becker (1995) reared *O. rivicola* (s.l., see above) from gravel and stones and also from a sandy bottom. Schmid (1993) found the larvae up to 10 cm deep in gravel sediments. The larvae usually live in gelatinous tubes covered with sand grains and detritus; the pupal tubes have holes at both ends for the current (Soponis, 1990).

WATER TYPE

Kownacka & Kownacki (1972), Laville & Vinçon (1991) and Bitušik (2000) found *O. rivicola* mainly at very fast and fast flows (rhithral zone). However, these authors have not mentioned *O. luteipes* and most probably they have not distinguished both species. Michiels (1999, 2004) found both species together in two fast-flowing streams in Germany. Ringe (1974: 255) attribute the low abundance in the Rohrwiesenbach to a combination of a slow current and a high temperature.
O. rivicola lives in goodish numbers also further downstream, even in large rivers such as the Garonne and the Rhine (Soponis, 1990; Schmid, 1993; Garcia & Laville, 2001; Móra, 2008; reference coll. Het Waterlaboratorium). There are no reliable records from typical lowland brooks. Reiff (1994) collected the exuviae rarely in lakes in Bavaria.
Kownacka & Kownacki (1972) suggest that *O. rivicola* is more or less characteristic of granite substratum and scarce on calcareous substratum. According to Braukmann (1984: 221, 455) this is probably incorrect.

SAPROBITY

Many data in the literature indicate that *O. rivicola* is a species of hardly polluted water (see also Moog, 1995). Exact data about the influence of organic pollution are rare. Kownacka & Kownacki (1972) collected *O. rivicola* mainly in non-polluted glacier streams in the Caucasus. However, Bitušik (2000) writes that the numbers in mountain streams increase when the stream is organically polluted. For further information see the introduction to the genus.

Orthocladius (Orthocladius) rivinus Potthast, 1914

IDENTIFICATION

Identification of adult males and exuviae of *O. rivinus* is sometimes very difficult and the larvae are still unknown. In their key to exuviae Langton & Cranston (1991) give as the main character that the posterior band on tergite III extends further laterally than the apical band, but fig. 2a shows that it can be more or less the same as sometimes found in other species. The forked or branched setae on the sternites are the most important character, but this is variable as well. Therefore, not all published records are reliable (e.g. Orendt, 2002: 114).

DISTRIBUTION IN EUROPE AND THE NETHERLANDS

O. rivinus is only known from a limited number of countries, possibly owing to identification problems. Bitušik (2000) called the species rare in Slowakia. We found it only in two localities in the Netherlands, one in Germany and one in Belarus. Klink (pers. comm.) collected it only once, in Hungary.

LIFE CYCLE

Rossaro et al. (2003) stated emergence in Italy in winter and spring. We found the exuviae in the Netherlands and Germany in April and May. See further the Life cycle section in the introduction to the genus.

WATER TYPE

The species has been collected in cold springs, small brooks, large rivers and lakes (Langton, 1991; Ruse, 2002; Rossaro et al., 2003; own data). Our own records do not indicate a preference for low temperature. It is not clear why the species is so scarce. We found a larger population only in a rather fast-flowing woodland brook.

Orthocladius rivulorum Kieffer 1909

IDENTIFICATION

Descriptions and many remarks are given by Soponis (1990). Identification of the adult male and pupa do not present any problems. The larva is characteristic because of the 8 to 11 pairs of lateral mental teeth; however, larvae with 8 pairs of teeth can belong to spec. Aalten, see p. 157.

DISTRIBUTION IN EUROPE AND THE NETHERLANDS

O. rivulorum has been recorded in almost the whole of Europe except for Russia (Saether & Spies, 2010). In the Netherlands Klink (1985) collected exuviae in the river Meuse in southern Limburg (Grensmaas), possibly originating from Belgium. From palaeolimnological investigations it appears that the species was previously present in this river and also in the Rhine and Waal (Klink, 1985).

LIFE CYCLE

Pinder (1977) found *O. rivulorum* to be univoltine in an English brook, emerging in late spring. In an alpine gravel brook Schmid (1992) stated that (in addition to the emergence in May/June) small numbers emerged in late winter, late summer and autumn. He found a high proportion of first and second instars between September and November, and towards the end of February the majority of the larvae developed into the third instar. Variations on these life cycles are reported by Drake (1982) in England: emergence in early spring; and by Michiels (1999) in Germany: emergence mainly in February/March and in May. From these data it is not possible to conclude whether there can be a quiescence in the egg stage in summer, as in some other *Orthocladius* species. See the Life cycle section in the introduction to this genus.

FEEDING

The larvae feed exclusively on diatoms and filamentous algae (Thienemann, 1954:108). These algae grow on the tubes in which the larvae live and which are fixed to the substrate.

MICROHABITAT

The larvae live on stones (Thienemann, 1954), but also on gravel (Pinder, 1980) and plants (Pinder, 1977). Schmid (1992) found the larvae in large numbers in the sediment of a gravel stream, even at a depth of 30 cm. Third and fourth instars preferred the surface layer.

WATER TYPE

O. rivulorum inhabits almost only more or less fast flowing streams (meta- and hyporhithral zone) (Lehmann, 1971; Laville & Vinçon, 1991; Bitušik, 2000; Michiels, 1999, 2004). The species has been collected rarely in typical lowland streams, such as Linding Å in Denmark (Lindegaard-Petersen, 1972). The low water temperature in summer (11–12 °C.) may have played a role here. In Italy Rossaro (1991) found a remarkable low mean temperature for this species of 6.8 °C. Reiff (1994) collected the exuviae rarely in lakes in Bavaria.
Laville & Vinçon (1991) collected the species in narrow (2 m) to wide (18 m) Pyrenean streams; it also lived in very low numbers in a larger river (Middle Garonne: Garcia & Laville, 2001). In recent times *O. rivulorum* seems to be absent from the river Rhine (Caspers, 1991; Becker, 1995), but it has lived there in the past (see Distribution). Its absence from many large rivers and also from the fast-flowing brooks and streams in southern Limburg (the Netherlands) seems to be attributable to morphological changes and/or water quality (see under Saprobity).

OR

SAPROBITY

According to Wilson & Ruse (2005), all species of the subgenus *Euorthocladius* are intolerant of organic pollution. Moog (1995) called *O. rivulorum* not very intolerant, but less tolerant than *O. thienemanni* or *O. rubicundus*. Therefore, the absence of this species from the brooks and streams in southern Limburg and from the lower courses of most European rivers could be partly explained by the decreasing water quality in recent times. However, the structure of the river beds has also changed and water temperature has increased. Because we have not enough information about the influence of water quality parameters on this species we cannot attribute the absence of *O. rivulorum* to saprobity alone. The figures in table 5 in Chapter 9 have to be used with caution.

Orthocladius (Orthocladius) rubicundus (Meigen, 1818)

Orthocladius saxicola Lehmann, 1971: 487; Schmid, 1993: 136, fig. 88

IDENTIFICATION

Identification of the **adult male** using older keys (before 2003) was not reliable. It remains rather difficult to separate *O. rubicundus* and *O. oblidens*. According to Langton & Pinder (2007), the antennal ratio of *O. oblidens* is 2.0 or more, but Rossaro et al. (2003) give for this species an AR of 1.3–1.9 (in *rubicundus* 1.3–1.6; so specimens with AR 2.0 seem to be always *oblidens*). The most reliable character seems to be the inconspicuous, short dorsocentral setae (< 75 μm), not inserted on pale spots, in *O. rubicundus* (see Rossaro et al., 2003: 225, 232 and Langton & Pinder, 2007: 120).
Identification of the pupal **exuviae** was already possible using Langton (1984) and is very easy in all keys published since then (see also Lehmann, 1971: 487, fig. 14).
The **larva** can be identified by the insertion of the setae submenti and the short first antennal segment (see Schmid, 1993: 132, 136, fig. 88), although (at least in Austria) another species (sp. D) resembles *O. rubicundus*.

DISTRIBUTION IN EUROPE AND THE NETHERLANDS

O. rubicundus lives in the whole European mainland and has been found also on some surrounding islands (Saether & Spies, 2010). In the Netherlands it is one of the most common *Orthocladius* species in the large rivers (Klink, 1985; reference collection HWL). It also lives in fast-flowing brooks and streams in southern Limburg (own data). There are some records from lowland brooks on the Veluwe and near Winterswijk and from the Keersop in Noord-Brabant (Klink, pers. comm.; own data).

LIFE CYCLE

The life cycle of *O. rubicundus* resembles that of other species of the subgenus, such as *O. rhyacobius* and *O. oblidens*, but it has not been studied in detail. Most emergences have been stated in spring, from March to May, and in Italy already in winter (Drake, 1985; Gendron & Laville, 1992; Rossaro et al., 2003; own data). Most authors stated also a summer and/or autumn generation (June until October). Rossaro et al. (2003) wrote that the species is present in Italy in summer if the water temperature is under 15 °C. However, a summer generation in June to September (and sometimes also an autumn generation) has been stated also at higher temperatures, e.g. in brooks in Germany (Ringe, 1974), in England (Wilson, 1977), in the Meuse in southern Limburg (Klink, 1985), in the river Rhine (Wilson & Wilson, 1984: 131), in upper courses of the Seine (Klink, 2010) and in Bavarian lakes (Reiff, 1994: fig. 141).

FEEDING

See the introduction to this genus.

MICROHABITAT

According to Potthast (1914) the larvae live in short tubes on stones. Other workers collected the larvae mainly on stones, but also on plants in fast-flowing streams (Lehmann, 1971; Schmid, 1993; own data).

WATER TYPE

O. rubicundus lives from the uppermost upper courses in the mountains to brooks, streams and rivers far downstream (Lehmann, 1971; Laville & Vinçon, 1991). Laville & Vinçon found the species mainly at fast and very fast current velocities (cf. Ertlova, 1970: 296). This is consistent with our findings in the Netherlands. It is absent from

slow-flowing lowland brooks, but its numbers may increase in future because of the recent construction of fish ladders. See table 2 on p. 155.
The species is not rare in lakes in Bavaria (Reiss, 1984; Reiff, 1994).

SAPROBITY

According to Potthast (1914), Moog (1995), Bitušik (2000) and Orendt (2002), *O. rubicundus* is very tolerant of organic pollution if there is no serious oxygen deficit. Learner et al. (1971) found the species (if the adult males were correctly identified) in a stream in Wales after loading with coal particles. However, Wilson (1987) collected the exuviae mainly at the less polluted sites in the river Trent (BOD < 7). Wilson & Ruse (2005) consider *O. rubicundus* to be not very tolerant of organic pollution, which reflects our findings in brooks and streams in the Netherlands and Germany. Anyhow a good oxygen supply is important for this species. See further the introduction to this genus.

Orthocladius (Orthocladius) ruffoi Rossaro & Prato, 1991

Rheorthocladius sp. A Th ienemann, 1944: 598; Langton, 1991: 184, fig. 76e–h; Langton & Cranston, 1991: 241, fig. 1e, 2b

DISTRIBUTION IN EUROPE AND THE NETHERLANDS

The species appears to be rather scarce, but it has been collected in several countries scattered across Europe (Saether & Spies, 2010). In the Netherlands the exuviae have been found in four brooks and streams in the province of Limburg (in each of them only once).

WATER TYPE

Gendron & Laville (1997) collected *O. ruffoi* in a Pyrenean stream only in the montane stretch where the gradient was more than 10m/km. Other authors have also repor ted its occurrence in fast-flowing springs and small streams (rhithral zone) (Michiels, 1999, 2004; Bitušik, 2000; Rossaro et al., 2003). In the Netherlands the exuviae have been collected very rarely and only in rather fast-flowing streams in the province of Limburg. The current velocity, water temperature and/or water quality here probably represent marginal conditions.
Reiff (1994) collected one specimen in a lake in Bavaria.

Orthocladius saxosus (Tokunaga, 1939)

This species lives only in mountain streams and is not treated here (Soponis, 1990; Brabec, 1997; Bitušik, 2000).

Orthocladius (Symposiocladius) smolandicus Brundin, 1947

Saether (2003) considers *O. smolandicus* to be different from *O. holsatus*, in contrast to Langton & Cranston (1991) and Dettinger-Klemm (2001). Only the adult male has been described. It has been collected from lakes in Sweden and Finland and possibly also in France and Switzerland) (Saether & Spies, 2010). It is not treated here.

Orthocladius (Euorthocladius) thienemanni Kieffer, 1906

IDENTIFICATION

Identification of adult males is very difficult and is best achieved using Soponis (1990) together with Langton & Pinder (2007); attention has to be paid to the fact that *O. luteipes* is absent from the last key. The exuviae are easily identifiable. **Larvae** cannot be identified reliably to species level (Soponis, 1990: 13, 19; Schmid, 1993: 133, 144). The relatively narrow central mental tooth in combination with the ventromental plates and the insertion of the setae submenti (see fig. 10) are useful for recognising the subgenus *Euorthocladius*, also in young larvae. In many streams in the Netherlands *O. thienemanni* is the only species of this subgenus.

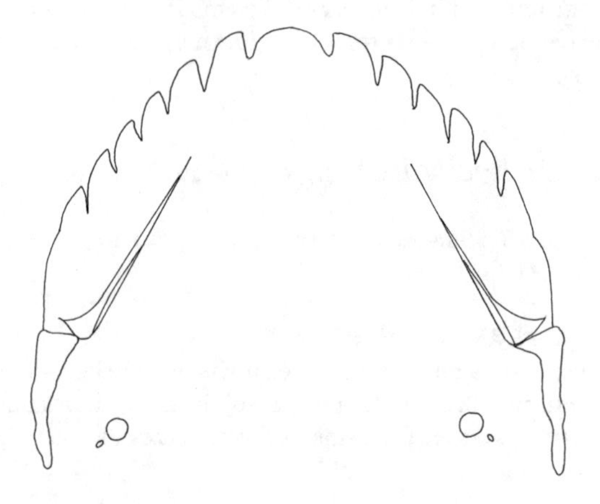

10 Orthocladius thienemanni mentum

DISTRIBUTION IN EUROPE AND THE NETHERLANDS

O. thienemanni probably lives in all parts of Europe (Saether & Spies, 2010). In the Netherlands the species is only known from the large rivers, the whole Pleistocene part of the country and southern Limburg. It is common only in southern Limburg; in the provinces Drenthe and Noord-Brabant it is a rare species.

LIFE CYCLE

Many authors stated a spring generation with most emergences in February, March or April, and a smaller peak of emergences in summer and/or autumn (Lehmann, 1971: 486, 511; Pinder, 1980; Michiels, 1999). Drake (1982) suggested two generations within the spring period. This seems to be probable, because we also found second instar larvae in early April. Sometimes (especially in lowland streams?) the summer and autumn generation are missing (Lindegaard & Mortensen, 1988; own data). The autumn generation can be large, as we found in a fast-flowing brook near Aachen.
With the life cycle of *O. rhyacobius* in mind, these data suggest a **dormancy** in the egg stage in lowland streams in summer and hatching of eggs early in winter. In the other case (at lower temperatures?) most or all eggs hatch immediately and give a (sometimes small) summer generation. See also the Life cycle section in the introduction of the genus.

FEEDING

In glacial streams the larvae of *O. thienemanni/rivicola* fed mainly on blue green algae (Kawecka et al., 1978). In the guts of *thienemanni* larvae from Germany and the Netherlands we sometimes found mainly diatoms, but in one small stream almost only detritus. See further the introduction to this genus.

MICROHABITAT AND DENSITY

The larvae live in clear gelatinous tubes encrusted with sand grains, often in fissures and depressions of stones (Soponis, 1990). The larvae are found in large numbers on gravel and stones, relatively less on plants and only scarcely on the sandy bottom of streams (Pinder, 1980; Kretzschmar & Böttger, 1994; own data).
On gravel Pinder (1980) found densities up to 900 larvae/m^2 .

WATER TYPE

Current

O. thienemanni is absent from mountain streams and is characteristic of secondary mountains and lowland streams (Lehmann, 1971; cf. Laville & Vinçon, 1991; Bitušik, 2000; Orendt, 2002a; Michiels, 2004). Drake (1982: 231) did not collect the larvae during a period of low flow in a chalk stream in England. In the Netherlands the species is almost confined to brooks with a gradient of more than 2 m/km, which can be found mainly in southern Limburg and here and there in the eastern part of the country.

Dimensions

The species is rather common in brooks and streams with an appropriate microhabitat and current, but is rare or absent in larger streams and rivers (e.g. Klink & Moller Pillot, 1982; Becker, 1994; Garcia & Laville, 2001; Orendt, 2002a; Klink, 2010). The larvae can be found abundantly in narrow brooks no more than 1 m wide.

Permanence

The species can live in brooks without any surface water in summer, probably because of the possibility of a summer diapause; see Life cycle.

SAPROBITY

According to Wilson & Ruse (2005), the species is intolerant of organic pollution. Peters et al. (1988) called the species characteristic of hardly polluted water in lowland brooks in the Netherlands. In southern Limburg the species is absent from seriously polluted streams and the larvae need well oxygenated water (Cuijpers & Damoiseaux, 1981; own data). In alpine streams, however, the species appears to live also in α-mesosaprobic water (Moog, 1995). At least the older larvae are able to survive on plants and stones in polluted environments (see also Klink, 1985).

Orthocladius (Orthocladius) wetterensis Brundin, 1956

IDENTIFICATION

Identification of the adult male and pupa present no problems. The records of adult and pupae are probably always reliable. The larvae can be identified by the shape of the premandible (Schmid 1993: 132, fig. 92), but it is not sure if this character excludes all other species.

DISTRIBUTION IN EUROPE AND THE NETHERLANDS

O. wetterensis has been recorded from several countries scattered across Europe (Saether & Spies, 2010). In the Netherlands the exuviae have been collected in the river Waal (reference coll. Het Waterlaboratorium).

LIFE CYCLE

The life cycle seems to be the same as in *O. rhyacobius*. Almost all data from Europe exhibit emergence from February until May and rarely in autumn (Lehmann, 1971: 488, 511; Becker, 1995; Michiels, 1999; Rossaro et al., 2003; Krasheninnikov, 2011). See also the Life cycle section in the introduction to the genus.

MICROHABITAT

Becker (1995) reared the species from stones in the river Rhine.

WATER TYPE

O. wetterensis has been found in large rivers such as the Danube (Schmid, 1993), the Rhine (Becker, 1994) and the Oder (Steinhart, 1998: 37) as well as in many rather fast flowing streams in central Europe (Bitušik, 2000; Orendt, 2002a; Rossaro et al., 2003; Michiels, 2004). There are no records from lowland brooks.
The type specimen was collected in a swarm near the shore of lake Vättern in Sweden (Brundin, 1956: 106).

Paracladius conversus (Walker, 1856)

Cricotopus inserpens Reiss, 1968: 237
Paratrichocladius inserpens Pankratova, 1970: 206
Cricotopus conversus Lehmann, 1971: 479

DISTRIBUTION IN EUROPE AND THE NETHERLANDS

P. conversus is known from many countries throughout Europe (Saether & Spies, 2010). It seems to be comparatively more common in large lowland regions, as in Russia (see Zinchenko, 2002: 63, fig. 42), and scarce in upland and montane regions (e.g. Italy: Rossaro, 1982). In the Netherlands the species is most common in the central and eastern part of the country. It is not confined to the Pleistocene regions, but records from the Holocene part of the country are mainly from regions with groundwater feeding (Steenbergen, 1993; Nijboer & Verdonschot, unpublished; Limnodata.nl; own data).

LIFE CYCLE

In Germany and the Netherlands *P. conversus* flies from March until October (Reiff, 1995; own data). In some cases emergence has been stated to be only or mainly in spring (e.g. Reiss, 1968; Wilson, 1977; Orendt, 1993: 140), but a large generation in summer and/or autumn is also possible (Klink & Mulder, 1992; Reiff, 1995; own data). It is not clear whether developing larvae can be totally absent in summer. Nothing is known about hibernation. Tolkamp (unpublished) collected some fourth instar larvae in January.

FEEDING

The gut usually contains almost only detritus, but sometimes also a considerable amount of diatoms (own data).

MICROHABITAT

P. conversus is one of the few Orthocladiinae which are exclusively bottom-living. It has been collected usually from the upper layers of bottoms without vegetation or with sparse vegetation, in silt or muddy sand (Reiss, 1968; Srokosz, 1980; Zinchenko, 2002; own data), but the larvae also live (more rarely) on stones and wood (Hirvenoja, 1973; Cranston, 1982).

WATER TYPE

It is worth noting that most (or all) records of *P. conversus* are from water bodies (flowing or stagnant) fed by groundwater (own data; see also Pankratova, 1970).

Current

In the Netherlands the larvae are most often collected in slow or very slow-flowing brooks and ditches, but also in stagnant water of lakes, canals, pools and ditches. Elsewhere they are also usually found in slow-flowing streams; rarely in fast currents, and in such cases they have most probably descended from still-water sites (Laville & Vinçon, 1991; Michiels, 1999; Zinchenko, 2002; Orendt, 2002a). The literature contains many data from lakes (Reiss, 1968, 1984; Hirvenoja, 1973; Reiff, 1995).

Dimensions

In flowing water the larvae are most common in brooks and ditches and rare in large rivers (Klink & Moller Pillot, 1982; Schmid, 1993; Becker, 1995; Limnodata.nl; own data). Smit et al. (1996) sometimes collected the larvae on a tidal sandy flat of an estuary in the delta region. In stagnant water there seems to be no clear preference for small or large water bodies.

Permanence

There are no records from temporary water. This may be connected with the fact that the occurrence of larvae seems to be confined to water bodies fed by groundwater (see above).

pH

P. conversus is absent from acid water and most common at pH > 7.5 (Limnodata.nl). See under Water type.

SAPROBITY AND OXYGEN

Although the larvae often live in bottoms with much silt, the main records are from hardly polluted water (see table 5). Limnodata.nl gives rather stable oxygen conditions, a mean BOD of 1.9 mg O_2/l and a low ammonium content. Peters et al. (1988) found the larvae only in hardly polluted brooks in the Netherlands. However, Srokosz (1980) collected the larvae in at least equal numbers after organic pollution in the river Nida in Poland (at BOD 6.5 and an ammonium-N content of 2.68 mg/l). We also have some records from organically polluted water with less stable, although probably never very low, oxygen content. According to Moog (1995), the larvae can live in α-mesosaprobic water. The key factors for the occurrence or absence of the species are hitherto unknown.

SALINITY

The species is almost only known from fresh water (Steenbergen, 1993; Limnodata.nl). Krebs (1981) reared the adults from slightly brackish water.

Paracricotopus niger Kieffer, 1913

SYSTEMATICS AND IDENTIFICATION

P. niger is probably the only species of this genus in the western European lowland. A second species, still undescribed as adult, may live in mountain brooks (Schmid, 1993: 151). Baranov (2011) described the adult male of a new species (*P. korneyevi*) of this genus from the Crimean Peninsula. Most probably *P. uliginosus* (Brundin, 1947) is not a separate species (see Saether, 1980: 132). The adult material from different localities is very heterogeneous.

The female, pupa and larva have been described by Saether (1980). The adult male is absent from the key by Pinder (1978), but can be identified using Langton & Pinder (2007). The **larvae** are absent from the keys by Cranston (1982) and Moller Pillot (1984), but they have been keyed and figured by Schmid (1993). The rather long abdominal setae are characteristic. The height of the central mental tooth of the larva is dependent on wear and is not a species character. The larvae collected by Schmid (1993) and the larvae collected in the Netherlands differ from those described by Saether (1980) on that the setae submenti are inserted more distally (see figure 11). This character is also useful for identifying younger larvae in comparison with other genera.

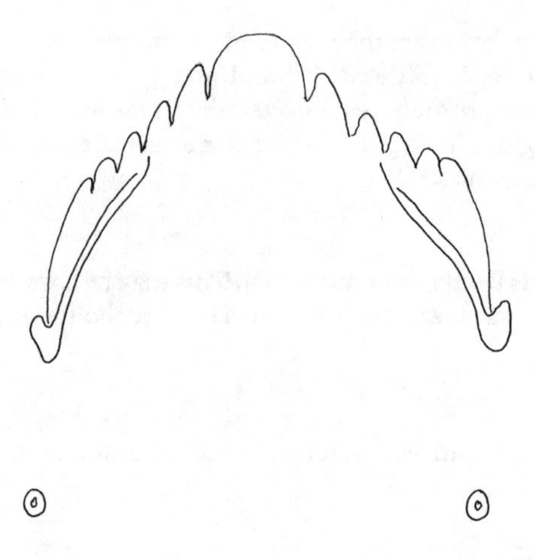

11 *Paracricotopus niger mentum*

DISTRIBUTION IN EUROPE AND THE NETHERLANDS

P. niger has been collected in many countries scattered across Europe (Saether & Spies, 2010). In the Netherlands the species is rather common in southern Limburg and very scarce in the other parts of this province (own data). Elsewhere there are only records from the Veluwe and Winterswijk (province of Gelderland).

LIFE CYCLE

Lehmann (1971) stated emergence in the river Fulda only from June to August. In the Netherlands we found pupae and exuviae from April until October. In winter we collected only larvae in second and third instar (see also Prat et al., 1983).

MICROHABITAT

We collected the larvae from stones, gravel bottoms and water plants in brooks; also from mosses in hygropetric environments (mentioned by Thienemann, 1944, as var. *muscicola*). During rearing the pupae crept upwards (partly backwards) against the mosses or the wall of the vessel to above the water level.

WATER TYPE

Current

Laville & Vinçon (1991) found the species in fast and very fast flowing Pyrenean streams. Larvae and exuviae are rather common in pre-Alpine streams in Bavaria (Orendt, 2002a), but are not found in more or less slow-flowing lowland streams (e.g. Pinder, 1974; Lindegaard & Mortensen, 1988). In the Netherlands and adjacent parts of Germany *P. niger* seems to be confined to fast-flowing stretches of foothill brooks or weirs and fish ladders.

The exuviae have been collected rarely in lakes in Bavaria (Reiff, 1994). *Paracricotopus uliginosus*, which is possibly a separate species, lives in Sweden in dystrophic stagnant water (Brundin, 1956: 120).

Dimensions

As a rule, *P. niger* is absent from large rivers, although in some cases a few larvae or exuviae have been collected (Klink & Moller Pillot, 1982; Caspers, 1991; Móra, 2008; Janecek, 2010; Klink, 2010). Laville & Vinçon (1991) found the species in stream stretches up to a width of 18 m.

Temperature

Rossaro (1982) called the species thermophilous. The mean temperature where *P. niger* had been collected in Italian running waters was 17.7 °C (Rossaro, 1991). Gendron & Laville (2000) called *P. niger* a foothill species and it was found in the Pyrenees not higher than 1600 m (Laville & Lavandier, 1977; Laville & Vinçon, 1991). Its absence from slow-flowing brooks and streams can be caused by a high oxygen demand.

SAPROBITY

Very little is known about the ability of *P. niger* to live in polluted water. The species has been found often in oligosaprobic and β-mesosaprobic conditions (Moog, 1995; Michiels, 1999; own data). The larvae probably need a good oxygen supply (see under Temperature above). The figures given in table 5 in chapter 9 are indicative.

Parakiefferiella Thienemann, 1936

IDENTIFICATION

This genus is still incompletely studied and in this book we discuss only the species to be expected in and around the Netherlands. A list of European species can be found in Langton & Visser (2003). Adult males and exuviae have been described and figured in detail by Tuiskunen (1986). However, some species (e.g. *P. gracillima*) are not keyed here; see under the species names below. The most complete key of exuviae can be found in Langton & Visser (2003). For problems with *P. bathophila* and *smolandica*, see under the former species. There is no reliable key of the larvae; see under the species.

MICROHABITAT AND FEEDING

Whitman & Clark (1984) investigated the sand-dwelling community of a stream in

Texas. They stated that an undescribed species of *Parakiefferiella* lived in open sandy areas and suggested that larvae of the whole genus are psammobiotes and can be absent from streams where this habitat is absent. Most larvae lived in the upper 2–3 cm, but in spring most larvae were found more than 20 cm deep. They found only low densities: less than one larva per dm^3 . They called the genus detritivorous.
The European species have been found in sandy bottoms, but also on stones and on organic bottoms with or without vegetation (see under *P. bathophila*). We also collected larvae of *P. smolandica* on grasses and other water plants with much silt on the leaves.

Parakiefferiella bathophila (Kieffer, 1912)

IDENTIFICATION

Identification of the **adult males** of *P. bathophila* and *P. smolandica* can be difficult (see Tuiskunen, 1986; Langton & Pinder, 2007). In particular, the anal point of *smolandica* can resemble that of *bathophila* (Tuiskunen, p. 194; own material). **Exuviae** of these two species can be distinguished using Langton (1991) or Langton & Visser (2003); however, the anterior transverse band of tergite II can be widely broken in both species and is very variable in *bathophila* (Tuiskunen p. 178; see further the descriptions in Langton & Visser). Both species are easy distinguishable in the **larval** stage (see under *S. smolandica*). A short description of the larva with good figures can be found in Schmid (1993). The differences between the larvae of *bathophila* and *coronata* as given in the existing keys are not reliable; see under *coronata*. See further under the genus.

DISTRIBUTION IN EUROPE AND THE NETHERLANDS

P. bathophila has been collected throughout Europe. In the Netherlands it occurs throughout the whole Pleistocene area and has been found in a few sand pits elsewhere in the country (Moller Pillot & Buskens, 1990; Limnodata.nl).

LIFE CYCLE

In Germany and the Netherlands, emergence has been stated from March to October (Orendt, 1993; Reiff, 1994; own data). In Bavarian lakes there are three or four generations a year, but the numbers are low in summer and autumn. In the Netherlands larger numbers have been stated also in summer (Kouwets & Davids, 1984; van Kleef, unpublished).

MICROHABITAT

As described under the genus, *P. bathophila* is primarily an inhabitant of sandy and muddy-sandy bottoms (Srokosz, 1980; Arts, 2000: 63; Buskens, unpublished). However, the species has also been found on stones (Meuche, 1938; Brundin, 1949; Brodersen et al., 1998; Buskens, unpublished) and on or between submerged plants (Brundin, 1949; Janecek, 2000: 453; Buskens, unpublished). Lindegaard et al. (1975) collected the larvae rarely among mosses in a Danish spring.
Possibly the larvae are more common among the root systems of water plants (own data).

WATER TYPE

P. bathophila is widely distributed in lakes (e.g. Brundin, 1949; Orendt, 1993) up to a depth of 30 or 40 m (Tuiskunen, 1986). In the Netherlands it is mainly found in sand pits (Buskens & Verwijmeren, 1989: 56, 59), rarely in moorland pools and never in ditches.

The species is much less common in flowing water. However, it is sometimes found in rather fast-flowing streams (Gendron & Laville, 1997; Brabec, 2000; Orendt, 2002a) and somewhat more often in lowland brooks and streams (Pinder, 1974; Limnodata.nl.). Nevertheless, it is a scarce species in all types of streams. Klink (2010) did not found it in the upper courses of the Seine and Becker (1995) collected it rarely in the river Rhine. The species is very rare in springs (Lindegaard et al., 1975).

pH

Raddum & Saether (1981) stated occurrence of the species in acid Norwegian lakes even at pHs of about 4.5, but in another lake Raddum et al. (1984) stated an obvious increase in abundance after liming, which raised the pH from 4.8 to 6.7. However, Brodin & Gransberg (1993) found larger numbers of the species after acidification of a Scottish lake in the twentieth century (pH falling to below 5).
Buskens (unpublished) collected *P. bathophila* rarely in acid sand pits in the Netherlands and often in other sand pits with soft and hard water. The species is also rare in moorland pools with a pH of about 5 (Leuven et al., 1987; Arts, 2000: 63; van Kleef, unpublished; own data). It is not clear whether an important difference in pH preference exists between waters in the Netherlands and those in northern countries; see section 2.12.

TROPHIC CONDITIONS AND SAPROBITY

Ruse (2002) collected the species only in water with a rather low conductivity. Most authors stated an obvious preference for water with a low or moderate nutrient content (Särkkä, 1983; Klink, 1986; Buskens, 1989, 1989a; Brodersen et al., 1998; Marziali et al., 2008) and in acid lakes (see under pH). However, it also lives in eutrophic lakes (Tuiskunen, 1986; Orendt, 1993). The last author found three to four generations of this species in eutrophic water and only one in mesotrophic water.
Moog (1995) gives a preference for β-mesosaprobic water in streams. Tuiskunen (1986) mentioned a record from a rapid in a polluted river in Finland. Whitman & Clark (1984) and Arts (2000: 63) suggested that an aerobic sandy bottom is important. This may be the key factor in many lakes and streams (see also Microhabitat).

SALINITY

In northern Europe the species also lives in brackish environments. According to Izvekova et al. (1996), the larvae have been found at a chloride content of 900 mg/l; Paasivirta (2000) found the species where the chloride content was more than 1000 mg/l.

Parakiefferiella coronata (Edwards, 1929)

IDENTIFICATION

Adult males and pupae can be identified without much difficulty. For the difference between the pupae of *coronata* and *gracillima* see Wülker (1957). The characters given for larvae in the keys by Cranston (1982), Moller Pillot (1984) and Schmid (1993) are not reliable. The keys are based on unreared material and the characters used are variable. The larva resembles *P. bathophila* (see also Pagast, 1931: 206, fig. 1). Mentum, antenna and S I are figured by Epler (2001).

DISTRIBUTION IN EUROPE

The species has been recorded in many countries throughout Europe (Saether & Spies, 2010). There are no records from the Netherlands or adjacent lowlands.

LIFE CYCLE

Pagast (1931) stated emergence in Latvia from early May until the end of July. In Sweden the adults were caught by Brundin (1949) only in summer. In Bavaria Orendt (1993) found two generations, flying in May–June and August–September.

MICROHABITAT

The larvae live on and in the bottom, often among vegetation of Characeae or other plants (Pagast, 1931; Brundin, 1949; Tuiskunen, 1986).

WATER TYPE AND WATER QUALITY

The larvae inhabit oligotrophic lakes, ponds and even marsh systems (e.g. Tuiskunen, 1986; Orendt, 1993; Reiff, 1994), but there are several records from fast-flowing streams (e.g. Laville & Lavandier, 1977; ? Schmid, 1993; ? Brabec, 2000: identified as larvae).

Koskenniemi & Paasivirta (1987) found the species at pH 5.5–6, but in Bavarian lakes the pH was above 8 (Orendt, 1993).

Parakiefferiella gracillima (Kieffer, 1922)

IDENTIFICATION

The adult males are not keyed by Tuiskunen (1986) and Langton & Pinder (2007) and can be identified only using Brundin (1956: 152); the hypopygium is very characteristic. The exuviae are absent from the key by Tuiskunen, but can be identified using Langton (1991); they are described in Wülker (1957) and Langton & Visser (2003). The larval descriptions by Wülker (1957) and Schmid (1993, as cf. *gracillima*) are different; see also Rossaro (1982: 68). Most probably the central mental tooth had been damaged in Wülker's specimen.

DISTRIBUTION IN EUROPE

P. gracillima has been collected in few European countries, most of them in central Europe. The species is possibly absent from Scandinavia and large parts of the Mediterranean area (Saether & Spies, 2010). Its occurrence in Dutch southern Limburg cannot be excluded.

ECOLOGY

Emergence has been reported only in (early) spring (Lehmann, 1971; Stur et al., 2005) and in June (Wülker, 1957). All records apply to springs or spring brooks (see also Lindegaard, 1995); Langton (1991) mentioned its occurrence in hygropetric conditions in waterfalls. According to Moog (1995), the species is restricted to xenosaprobic and oligosaprobic water.

Parakiefferiella nigra Brundin, 1949

A boreal and arctic species, in Europe only known from Scandinavia. The larva and ecology are described in Walker et al. (1992). The larva resembles that of *P. gracillima* in Wülker (1957) and Schmid (1993). The species is not treated in this book.

Parakiefferiella scandica Brundin, 1956

DISTRIBUTION IN EUROPE

P. scandica has been recorded in northern and central Europe and on the British Isles (Saether & Spies, 2010).

ECOLOGY

The species has been found in many, mainly oligotrophic lakes and in some brooks in Scandinavia (Tuiskunen, 1986; Paasivirta, 2012). In Bavaria Michiels (1999) collected it occasionally in the lower course of the fast-flowing river Salzach, and two exuviae have been collected in a mesotrophic part of the Chiemsee (Reiff, 1994).

Parakiefferiella smolandica (Brundin, 1947)

Paratrichocladius triquetra Pankratova, 1970: 206–207, fig. 126

IDENTIFICATION

For identification of adults and exuviae see the introduction to the genus and under *P. bathophila*. It is important to be aware of mistakes in identification of exuviae. The larvae of *P. smolandica* have been described and figured by Chernovskij (1949: 130, fig. 120) and Pankratova (1970: 206–207, fig. 126). Based on many associated larvae, pupae, exuviae and adult males from two different brooks, we have stated that *triquetra* is the same species as *smolandica*. This synonymy has been confirmed by rearings by Makarchenko and Pozdeev; a publication is in preparation (A. Przhiboro, pers. comm.).

DISTRIBUTION IN EUROPE AND THE NETHERLANDS

The species has been recorded in a small number of countries scattered across Europe, but not yet in the Mediterranean area (Saether & Spies, 2010). In the Netherlands it has been found only in the southern part of Limburg.

LIFE CYCLE

Reiff (1994) stated emergence in Bavarian lakes from April to October in two (possibly three?) generations. From brooks in the Netherlands and Westphalia (Germany) we collected pupae, exuviae and adults, also from early April until October.

MICROHABITAT

Pankratova (1970: 207) collected the larvae from detritus bottoms of lakes and rivers, Särkkä (1983) on soft bottoms in lakes. We found them in brooks in the Netherlands and Germany in sandy-silty bottoms, but also on grasses and other water plants. See further under the genus.

WATER TYPE

In Russia the larvae have been found up to 10 m deep in lakes and also in rivers (Pankratova, 1970). Elsewhere the species has been recorded mainly from lakes, e.g. in Finland (Särkkä, 1983; Tuiskunen, 1986; Koskenniemi & Paasivirta, 1987) and in Bavaria (Reiff, 1994). We found it in the Netherlands only in small lowland brooks and in adjacent Westphalia (near Coesfeld) in lowland brooks and streams. Reiff (1994) found the species also in a brook in central Germany and Milošević (unpublished) in a small stream in Serbia.

pH

Koskenniemi & Paasivirta (1987) found the species at pH 5.5–6, but in Bavarian lakes the pH was above 8 (Reiff, 1994).

TROPHIC CONDITIONS AND SAPROBITY

In Finland *P. smolandica* lives mainly in oligotrophic oligohumic lakes (Särkkä, 1983; Tuiskunen, 1986), but it has been found also in somewhat loaded mesohumic lakes and even in brackish water (Tuiskunen, 1986; Paasivirta, 2000). In Bavaria the species lives in oligotrophic as well as mesotrophic lakes (Reiff, 1994). The brooks where we collected the species in the Netherlands and Germany were not or hardly polluted.

Paralimnophyes longiseta (Thienemann, 1919)

Limnophyes longiseta Kreuzer, 1940: 465, 467
Paralimnophyes hydrophilus Cranston, 1982: 112; Moller Pillot, 1984: 121, fig. VI.33

SYSTEMATICS AND IDENTIFICATION

The genus is closely related to *Limnophyes*. All the usual keys of adult males, pupae and larvae can be used for identification. The female can be distinguished from *Limnophyes* because the last antennal segment has no long verticillate setae (Edwards, 1929: 355; Goetghebuer, 1940–50: fig. 82, p. 135).

DISTRIBUTION IN EUROPE AND THE NETHERLANDS

The species seems to be absent from southern Europe, but distributed throughout the other parts of Europe (Saether & Spies, 2010). It seems to be scarce in the British Isles and Ireland (Cranston, 1982). In the Netherlands it is very common in the whole Pleistocene part of the country and on the Frisian Islands, but scarce in the other parts of the Holocene and in southern Limburg (Moller Pillot & Buskens, 1990; Limnodata.nl).

LIFE CYCLE

Emergence has been stated from the end of March until the autumn. The species is polyvoltine (Dettinger-Klemm, 2003). Fourth instar larvae are present all year round (own data). Dettinger-Klemm (2000) did rearing experiments and stated that 310–340 degree-days were necessary for total development. One egg mass contained 10 to 100 (mean 37) eggs, laid by females reared at 20 °C (at lower temperature higher egg numbers are expected).

FEEDING

P. longiseta is without doubt a detritus feeder. However, we did not find any further studies and it cannot be excluded that fungi and/or algae are also consumed.

MICROHABITAT

The larvae live mainly at the water's edge and among vegetation.

WATER TYPE

The larvae live in aquatic and semiterrestrial habitats, most commonly in small water bodies in woodland and bogs (including ditches and small upper courses of streams), less often in moorland pools, marshes, marshy vegetation along lakes and very wet grassland. We found the species in optimal habitats (see also pH) almost everywhere in the Netherlands and in Belarus. Most probably the eggs are deposited only when the females find at least a small patch of open water (own data). Waajen (1982) and Werkgroep Hydrobiologie (1993) found the larvae in peat cuttings mainly after advanced succession.

Temperature
Dettinger-Klemm (2000) found in rearing experiments an upper lethal limit for total development between 25 and 29 °C.

Permanence
The larvae are only numerous in temporary water bodies; they are able to survive a period of drought in relatively dry mud without a true diapause (Dettinger-Klemm & Bohle, 1996; Dettinger-Klemm, 2003). However, there is no survival in soils with very low water content (own data; see also under *Limnophyes asquamatus*).

pH

We found the species mainly in acid water (pH 3.1 to 6), but also in other soft water habitats and sometimes in very hard water (alkalinity up to 4 and higher, pH rarely > 8) (cf. Leuven et al., 1987; Werkgroep Hydrobiologie, 1993; Steenbergen, 1993; Duursema, 1996; Dettinger-Klemm & Bohle, 1996: 406–407). However in the water bodies regularly investigated by Dutch water authorities, such as ditches, lakes and canals, a mean pH of about 7 has been stated, obviously because relatively few of these water bodies are acid (Limnodata.nl). The low presence at pH 7, as given in table 6 on p. 284, is because the species is common at this pH only in certain dune and woodland pools, at least in the Netherlands. In Belarus the larvae appeared to be common in circumneutral marshes and temporary pools as well (own data).

TROPHIC CONDITIONS AND SAPROBITY

According to Cranston et al. (1983), the larvae inhabit eutrophic pools and ditches. Nevertheless, they are found mainly in water with low phosphate content and seem to be absent from fertilised grassland. They live mainly in water where (slow) decomposition predominates, also in polyhumic peat cuttings and marshes. Low oxygen contents (even < 5% saturation) are endured and can occur regularly during the night (cf. Waajen, 1982; Werkgroep Hydrobiologie, 1993; Limnodata.nl), probably because the larvae often live near the water surface. In the Netherlands high ammonium-N contents have also been stated (e.g. Limnodata.nl).

SALINITY

There is one record from slightly brackish water (Steenbergen, 1993).

Parametriocnemus Goetghebuer, 1932

SYSTEMATICS AND IDENTIFICATION

Parametriocnemus is most closely related to *Paraphaenocladius* and belongs to the *Heterotrissocladius* group of the Orthocladiinae (Saether, 1975, 1977a). Within the genus *Parametriocnemus* six European species have been described as adult or as pupal exuviae (Langton & Visser, 2003). Several species are known only from southern Europe. As far as is known just one species lives in the lowland of north-west Europe: *P. stylatus*. *P. boreoalpinus* and *P.* Pe1 Langton can also be found in adjacent montane streams.
There is no key to the adult males of all European species. Pupal exuviae can be identified using Langton & Visser (2003). Most larval keys describe only *P. stylatus*; Schmid

(1993) also gives *P. boreoalpinus*. The sizes of the larval head and antenna within *P. stylatus* are very variable (Thienemann, 1937: 28; Moller Pillot, 1984); however, all this material seems to belong to one species.
Only *P. stylatus* is treated in this book.

Parametriocnemus stylatus (Kieffer,1924)

Limnophyes transcaucasicus Chernovskij, 1949: 146, fig. 156; Pankratova, 1970: 248-249, fig. 156

DISTRIBUTION IN EUROPE AND THE NETHERLANDS

P. stylatus is distributed throughout Europe (Saether & Spies, 2010). In the Netherlands it has been collected mainly in southern Limburg and also scattered over the most eastern and southeastern parts of the country, on the Veluwe and in the Gelderse Vallei. In the past the species was more common, probably in a large part of the Pleistocene region, as indicated by palaeoecological investigations (Klink, 2010a).

LIFE CYCLE

Ringe (1974), Caspers (1980a) and Michiels (1999) collected adults or exuviae from March or April until October. Their data indicate two or three generations a year. However, Lindegaard et al. (1975) and Singh & Harrison (1984) stated only a spring generation (in May). In these cases there was most probably also a small second (and third?) generation, which can be easily overlooked, as found by Wilson (1977: 12). In the Netherlands two or three generations seem to be the rule, and fourth instar larvae have been collected almost in every month of the year.
Caspers (1980a) points out that the abundance of the different generations most probably depends on environmental factors; in one year he found a small spring generation and a larger summer generation.

FEEDING

Crisp & Lloyd (1954) found the larvae in a patch of woodland mud, where the only food consisted of decaying leaves and twigs. We also collected them often in mud and on wood, and the guts contained mainly detritus, bacteria and small remains of plant material (see also Microhabitat). The preferred habitat (woodland springs and brooks) also suggest that rather slowly decaying leaves and wood are the main food source and not the rapidly decaying algae and water plants of open landscapes.

MICROHABITAT

Lindegaard et al. (1975) collected large numbers of larvae in the interior part of the moss carpet in the Danish spring Ravnkilde and only low numbers in the border of the carpet, where current velocity was higher and detritus deposits smaller. Most other authors found the larvae on silty/sandy bottom or in detritus, sometimes more along the water's edge. Larvae have also been collected on submerged wood, stones, gravel and plants (Caspers, 1980a; Pinder, 1980; Rossaro, 1982; Spänhoff et al., 2000; own data).

WATER TYPE

Current

Lindegaard (1995) called *P. stylatus* more a lotic than a crenophilous species, although there are clear indications that the larvae prefer a low current velocity (Lindegaard et al., 1975; Janzen, 2003). It is found in many springs and even in a patch of woodland

mud with seepage (Crisp & Lloyd, 1954). Nevertheless, Laville & Vinçon (1991) stated a preference for a fast flow in Pyrenean streams, and almost all records from the Netherlands are in faster flowing brooks, mainly in hilly regions. These data together suggest that the larvae live mainly in sheltered places in fast-flowing streams (cf. Feeding and Microhabitat).
There are few records from stagnant water (Thienemann, 1944). Possibly in all such cases there is seepage of groundwater. Klink (2010a) suggested that the disappearance of the species in several Dutch regions has been caused by decline of groundwater seepage.

Dimensions
Most records of the species apply to springs and narrow upper courses of brooks. However, the larvae can be found frequently in downstream parts of brooks and streams and even in large rivers (Becker, 1995; Gendron & Laville, 1997; Klink, 2010; Waterschap Roer en Overmaas, unpublished). It is not known if they spend their whole life cycle there; possibly such specimens have been brought down by drift.

Shade
P. stylatus is especially common in wooded landscapes, even when the whole brook is shaded around the year. Nevertheless, larvae have also been recorded from lower courses in open country (see under Dimensions).

pH

P. stylatus seems to be absent from acid streams (Orendt, 1999; Janzen, 2003; Syrovátka et al., 2012; Limnodata.nl; own data). This may be the reason why the species is almost absent from upper courses of brooks in the Dutch province of Noord-Brabant. Verdonschot & Schot (1987) collected the larvae in springs with a pH of about 6.5.

SAPROBITY

According to Moog (1995), *P. stylatus* is a species of mainly oligosaprobic and β-mesosaprobic water. The more than 100 records in the Netherlands apply to brooks and springs with a rather low BOD and high oxygen content (Limnodata.nl; Waterschap Roer & Overmaas, unpublished; own data). Tang et al. (2010) found it subdominant in streams in South Korea with low pollution levels (mean BOD 1–3 mg/l). However, the larvae are found especially in woodland brooks in detritus (e.g. in mud of decaying leaves) and on decaying wood (see Microhabitat and Feeding). It is possible that they avoid anthropogenic organic matter and rapidly decaying algae and water plants. In this context it is noteworthy that the species prefers basic conditions (see above).

SALINITY

Paasivirta (2000) collected the species at two localities in the Baltic Sea with a chloride concentration of about 2000 mg/l.

SUMMARY

The ecological data can be summarised as follows. The species can be found mainly in more or less basic conditions in brooks, springs and other places with groundwater seepage in places sheltered from fast flows, especially in wooded landscapes. Slowly decaying organic matter seems to be the main food source and anthropogenic polluted silt seems to be unsuitable. In northern countries the tolerance spectrum can be different (see under Salinity).

Paraphaenocladius Thienemann, 1924

SYSTEMATICS

In the revision of the genus by Saether & Wang (1995) the authors described several new taxa which they regarded as subspecies. However, Ekrem et al. (2010) stated a considerable genetic distance between the subspecies, so that these probably should be raised to species. Because even the species has not been stated in much of the literature and in most of our data (often identified as larvae) and because *Paraphaenocladius* lives mainly in semiterrestrial habitats the genus is treated here only very briefly. We confine ourselves to the species level as used in Saether & Wang (1995).
Within the genus three species groups are represented in northwest Europe: the *irritus* group, the *penerasus* group and the *impensus* group.

IDENTIFICATION

Adult males can be identified using Saether & Wang (1995) or Langton & Pinder (2007). Identification of exuviae is only possible using Saether & Wang (1995) or Langton & Visser (2003). All keys to larvae are incomplete; at present only the three groups can be distinguished. We give here the differences between the larvae of *pseudirritus* (*irritus* group) and *penerasus*, based on specimens reared by H. van den Elsen (unpublished):

a Base of anterior parapods with many sclerotised plates with single up to ten or more narrow spines. Ventromental plates partly duplicated, with turned 'wings'. Mostly one median tooth. First antennal segment about three times longer than wide *pseudirritus*

b Base of anterior parapods without striking sclerotised plates, only with groups of narrow spines. Ventromental plates without turned wings. Median mental tooth (usually?) excavated. First antennal segment two times longer than wide *penerasus*

ECOLOGY

Probably all species of the genus are more or less semiterrestrial, living in wet bottoms and in springs. The larvae seem to avoid acid environments. We found no clear difference in ecology between the species. The following descriptions focus mainly on the more or less aquatic ecology of the species.

Paraphaenocladius cuneatus (Edwards, 1929)

This species is described from two females only and these females could not be found (Saether & Wang, 1995: 65). Because of the cuneiform wing and the long apical seta it could be a synonym of *Molleriella calcarella*. However this cannot be verified.

Paraphaenocladius exagitans (Johannsen, 1905)

Paraphaenocladius impensus monticola Strenzke, 1950: 211–215, fig. 2, 4

Only the subspecies *P. exagitans monticola* Strenzke is known from Europe. It is distributed in the alpine region, Scandinavia and Luxembourg (Saether & Spies, 2010; Stur et al., 2005).
The morphological differences from other species of the *impensus* group are small (see Saether & Wang, 1995). Its ecology is probably not very different from that of *impensus* (Strenzke, 1950: 215, 370).

Paraphaenocladius impensus (Walker, 1856)

SYSTEMATICS

Within the current definition of *P. impensus* Ekrem et al. (2010) found five separated clusters which possibly merit species status.

DISTRIBUTION IN EUROPE AND THE NETHERLANDS

The species has been found in many countries throughout Europe, even far from the continent on Madeira, Iceland and the Faeroe Islands (Saether & Spies, 2010). In the Netherlands it is a common species everywhere, even on the West Frisian Islands (Moller Pillot & Buskens, 1990).

LIFE CYCLE

In early winter almost all larvae are in second and third instar, suggesting a diapause. During winter many larvae develop into fourth instar. Adults have been collected throughout spring and summer, from April until autumn (own data; cf. Strenzke, 1950: 397; Pinder, 1974). Lindegaard et al. (1975) found most adults in May and suggested slow growth of the larvae.

FEEDING

There is hardly any information about feeding. In the guts we found mainly detritus.

MICROHABITAT AND ECOLOGY

The larvae live almost everywhere in wet bottoms: on banks of brooks and streams, ditches and pools, in wet grassland, reedbeds and woods. They are common in springs and can be collected also below the water level, but they do not swim and therefore they are rarely observed in aquatic samples. Nevertheless, they appeared to be locally common in the lower courses of large rivers, such as Hollands Diep and Oude Maas (Limnodata.nl).

The larvae can often be found in dense marshy vegetation, sometimes together with *Pseudorthocladius*, but they are absent from very acid environments, where *Pseudorthocladius* species are not rare (Strenzke, 1950: 350–358; own data). Lindegaard et al. (1975) collected the larvae in the moss carpet in the Danish spring Ravnkilde, mainly in the madicolous zone just above the water surface where the moss was constantly moistened by capillary water. We reared the species sometimes from submerged wood and stones.

SALINITY

We also collected the larvae in slightly brackish environments (Moller Pillot & Buskens, 1990: 14). It is a common inhabitant of the Baltic Sea (Paasivirta, 2000).

Paraphaenocladius intercedens Brundin, 1947

P. intercedens is only known from Scandinavia and the Pyrenees (Moubayed-Breil, 2008; Saether & Spies, 2010). The first author collected the exuviae in the upper courses of streams in the Pyrenees. Lundström et al. (2010) collected the species in temporary flooded wetlands in Sweden, sometimes in large numbers.

Paraphaenocladius irritus (Walker, 1856)

nec *Paraphaenocladius irritus* Langton, 1991: 160 (misidentification)

P. irritus has been recorded throughout Europe (Saether & Spies, 2010), but not all records have been verified. There are no data from the Netherlands; the record in Moller Pillot (1984: 127) is based on a misidentification. The data mentioned in Strenzke (1950), Lehmann (1971) and Saether & Wang (1995) suggest about the same ecology as in other species of the genus.

Paraphaenocladius penerasus (Edwards, 1929)

Paraphaenocladius irritus Langton, 1991: 160, fig. 65i (misidentification)

SYSTEMATICS AND IDENTIFICATION

P. penerasus is the only representative of the *penerasus* group in Europe (Saether & Wang, 1995). The adult male can be identified using Saether & Wang (1995) or Langton & Pinder (2007); the pupa has been keyed in all keys, but in Langton (1991) incorrectly as *P. irritus*. The larva resembles that of *pseudirritus* but can be identified as described under the genus above.

DISTRIBUTION

Most records of the species are from western and central Europe, but it has been collected also in the East Palaearctic (Saether & Spies, 2010). In the Netherlands it has been found only in the central part of the province Noord-Brabant (Moller Pillot & Buskens, 1990).

LIFE CYCLE

Adults have been collected in spring, summer and autumn (Saether & Wang, 1995; Laville & Lavandier, 1977). We found larvae in December and January in third and fourth instar.

ECOLOGY

Larvae have been found among mosses and *Cladophora* cushions in seepages, springs and upper courses of streams, mainly in mountain regions (Geijskes, 1935; Laville & Lavandier, 1977; Laville & Vinçon, 1991; Langton, 1991). In the Netherlands the larvae (some reared to adults) were rather common on the banks of the Wilhelminakanaal overgrown with herbs and grasses. The larvae lived in the wet to moist zone from 0 to about 50 cm (mainly 5–12 cm) above the water level (H. van den Elsen, unpublished). Because of the rather stable water level in the canal and possibly a little inward seepage of water, this zone had a constant humidity. The presence or absence of mosses had no influence on the occurrence of the species.

Paraphaenocladius pseudirritus Strenzke, 1950

SYSTEMATICS AND IDENTIFICATION

P. pseudirritus belongs to the *irritus* group. For identification see the introduction to the genus. Within the species Ekrem et al. (2010) found several clusters with a high degree of genetic divergence (> 10%).

DISTRIBUTION IN EUROPE AND THE NETHERLANDS

The species lives throughout Europe (Saether & Spies, 2010). In the Netherlands there are about 20 records of larvae of the *irritus* group scattered across the whole country. Only from two localities adult males of *P. pseudirritus* have been reared.

ECOLOGY

The larvae live in springs and in the banks of brooks, streams, canals and ditches around the water level and also on wet to moist soil near lakes, in marshes and in grassland with seepage (Strenzke, 1950; Ekrem et al., 2010; own data, mainly based only on identification of larvae). We saw one larva in a tube.

Parasmittia Strenzke, 1950

As far as is known the larvae of *Parasmittia* are strictly terrestrial and therefore the ecology will not be described in this book. For description of all stages and the ecology of the larvae, see Strenzke (1950). Most probably there are still several undescribed species (Strenzke, 1950; Paasivirta, 2012; own data).

Paratrichocladius Santos Abreu, 1918

Many mistakes have been made in interpreting the genus *Paratrichocladius* and literature older than Hirvenoja (1973) cannot be used at all. The species placed under this genus by Pankratova (1970) belong to other genera. Hirvenoja (1973) keyed the adults and pupae of two species without figures or detailed description; his short description of a larva does not match *P. rufiventris*.
At least six species of the genus live in Europe. There is no key to all these species as adult, pupa or larva. In the Netherlands and adjacent lowlands only *P. rufiventris* and possibly *P. skirwithensis* can be found. These species can be distinguished from each other as adult male using the key by Hirvenoja (1973) or Langton & Pinder (2007), as exuviae using Langton (1991) and as larva using Cranston (1982), Schmid (1993) or Bitušík (2000). However, as pointed out by Cranston et al. (1983), it is not always easy to separate larvae of *Paratrichocladius*, *Cricotopus* and *Orthocladius*.

Paratrichocladius rufiventris (Meigen, 1830)

Syncricotopus rufiventris Lehmann, 1971: 490; Ringe, 1974: 239

IDENTIFICATION

For identification of adults and pupae see the introduction to the genus. Because of the difficulties in identifying the larvae, it can be useful to look for the usually yellow colour of the head, the often violet-blue thorax and blue-greenish abdomen. In addition, the base of the ventromental plates and the position of the bases of the setae submenti can help to distinguish the species from most species of *Orthocladius* (also in third and second instar). The first antennal segment is relatively short. Very good figures are given by Schmid (1993: 421).

DISTRIBUTION IN EUROPE AND THE NETHERLANDS

P. rufiventris has been recorded in the whole of Europe (Saether & Spies, 2010). In the Netherlands the species is widely distributed in southern Limburg, in the Pleistocene part of the country and in the large rivers, and is scarce elsewhere in the Holocene parts (Limnodata.nl; Nijboer & Verdonschot, unpublished). It is almost absent from the province of Zeeland. Other blank areas in distribution maps are probably due to some researchers not recognising the larvae.

LIFE CYCLE

There are probably two to four generations a year. Usually emergence has been stated from March or April to October. Locally (possibly dependent on water temperature and food) small numbers emerge as early as February or even January and as late as November or December (Drake, 1985; Becker, 1995; Michiels, 1999). In some cases, but not always, the highest numbers were stated in spring.
In winter we collected only larvae in second and third instar, suggesting a diapause as in *Cricotopus*. As appears from the emergence data above, further development is possible also in winter.

FEEDING

According to Bitušik (2000) the larvae feed on epilithic algae. Moog (1995) also mentioned feeding on detritus.

MICROHABITAT

The larvae live on stones, wood and plants, less abundantly also on the bottom (Lehmann, 1971; Becker, 1995; own data). They build tubes cemented to the substrate.

WATER TYPE

Apart from the factors described below and in the section on Saprobity, *P. rufiventris* seems to be absent in some streams without a clear cause (e.g. in the river Inn: Orendt, 2002a; Tadnoll Brook: Pinder, 1974). In other streams the densities are much lower than expected (e.g. Geul and Jeker in southern Limburg). Factors like concurrence with other species cannot be excluded, but also chemical factors may be the cause.

Current

Laville & Vinçon (1991) collected *P. rufiventris* equally numerously in stretches of Pyrenean streams with fast currents and (relatively) slow currents. However, in the Netherlands and adjacent parts of Germany we found the species to be much more abundant in brooks and streams with a current velocity higher than 75 cm/s and on fish ladders in lowland streams. It seems probable that the larvae in fast-flowing brooks and streams settle in slower flowing stretches as well. In the faster flowing streams in Bavaria the species is rarely absent (Orendt, 2002a). Gendron & Laville (2000) stated that the larvae in the river Aude in the Pyrenees sharply decreased in numbers after a catastrophic flood.
In stagnant water the species can be found scarcely, mainly in lakes and canals (Kouwets & Davids, 1984; Limnodata.nl; own data), but sometimes also in small water bodies, such as city fountains (Hamerlik & Brodersen, 2010).

Dimensions

The larvae live in small upper courses as well as in larger streams and rivers (Lehmann, 1971; Becker, 1995; own data). Highest densities are found further downstream (e.g. Gendron & Laville, 1997), possibly because of increasing primary production (Bitušik, 2000). In stagnant water most records are from larger water bodies, probably because of better oxygen conditions.

Shade
We found the species locally in a brook in deciduous woodland in spring. Elsewhere it was absent in such brooks in summer; it is not clear if this can be caused by shading. Klink (2011) found the larvae in the river Meuse substantially less on shaded trees, possibly because of lower overgrowth with periphyton.

Permanence
P. rufiventris has been collected in brooks falling dry in summer where the larvae could survive deeper in the bottom (own data). Hamerlik & Brodersen (2010) found them also in city fountains which contained no water in winter.

pH

All data indicate a preference for more or less alkaline water. There are no records from acid brooks or streams.

TROPHIC CONDITIONS AND SAPROBITY

Stagnant water
Ruse (2002) collected *P. rufiventris* only in lakes with low conductivity. However, the species seems to be rare or absent in lakes with low productivity (e.g. Reiss, 1984; Buskens & Verwijmeren, 1989). Records from stagnant water bodies in the Netherlands are from large (usually eutrophic) lakes or canals with good oxygen conditions (Limnodata.nl; own data). The species has also been found in city fountains (Hamerlik & Brodersen, 2010), where enough food is available and the oxygen content is usually high because of the turbulent water.

Running water
Many investigations have demonstrated a positive correlation with increasing primary production, nitrogen content, ammonium-N and BOD. The larvae are especially numerous in streams with much organic pollution, even at BOD levels above 30 mg/l (Davies & Hawkes, 1981; Bazzanti & Bambacigno, 1987; Bitušik, 2000; Tang et al., 2010; Ruse, 2012). The larvae appeared in large numbers after loading with coal particles in a stream in Wales (Learner et al., 1971). On the other hand, the species also lives in unpolluted or hardly polluted streams (Wilson, 1987; Michiels, 1999; Limnodata.nl; own data). Davies & Hawkes (1981) stated that the species was absent in water with an oxygen content of less than 2 mg/l (during daytime?). Also the preference for fish ladders in Dutch lowland streams is an indication of their oxygen demand. This is summarised in table 5 on p. 283.

SALINITY

The species has often been found in brackish water in Finland (Hirvenoja, 1973: 74; Paasivirta, 2000). In the Netherlands all records are from fresh water with a very low chloride content (Limnodata.nl).

Paratrichocladius skirwithensis (Edwards, 1929)

Syncricotopus nivalis Lehmann, 1971: 489, 518 (nec aliis)

DISTRIBUTION IN EUROPE

Due to problems in identification (see the introduction to the genus) this species has been stated definitely in few European countries, mainly in central Europe, but also in the British Isles and in Scandinavia (Saether & Spies, 2010). It is absent from the Netherlands.

WATER TYPE

P. skirwithensis lives mainly in fast-flowing brooks and streams in the epirhithral zone (Lehmann, 1971; Gendron & Laville, 1997; Michiels, 1999; Bitušík, 2000). However, it has been found sometimes in lower parts of such streams and in lakes in hilly regions (Reiff, 1994; Michiels, 2004). The species is absent from typical lowland streams.

Paratrissocladius excerptus (Walker, 1856)

Trissocladius fluviatilis Pankratova, 1970: 139–140, fig. 76; Lehmann, 1971: 490
Chaetocladius excerptus Pinder, 1974: 198

SYSTEMATICS

In Europe only one species of the genus has been described. For systematics and synonyms see Saether (1976: 253–265).

DISTRIBUTION IN EUROPE

P. excerptus has been collected in most of Europe (Saether & Spies, 2010). There is only one record from the Netherlands: the exuviae of one specimen are collected in the Zieversbeek near Vaals (May 2012, leg. B. van Maanen & M. Korsten).

LIFE CYCLE

Pinder (1974) and Ringe (1974) stated emergence in summer and autumn in southern England and central Germany. Elsewhere emergence has also been reported in spring (Lehmann, 1971; Reiff, 1994). In Ireland the adults were flying in winter (Fahy, 1973).

MICROHABITAT

The larvae live in sand and mud in streams, where they build long tubes of sand and detritus (Pankratova, 1970; cf. Kretzschmar & Böttger, 1994), and sometimes also in more course material (Schmid, 1993).

WATER TYPE

Current and dimensions

There are very few records from typical lowland streams (e.g. Pinder, 1974), but the species has often been collected in somewhat faster flowing streams, mainly in upper courses, in hilly or mountainous environments (Lehmann, 1971; Ringe, 1974; Bazerque et al., 1989; Kretzschmar & Böttger, 1994; Gendron & Laville, 1997; Janzen, 2003; Michiels, 2004). The larvae seem to be scarce in small rivers, such as the upper courses of the Seine (Klink, 2010), and rare in large rivers (e.g. Becker, 1995).
The species is very rarely collected in lakes (Reiff, 1995).

Shade

According to Cranston (1982), all larval sites in Britain were cool, tree-shaded streams. Kretzschmar & Böttger (1994) collected the species in a heavily shaded brook in northern Germany.

SAPROBITY AND OXYGEN

Almost all data apply to unpolluted or hardly polluted upper courses. Bazerque et al. (1989) collected the species in the upper courses of the small rivers Selle and Somme only at a BOD of 2 mg O_2/l and not at BOD 3 or more. Kretzschmar & Böttger (1994) stated emergence at BOD up to 5.6 in a woodland brook with minimum O_2 saturation of 65%. Wilson & Ruse (2005) called *P. excerptus* intolerant to organic pollution.

Parorthocladius Thienemann, 1935

Synorthocladius Pankratova, 1970: 169–172 (pro parte)

SYSTEMATICS AND IDENTIFICATION

The genus is related to *Synorthocladius* and the larvae of both genera are very similar. Of the four European species, only the larva of *P. nudipennis* is known. For larval descriptions and figures, see Cranston et al. (1983) and Schmid (1993).

DISTRIBUTION IN EUROPE

Parorthocladius nudipennis seems to be the most common species in Europe. It has been collected in many countries throughout Europe (Saether & Spies, 2010). The genus is not known from the Netherlands and is not to be expected there.

MICROHABITAT

Larvae and pupae have been found on stones in mountain brooks (Potthast, 1914: 298; Geijskes, 1935).

WATER TYPE

The genus lives mainly in cold mountain brooks (Lehmann, 1971; Braukmann, 1984; Rossaro, 1991; Schmid, 1993). Langton (1991) gives rivers as habitat for *Parorthocladius nigritus*. According to Braukmann (1984: 221), *P. nudipennis* is not confined to carbonate streams, as suggested by Kownacka & Kownacki (1972).

Propsilocerus lacustris Kieffer, 1923

Based on molecular phylogenetic investigations, the genus *Propsilocerus* belongs to the subfamily Prodiamesinae and is treated on p. 261.

Psectrocladius Kieffer, 1906

SYSTEMATICS

The genus is divided into four subgenera: *Psectrocladius*, *Monopsectrocladius*, *Mesopsectrocladius* and *Allopsectrocladius*. In this book the subgenus is included between brackets in the names of the species of the last three subgenera. Within the subgenus *Psectrocladius* three groups have been distinguished based on Wülker (1956:8): the *limbatellus* group, the *sordidellus* group and the *psilopterus* group.
Until about 1990 many authors used the names *psilopterus* group and/or *limbatellus* incorrectly; see below under these species.

IDENTIFICATION OF THE LARVAE

A **description** of the larvae of subgenus *Psectrocladius* in all four instars can be found in Zelentsov (1980).
A **key** to the subgenera is given by Cranston et al. (1983: 192) and Bitušik (2000); these authors have figured the mentum of the four subgenera. Their division into species groups based on occipital margin colour is not reliable. The keys to species by Pankratova (1970), Cranston (1982) and Moller Pillot (1984) are incomplete: see the comments below under the species names. Identification of all larvae to species level is still impossible.

Mistakes

Because Thienemann accidentally mixed up material, the name *P. psilopterus* has been used for larvae of other species; this mistake is still present in Pankratova (1970) and Rossaro (1982) and applies to most of the older literature. Mistakes can also be found in Wülker (1956): they may also have been caused by the accidental mix-up of the Thienemann material. For this reason this publication cannot be used to identify larvae (Wülker, in litt. d.d. 21 January 1986). A mistake concerning *P. oligosetus* has been copied in Moller Pillot (1984: 136, 137).

Adults and exuviae can be identified in reliable manner using Langton & Pinder (2007) and Langton (1991) respectively.

LIFE CYCLE

The life cycle of the species of this genus is not fully known. In many cases there are more than two generations a year. Goddeeris (pers. comm.) found a sharp decrease in the numbers of fourth instar larvae of a species of the *limbatellus* group during August and an accumulation of second and third instar larvae, suggesting a diapause induced by short-day conditions. We also stated an accumulation of second and third instars and a gradual absence of fourth instars in *P. platypus* in autumn (see under this species). In *P. bisetus* we found at the end of November only larvae in first, second and third instar (especially third instar in large numbers). In *P. ? oligosetus* we found in the same sample most larvae in early fourth instar (imaginal discs up to subphase 4). Probably an autumnal diapause is widespread in this genus, but the species can be specialised in their time niche.

In the laboratory Shilova & Zelentsov (1972) stated pupation throughout autumn.

FEEDING

Most species of *Psectrocladius* feed on algae and/or detritus. However, there are large differences between the species. All or most species also feed on filamentous algae, but some algal species, for example species of *Bulbochaete*, are only incidentally consumed or even avoided (Botts & Cowell, 1992; own observations). Pankratova (1970:31) observed the consumption of small chironomids and Oligochaeta.

WATER TYPE

The genus as a whole is characteristic of stagnant water. It is often absent or only incidentally mentioned in articles about brooks, streams and rivers (e.g. Lehmann, 1971; Caspers, 1980; Lindegaard & Mortensen, 1988; Laville & Vinçon, 1991; Orendt, 2002a). Some species are not rare in upper courses of lowland brooks (e.g. *P. platypus* and *P. psilopterus*); other species probably arrive in flowing water from surrounding pools or dead river arms.

The occurrence of certain species of this genus in temporary water seems to be very characteristic. It is dependent on the dispersal power of the species (see under Dispersal) and on the preference for large or small water bodies.

More detailed information about current and permanence in linear water bodies is given for some species in table 7 on p. 287.

pH

The species of the genus *Psectrocladius* display conspicuously large differences in their preference for acid or calcareous water. Some species are only found in calcareous water (e.g. *P. barbimanus*), other species live more or almost only in very acid water (*P. platypus*, *P. oligosetus*, *P. bisetus*). See fig. 12 and table 6 in Chapter 9.

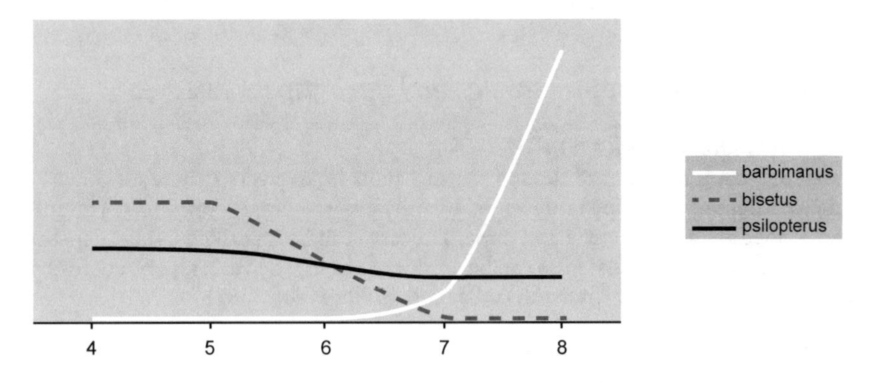

12 *Some examples of pH preference in species of Psectrocladius*

TROPHIC CONDITIONS AND SAPROBITY

The genus *Psectrocladius* is often considered to be more or less characteristic of less eutrophic conditions (e.g. Moog, 1995). Brodersen et al. (1998) collected the larvae at a relatively low chlorophyll-a content and Langdon et al. (2006) found them scarcely at chlorophyll-a concentrations above 50 µg/l. Steenbergen (1993) stated for all species of the genus significantly higher occurrence at low orthophosphate contents (< 0.05 mg P/l) and chlorophyll-a concentrations below 20 µg/l. However, Orendt (1993) collected *P. sordidellus* and *P. brehmi* significantly more frequently in very eutrophic water and Wotton et al. (1992) found *P. limbatellus* numerously in sand filter beds with intensive decomposition of organic matter. Limnodata.nl gives as a mean a low trophic level for the *sordidellus* group, but the larvae have sometimes been found in water with a high BOD (up to 9 mg O_2/l) and/or very low oxygen content (less than 1 mg/l in daytime). Several species of *Psectrocladius* are common in polyhumic (dystrophic) water, where phytoplankton is scarce and organic matter washes in from the surrounding peat (see also par. 2.17).

In many of the above-mentioned studies the genus name has been used because the larvae could not be identified to species level. The species inhabit very different water types and these are usually not all covered in each of the studies. It seems highly likely that *P. oxyura* and *P. bisetus* are much less tolerant to low oxygen concentrations than *P. sordidellus* and *P. platypus*. Nevertheless, it appears that even species like *P. sordidellus*, which are able to tolerate higher trophic and saprobic conditions, are found only relatively scarcely in hypertrophic or polluted water. Little is known about the combined effect of factors likes oxygen supply, water movement, decomposition products, pH and temperature. For this reason only few species have been included in table 5 in Chapter 9. It is necessary to consult the descriptions of the ecology of the species.

DISPERSAL

Some species of *Psectrocladius*, especially *P. limbatellus* and *P. platypus*, are found very often in temporary water (see above under Water type). Without doubt these species have a high dispersal power. From investigations by van Kleef & Esselink (2005) it appears that the settlement of three other species of the genus (*P. psilopterus*, *P. oligosetus* and *P. bisetus*) is highly dependent on the presence of populations of these species within a radius of one or two kilometres, which is an indication of a very limited dispersal power.

Psectrocladius (Mesopsectrocladius) barbatipes Kieffer, 1923

SYSTEMATICS AND IDENTIFICATION

Based on a comparison of adults and pupae, Laville (1971) erected a new subgenus *Mesopsectrocladius*, in Europe represented by only one species: *P. barbatipes*. The author redescribed the male, female and pupa. The larva can be identified using Cranston et al. (1983), but the antennal ratio can be higher, up to 2.9 (J. Mulder, pers. comm.). It is absent from the keys by Cranston (1982) and Moller Pillot (1984).

DISTRIBUTION IN EUROPE AND THE NETHERLANDS

P. barbatipes has been found in only a few European countries and is not known from eastern Europe (Saether & Spies, 2010). In the Netherlands the species has been collected from locations scattered across almost the whole Pleistocene part of the country, but is nevertheless scarce (unpublished data).

LIFE CYCLE

Exuviae have been found in Bavaria from the end of April until October (Reiff, 1994), but Reiss (1984) stated emergence almost only in spring and Laville (1971) in the Pyrenees only in summer. In the Netherlands exuviae have been collected from early June to the end of August. Reiff's data indicate two or three generations a year.

pH

Reiff (1994) collected the exuviae in large numbers in the Lustsee in Bavaria at pH 8.1. In the Netherlands the species has been found only in acid moorland pools, with a pH often not higher than 4. Obviously pH is not an important factor (see under Trophic conditions).

WATER TYPE

The species has been found only in stagnant water bodies, such as lakes and moorland pools (Laville, 1971; Langton, 1980; Reiss, 1984; Reiff, 1994; van Kleef, unpublished). The Dutch data demonstrate that the species is not characteristic of cold water, as suggested by Reiss (1984).

TROPHIC CONDITIONS

All records of *P. barbatipes* are from oligotrophic lakes and pools, often in northern or submontane areas, but also in the Netherlands (see under pH and Water type).

Psectrocladius barbimanus Edwards, 1929

SYSTEMATICS AND IDENTIFICATION

P. barbimanus belongs to the *limbatellus* group of the subgenus *Psectrocladius*. It is the only species within this group which can be identified easily in the larval stage: see Thienemann (1944: 629) and Moller Pillot (1984: 139, 143). The larva is described and figured by Zelentsov (1980b). The larva and pupa are larger than other species of the group (see, however, *P. ventricosus*).

DISTRIBUTION IN EUROPE AND THE NETHERLANDS

The species has been found in many countries scattered across the whole of Europe (Saether & Spies 2004). In the Netherlands it has been collected mainly in the Holocene western part of the country, especially in the dune region, and in some places around the large rivers (Limnodata.nl; own data).

LIFE CYCLE

Wotton et al. (1992) stated a total development time of 28 to 46 days in sand filter beds under favourable conditions (abundant food). The much smaller *P. limbatellus* needed about half of this development time. Emergence was stated until the end of October. Zelentsov (1980b) found in deeper parts of the lake littoral (more than 1 m) only one generation a year. Dutch records of pupae and exuviae are from May until September.

MICROHABITAT

Most records of *P. barbimanus* in the Netherlands are from lakes and pools with no or very few water plants, and in some cases it has been stated that the larvae lived on bare bottoms of silt, sand or gravel. This is consistent with the literature (e.g. Cannings & Scudder, 1978; Learner et al., 1989; Wotton et al., 1992; Ruse, 2002a). However, the larvae have also been collected on or between filamentous algae or water plants like *Myriophyllum* (Driver, 1977; Zelentsov, 1980b; Learner et al., 1989). Reiff (1994) even called the species phytophilous.

WATER TYPE

P. barbimanus lives mainly in stagnant water, but has been collected also in slow-flowing ditches and canals (Steenbergen. 1993; Limnodata.nl). Most records are from lakes and reservoirs, but in the same regions the species is also common in smaller pools (Ketelaars et al., 1992; Reiff, 1994; own data). It settles easily in temporary water (Driver, 1977; Wotton et al., 1992; Hamerlik & Brodersen, 2010).

pH

P. barbimanus is a characteristic inhabitant of water with high alkalinity and rather high conductivity (Ruse, 2002). In the Netherlands the species seems to be almost confined to the Holocene western part of the country (Nijboer & Verdonschot, unpublished; Moller Pillot & Buskens, 1990); elsewhere in Europe it can be found especially in the high alkalinity water bodies, as in Hungary and Neusiedler See in Austria (Berczik, 1967; Wolfram, 1996). Steenbergen (1993) stated increasing presence as pH increased within the range from 7 to 9 (see also Limnodata.nl).

TROPHIC CONDITIONS AND SAPROBITY

Although *P. barbimanus* occurs mainly in water bodies with a high pH and calcium content, it is most common in water bodies with low concentrations of phosphate, ammonium and chlorophyll-a (Steenbergen, 1993; Schmale, 1999). Nevertheless, the species was collected abundantly in sand filter beds with much decomposition of organic matter (Wotton et al., 1992) and it was sometimes found in hypertrophic conditions by Steenbergen (1993).

SALINITY

Cannings & Scudder (1978) collected *P. barbimanus* in Canada in all brackish water bodies they investigated. Paasivirta (2000) found it scarcely in the Baltic Sea. In western Europe the species appears to be rare in brackish environments (see table 6), although it lives especially in water with rather high conductivity (Limnodata.nl). Ruse (2002) did not find it in brackish water; in the province of Noord-Holland it was absent from water with chloride concentrations higher than 1000 mg/l and rare between 300 and 1000 mg/l (Steenbergen, 1993); in the province of Zeeland the species is very scarce (see Krebs, 1981, 1984, 1990). Tourenq (1975) did not collect *P. barbimanus* in brackish water in the Camargue (France).

Psectrocladius bisetus Goetghebuer, 1942

SYSTEMATICS AND IDENTIFICATION

Within the subgenus *Psectrocladius* the species belongs to the *psilopterus* group (Cranston et al., 1983; Coffman et al., 1986). The adult male can be identified using Langton & Pinder (2007). In older keys identification usually leads to the species *P. psilopterus*. Exuviae can be identified with all existing keys since Wülker (1956).
The larva is absent from almost all published keys and in fourth instar resembles a third instar larva of *P. psilopterus* (head length in *bisetus* fourth instar about 0.35 mm, in third instar about 0.23 mm). It has three long and two very short anal setae, in third instar three setae of different length and thickness (own data). The anal tubules can be longer or shorter than the posterior parapods (see also Wülker, 1956: p.29–30).

DISTRIBUTION IN EUROPE AND THE NETHERLANDS

P. bisetus has been recorded only from a small numbers of countries, but most probably it lives in almost the whole of Europe, possibly with the exception of parts of the Mediterranean area (Saether & Spies, 2010). In the Netherlands the species has been collected in more than 40 moorland pools, scattered over the whole Pleistocene area (Duursema, 1996; van Kleef, unpublished; own data).

LIFE CYCLE

The life cycle seems to be generally the same as in other species of the genus: three or more generations a year, flying from April until October. At the end of November we collected no fourth instar larvae and many larvae in third and second instar, suggesting a diapause (see the introduction of the genus).

MICROHABITAT

The larvae live among water plants and *Sphagnum*, but also on detritus bottoms (own data). In rearing experiments some of the larvae creep around freely, but most of them live in a silken tube more or less covered by algae or detritus. The tubes are attached to Sphagnum or other plants or to the bottom of the dish.

DENSITIES

In moorland pools in the Netherlands very high densities of *P. bisetus* have been stated more often than in other species. Exact data are lacking.

WATER TYPE

P. bisetus has been collected only in stagnant water and seems to be characteristic of small, polyhumic, more or less boggy lakes. Brundin (1949) does not mention the species, but Wülker (1956: 55) saw males collected by Brundin in the small lake O. Vontjärn. It is possible that also larvae of the *psilopterus* group, collected by Brundin in larger lakes, belong to this species.
In the Netherlands the species is very common in moorland pools (Duursema, 1996; van Kleef, unpublished). Duursema (1996: 81) did not find the species in temporary pools.

pH

In the Netherlands *P. bisetus* has been found only in acid water with a pH from 3.4 to 5.9, rarely higher (Buskens 1989a; van Kleef, unpublished).

TROPHIC CONDITIONS AND OXYGEN

The species has been found only in phosphate-poor conditions (Buskens, 1989a; Duursema, 1996: 81). In moorland pools with more or less increased phosphate content van Kleef (unpublished) collected no or very few exuviae of *P. bisetus*, in contrast to more oligotrophic pools. However, two of the inhabited Dutch moorland pools investigated by van Kleef had a high ammonium-N content (up to 61 µmol/l).
According to our own observations, the larvae seem to be intolerant of low oxygen concentrations. The figures in table 5 are tentative.

DISPERSAL

Duursema (1996) stated that *P. bisetus* occurs rarely or is absent in temporary moorland pools. Based on investigations by van Kleef & Esselink (2005), settlement is highly dependent on the presence of populations within one or two kilometres. Probably the species has a low dispersal power.

Psectrocladius brehmi Kieffer, 1923

SYSTEMATICS AND IDENTIFICATION

The species is absent from most keys to adult males and larvae. Wülker (1956: 6) distinguished males of *P. brehmi* and *P. sordidellus* by the antennal ratio: in *brehmi* 1.5, in *sordidellus* 1.5–2.0.
Pupae are a little easier to identify, but according to Langton & Visser (2003), *P. brehmi* appears to be no more than a small form of *P. sordidellus*. Identification is not very reliable in late summer when *P. sordidellus* also is smaller. However, Orendt (1993: 100) collected both small and large exuviae in the autumn; Reiff (1994) stated the same for other Bavarian lakes in spring. Most Dutch records are from July and August, but we collected the exuviae of one specimen at the end of May.

DISTRIBUTION IN EUROPE AND THE NETHERLANDS

The presence of *P. brehmi* has been stated in only a few countries scattered across Europe (Saether & Spies, 2010). Most probably the species will be found in many other countries too. In the Netherlands there are a small number of records of exuviae, almost exclusively from moorland pools (van Kleef, unpublished; own data).

LIFE CYCLE

The life cycle will be the same as that of *P. sordidellus*: two or three generations a year (see Orendt, 1993; Reiff, 1994).

TROPHIC CONDITIONS AND SAPROBITY

The species is often found in bogs and very acid moorland pools (Wülker, 1956; Reiss, 1984; van Kleef, unpublished). Orendt (1993) and Reiff (1994) collected it also in very eutrophic lakes in Bavaria. The ecology seems to be very similar to that of *sordidellus*. Possibly this small form occurs a little more in nutrition poor conditions (Bitušik, 2000).

Psectrocladius (Monopsectrocladius) calcaratus (Edwards, 1929)

nec *Psectrocladius calcaratus* Thienemann, 1944: 628

SYSTEMATICS AND IDENTIFICATION

P. calcaratus is probably the only species of the subgenus in Europe (Fittkau & Reiss, 1978).
The larva is not keyed in Cranston (1982) and Moller Pillot (1984), but the characteristic mentum is figured in Cranston et al. (1983); see also their key on p. 192.

DISTRIBUTION IN EUROPE AND THE NETHERLANDS

P. calcaratus is known from many countries in Europe, but is possibly absent from large parts of the Mediterranean area (Saether & Spies, 2010). In the Netherlands there are only three verified records from different parts of the country (two larvae, one adult male, coll. J. Meeuse and H. Moller Pillot).

WATER TYPE

The species has been found mainly in northern and montane and submontane lakes (Brundin, 1949; Langton, 1980; Serra-Tosio & Laville, 1991; Paasivirta, 2012). Ruse (2002) collected the exuviae in small acid lakes in the British Isles. The Dutch records are from a dystrophic moorland pool, a small non-acid lake in the fen peat region (Het Hol) and a non-acid pool in the valley of a lowland stream in Noord-Brabant.

Psectrocladius fennicus Storå, 1939

P. fennicus belongs to the *limbatellus* group and cannot be identified as a larva. It has been found in many lakes in Scandinavia and the British Isles and scarcely in lakes and pools in the alpine region (Wülker, 1956; Langton, 1980; Reiss, 1984). Records from the western and central European lowland are doubtful.
Brodin & Gransberg (1993) stated a strong increase during acidification in a Scottish lake, but it was not certain that the data applied to this species.

Psectrocladius limbatellus (Holmgren, 1869)

Psectrocladius limbatellus Pinder, 1978: 76, figs 38A, 116C (pro parte); Langton, 1984: 107, fig. 35C; Langton, 1991: 116, fig. 49 h–j
Psectrocladius edwardsi Brundin, 1949: 816–819, figs 180, 184; Langton, 1980: 79–81, 87, figs 1d, 2a, c, 5a
nec *P. limbatellus* Brundin, 1949: 818: fig. 185; Pankratova, 1970: 211–215; Langton, 1980: 86, figs 1e, 2d, 6d

SYSTEMATICS AND IDENTIFICATION

As has been mentioned in the introduction of the genus it is not possible to distinguish the larvae of the *sordidellus* and *limbatellus* groups as described in Cranston (1982) and Cranston et al. (1983).
For a long time there was much confusion about *P. limbatellus*. The name *limbatellus* was used correctly for the first time in the key to exuviae by Langton (1984). Reports of this species in the ecological literature up to 1990 have to be checked; usually the name applies to *P. oxyura*. Brundin (1949), who described the species as *P. edwardsi*, and Wülker (1956), in his revision of the genus, did not find reliable characters to

distinguish between this and other species of the group. Pinder (1978) also included *P. oxyura*, *P. oligosetus* and *P. octomaculus* in his *P. limbatellus*. Adults and exuviae can now be identified reliably using Langton & Pinder (2007) and Langton (1991) respectively, but identification of larvae to species level is still impossible.

DISTRIBUTION IN EUROPE AND THE NETHERLANDS

P. limbatellus has been collected in many European countries and probably lives almost everywhere in Europe (Saether & Spies, 2010). In the Netherlands the species is common, at least in the eastern and southern provinces. Most records given in Limnodata.nl seem to apply to the *limbatellus* group as a whole.

LIFE CYCLE

There can be many generations a year, depending on food and temperature. Wotton et al. (1992) stated mass emergence 18–20 days after colonisation of sand filter beds (some specimens already after 13 days). On a flat roof we stated total development within one month. In large water bodies with less food and lower temperature in summer, development will take more time.

In late winter larvae are in second, third and fourth instar. Pupae have been collected in nature from March until October. See further under the genus.

FEEDING

The larvae feed on green (probably also blue-green) algae and on fine particulate organic matter rich in bacteria (Wotton et al., 1992; own data). Hirabayashi & Wotton (1999) found about the same numbers of larvae in sand filter bed substrata with or without algae.

MICROHABITAT

The larvae often live on sandy substrata (Hirabayashi & Wotton, 1999), but we found them also in coarse organic matter. They build tubes consisting of filamentous algae or dead organic matter. Schmale & van der Hagen (1999) found the species in a dune lake almost only in a year in which the bottom was covered with filamentous algae. The same has been stated by Koskenniemi & Paasivirta (1987, as *P. edwardsi*) in a Finnish lake.

DENSITIES

In sand filter beds Hirabayashi & Wotton (1999) found densities up to 15 larvae/cm^2 . Such densities are not possible in natural water bodies.

WATER TYPE

P. limbatellus is common in small water bodies such as ponds and moorland pools and can live also in temporary or even ephemeral water bodies, in city fountains and in sand filter beds (Tourenq, 1975; Langton, 1991; Wotton et al., 1992; Hamerlik & Brodersen, 2010; van Kleef, unpublished; own data). The species has been found in smaller numbers in lakes and sand and gravel pits (Koskenniemi & Paasivirta, 1987, as *P. edwardsi*; Buskens, 1989 and unpublished data; Ruse, 2002a). Records from brooks and streams are scarce (Orendt, 2002a).

pH

P. limbatellus has been collected often in acid water bodies with a pH between 5 and 6 (Koskenniemi & Paasivirta, 1987; Buskens, 1989; van Kleef, unpublished). However, the densities seem to be lower than in less acid water containing more food in the form of algae and bacteria. The larvae can live in very alkaline water (e.g. Wolfram,

1996), but we have too few data about the occurrence in alkaline water to estimate their presence in such environments. Therefore the figures in the pH table in chapter 9 are doubtful.

TROPHIC CONDITIONS AND SAPROBITY

P. limbatellus has been collected in oligotrophic to very eutrophic lakes and pools. Probably the larvae live mainly in water where much food (detritus and algae) is available, e.g. ephemeral pools. Wotton et al. (1992) found the species abundantly in sand filter beds with decomposing organic matter. See also under pH.

SALINITY

From the Netherlands we have no records from brackish environments. Krebs (1981, 1984, 1990) never collected the species in the Dutch province of Zeeland. Tourenq (1975) found *P. limbatellus* (as *P. edwardsi*) sometimes in weakly brackish water in the Camargue (southern France). The species is widely distributed in the brackish Baltic Sea (Paasivirta, 2000).

DISPERSAL

Hamerlik & Brodersen (2010) collected the species in almost all investigated fountains in cities (only temporary filled with water) and elsewhere it is found very often in temporary or even ephemeral water, suggesting a very high dispersal power.

Psectrocladius (Allopsectrocladius) obvius (Walker, 1856)

Psectrocladius obvius agg. Moller Pillot, 1984: 140, 142; Limnodata.nl

IDENTIFICATION

The identification delivers no problems. Moller Pillot (1984) still used *P. obvius* agg., but most probably it is only one species.

DISTRIBUTION IN EUROPE AND THE NETHERLANDS

P. obvius has been collected in many countries in all parts of Europe (Saether & Spies, 2010). In the Netherlands the species is very common almost everywhere, but it is rare in the province of Zeeland and in southern Limburg (Krebs, 1981; Limnodata.nl).

LIFE CYCLE

P. obvius has at least two generations a year (Reiss, 1968; Shilova, 1976). Golobeva (1986) reared adults from eggs within 14 days. One egg mass contained about 500 eggs. In the Camargue it is one of the earliest species, emerging from February until November (Tourenq, 1975). In the Netherlands the species is not very early and some third instar larvae can be found until March. See further the introduction to the genus.

FEEDING

The larvae feed almost exclusively on filamentous algae; they die or leave their tubes when no algae are available (Gouin, 1936; Thienemann, 1954; Tourenq, 1975; Golubeva, 1986; Schmale & van der Hagen, 1999).

MICROHABITAT

The larvae live in places with many filamentous algae, between plants or on the bottom. They make a tube of mucous material with algal filaments and sometimes some silt. This tube is often transportable, but it can be attached to plants (Thienemann, 1954; Tourenq, 1975).

WATER TYPE

Current

The larvae live mainly in stagnant water, but they are not rare in slowly flowing brooks and streams (Langton, 1980; Peters et al., 1988; Steenbergen, 1993; Limnodata.nl). There are no records from faster flowing streams or rivers.

Dimensions

The species has been collected in relatively narrow ditches and brooks, but also in lakes (Langton, 1980; Steenbergen, 1993; Limnodata.nl). In larger lakes the larvae seem to be scarce and are only found in the sheltered zone near the shores.

Permanence

There are no data about temporary water. Most probably the species is very scarce there.

pH

P. obvius is very scarce in acid water, but there are several records from lakes and moorland pools with pHs between 4 and 6 (Koskenniemi & Paasivirta, 1987; Leuven et al., 1987; Buskens, 1989a; Hawczak et al., 2009; van Kleef, unpublished; Limnodata.nl). The larvae are also scarcely collected at pH > 8 (Steenbergen, 1993; Limnodata.nl).

TROPHIC CONDITIONS AND SAPROBITY

As described for the genus as a whole (see there), *P. obvius* lives mainly in water bodies which are not very eutrophicated or polluted. The species is rare in water with a chlorophyll-a concentration higher than 100 µg/l or orthophosphate concentration higher than 0.5 mg P/l (Steenbergen, 1993; Limnodata.nl). However, it has been found incidentally at very low oxygen contents during the day and probably it tolerates a short period of anoxia during the night, possibly at the water surface. *P. obvius* is probably absent from oligotrophic lakes (e.g. Brundin, 1949).

SALINITY

P. obvius is usually absent from brackish water (Tourenq, 1975: 347; Limnodata.nl). There are only a few records from water with chloride concentrations higher than 300 mg/l (Steenbergen, 1993; Ruse, 2002).

Psectrocladius octomaculatus Wülker, 1956

SYSTEMATICS

Wülker (1956: 16, 59) placed *P. octomaculatus* in the subgenus *Psectrocladius*, but not within one of the groups in this subgenus.

IDENTIFICATION

The adult male and pupa have been described by Wülker (1956: 15 et seq.). According to Cranston (1982), the larva can de identified by the long apical seta on the palpiger of the maxilla. The procercal setae do not deliver a reliable character. See also under the genus. The description of the larva by Schmid (1993: 173) is doubtful.

DISTRIBUTION IN EUROPE

P. octomaculatus has been recorded from many European countries, but is possibly absent from eastern Europe (Saether & Spies, 2010). The occurrence in the Netherlands is doubtful.

LIFE CYCLE

The species has two (sometimes three?) generations a year, emerging in spring and autumn (Reiss, 1984; Reiff, 1994), in montane areas from June to September (Laville, 1971).

WATER TYPE

Many records are from submontane or arctic pools and helocrene springs (Wülker, 1956). There are also some records from lakes and boggy seepages outside montane or arctic areas (Cranston, 1982; Reiss, 1984; Reiff, 1994), including the Lustsee in Bavaria, which is a lake with strong seepage. According to Reiss (1984) the species is cold-stenothermic.

TROPHIC CONDITIONS

The water types in which the species occurs are usually oligotrophic. This is also the case for the Lustsee in Bavaria (Reiff, 1994: 37).

Psectrocladius oligosetus Wülker, 1956

SYSTEMATICS AND IDENTIFICATION

P. oligosetus is a species of the *limbatellus* group. Identification of the adult male and the exuviae is sometimes difficult. Because the species is absent from Pinder (1978), it has been identified as *P. limbatellus* or *P. fennicus*. When in doubt it is necessary to compare the text and figures in Wülker (1956: 22) and Langton & Pinder (2007: 132–134, fig. 184C).

The description of the larva by Wülker (1956: 22, fig. 12 on p. 29) is probably based on a mistake (correspondence with the author, 1985/86). Currently, the larva cannot be distinguished from other species of the group; most or all larvae have five anal setae, not three as mentioned by Wülker (1956) and Moller Pillot (1984: 137). The bottle-shaped anal tubules (see Wülker, fig. 13, p. 30) may be characteristic.

DISTRIBUTION IN EUROPE AND THE NETHERLANDS

P. oligosetus has been recorded from a number of countries scattered across almost the whole of Europe (Saether & Spies, 2010). In the Netherlands the species is widely distributed in the Pleistocene part of the country (based on identification of exuviae).

LIFE CYCLE

Reiff (1994) stated emergence from April to October. Most probably the life cycle is not very different from the life cycle of other members of the genus, i.e. several generations between spring and autumn. However, possibly the larvae have an autumnal diapause in early fourth instar (see the introduction to this genus).

FEEDING

During rearing, larvae were seen feeding on leaves of *Sphagnum cuspidatum*, most probably mainly less vital (but not dead) parts of the leaves.

WATER TYPE

P. oligosetus has been collected in peat pools and subalpine and northern lakes (Wülker, 1956; Reiss, 1984; Langton, 1991; Reiff, 1994). In the Netherlands it is a common inhabitant of moorland pools (van Kleef, unpublished).

pH

Ruse (2002) found the species in small acid lakes with low conductivity. In Upper Bavaria it lives in lakes with pH > 8, but it is a scarce species there (Orendt, 1993: TAB XII.5; Reiff, 1994: 126). Most records from 28 moorland pools in the Netherlands apply to water with pH 3.4 to 5.5 (rarely up to 6.7) (van Kleef, unpublished).

TROPHIC CONDITIONS

Apart from one specimen mentioned by Orendt (1993) and one specimen collected in a Dutch moorland pool, all records are from mesotrophic to oligotrophic lakes and pools. However, some of the Dutch moorland pools investigated by van Kleef (unpublished) had a high ammonium-N content (up to 33.7 µmol/l).

DISPERSAL

According to investigations by van Kleef & Esselink (2005), settlement is highly dependent on the presence of populations within one or two kilometres. Probably the species has a much lower dispersal power than the related *P. limbatellus*.

Psectrocladius oxyura Langton, 1985

Psectrocladius oxyura Langton, 1991: 116, fig. 49l; Langton & Pinder, 2007: 132, figs 69A, C, 183B
Psectrocladius limbatellus Brundin, 1949: 818: fig. 185; Langton, 1980: 86, figs 1e, 2d, 6d

SYSTEMATICS AND IDENTIFICATION

P. oxyura is a species of the *P. limbatellus* group, which had not been closely examined for a long time. In the literature before 1985 the species is often named *P. limbatellus*, but this name was also used for other species of the group, for instance in Pinder (1978). See under *P. limbatellus*.
Identification of the larvae is still impossible.

DISTRIBUTION IN EUROPE AND THE NETHERLANDS

The species has been recorded from a small number of countries throughout Europe (Saether & Spies, 2010). In the Netherlands it most probably lives in all parts of the country, but there are only about thirty data.

LIFE CYCLE

See under the genus. Probably three (or four) generations a year are normal, emerging from April to October (Reiff, 1994; several Dutch data). In some lakes in Bavaria only one or two generations were numerous. In the dune lakes of Berkheide exuviae were rarely collected in late summer (Schmale, 1999). However, in other lakes exuviae were equally numerous in this period (Buskens, unpublished; own results).
Probably Macan (1949) found four generations a year, but it is not sure if his *P. limbatellus* applies to *P. oxyura*.

MICROHABITAT

Most records of *P. oxyura* in the Netherlands apply to lakes with no or very few water plants, and in some cases has been stated that the larvae lived on bare bottoms of silt, sand or gravel. This is consistent with the literature (e.g. Langton, 1980; Ruse, 2002a). However, in a Finnish reservoir the species settled in the first year, when the bottom was covered only with filamentous algae, but the larvae became more abundant when plants and mosses became dominant (Koskenniemi & Paasivirta, 1987, as *P. limbatellus*).

Klink (1991, app. 3) collected the species from artificial substrate in the river Meuse in southern Limburg (Grensmaas).

WATER TYPE

Current

P. oxyura lives almost exclusively in stagnant water. Records from running water (lowland streams) are rare and probably never concern a permanent population (Langton, 1980; own data). Langton supposed that exuviae in streams originate from stagnant water bodies, but this is probably not always the case. There are also records from rivers (e.g. Klink, 1991) and from slow-flowing canals and estuaries in the Netherlands (Limnodata.nl; own data).

Dimensions

The species is characteristic of large, deep water bodies (Langton, 1980: 81; Koskenniemi & Paasivirta, 1987, as *P. limbatellus*; Ruse, 2002a). Almost all Dutch records are from storage reservoirs, large sand pits and dune lakes (Ketelaars et al., 1992; Kuijpers et al., 1992; Buskens, unpublished; Schmale, 1999). However, there are several records from smaller pools and ditches (Limnodata.nl; unpublished Dutch data).

Temperature

Schmale (1999) stated the presence of the species in some infiltration lakes in the dunes of South Holland, where the water temperature in summer rose to values above 20 °C.

Permanence

In the Netherlands there is only one record from a temporary pool (Moller Pillot, 2003). However, Lundström et al. (2010) sometimes collected the species in large numbers in temporary flooded wetlands in Sweden.

pH

In a lake in southwestern Finland the species lived abundantly at pH 5.5–6 (Koskenniemi & Paasivirta, 1987, as *P. limbatellus*). Most other data apply to water bodies with a pH of about 7 or higher.

TROPHIC CONDITIONS AND SAPROBITY

The species is collected mainly in large water bodies without organic pollution, such as gravel pits, sand pits and storage reservoirs. Reiff (1994) and Bitušik (2000) found it mainly in mesotrophic to eutrophic lakes. Marziali et al. (2008) found it in the oligotrophic lake Mergozzo in Italia.

However, in the dune lakes of Berkheide (the Netherlands) *P. oxyura* was almost absent in the cleanest lakes. Moreover, the species disappeared after removing silt from one lake, whereas it remained in a similar lake where silt had not been removed. In another lake the species disappeared after supplying with cleaner water (Schmale, 1999). In the storage reservoir De Gijster (Netherlands) Kuijpers et al. (1992) found many more exuviae of *P. oxyura* than in the more oligotrophic reservoir Petrusplaat (cf. Ketelaars et al., 1992).

Anyhow all data suggest that the species makes much higher demands upon water quality and oxygen content than *P. sordidellus* (see table 5).

Psectrocladius (Allopsectrocladius) platypus (Edwards, 1929)

IDENTIFICATION OF LARVAE

Most keys give no differences between the larvae of *P. obvius* and *P. platypus*, or the latter is absent from the key. *P. platypus* can be distinguished because of the presence of a seta interna on the mandible and a head length (in fourth instar) of 0.5–0.6 mm. Procercus, antenna, mentum and mandible as in *P. obvius*. In third instar (head length about 0.35 mm) the larva resembles that of *P. psilopterus*, but can be identified because of the two obvious spurs on the procercus. It differs from *P. bisetus* by having five long anal setae, also in third instar.

DISTRIBUTION IN EUROPE AND THE NETHERLANDS

P. platypus is distributed throughout Europe (Saether & Spies, 2010). In the Netherlands the species is very common in the whole Pleistocene area and the dune region, and scarce elsewhere (Steenbergen, 1993; Limnodata.nl).

LIFE CYCLE

Macan (1949) stated emergence from May until October, in decreasing numbers. There are at least four generations. In the Netherlands the numbers are sometimes still high in late summer and emergence can be seen from April to October, rarely in early November. In November we found in several moorland pools large numbers of larvae, all in second and third instar, indicating a diapause (see the introduction to the genus). Some larvae collected by us in December emerged in the laboratory within 16 days.

FEEDING

The larvae live in tubes of filamentous algae and feed on these algae (own observations). In some cases, though, (possibly dependent on the algal species) we saw them feeding mainly on detritus (? bacteria) and other epiphyton. Possibly the larvae also feed on animal prey: Pankratova (1970: 31) stated that larvae of the *dilatatus* group ate Oligochaeta and pointed out that the mandibles of this group indicate carnivorous behaviour.

MICROHABITAT

The larvae live in tubes of filamentous algae or detritus and seem to occur only where these algae are present (Kreuzer, 1940; Thienemann, 1954). They can be found among plants or *Sphagnum* and in detritus, but can be absent from detritus bottoms without algae (own data).

WATER TYPE

Current

The larvae are common in temporary acid upper courses with slow currents in the Dutch province of Noord-Brabant (own data), but the species has not been mentioned as inhabitants of flowing water by Fittkau & Reiss (1978), Verdonschot et al. (1992) and Orendt (1999). The species seems to be absent from other brooks and streams.

Dimensions

The larvae live in small ephemeral puddles and peat cuttings, but also in moorland pools and more rarely in the littoral zone of larger lakes (Kreuzer, 1940; Macan, 1949; Reiss, 1984; Leuven et al., 1987; Buskens & Verwijmeren, 1989; Ruse, 2002).

Temperature
Kreuzer (1940) found the larvae even in a pool where the temperature could rise up to 28 °C.

Shade
Schleuter (1985) collected the species very rarely in small temporary water bodies in woodland. Possibly the species does not fly in woodland.

Permanence
Duursema (1996) collected the larvae abundantly in temporary moorland pools. We found the species in ephemeral water on a flat roof and in temporary upper courses (e.g. Moller Pillot, 2003).

Other aspects
Arts (2000) called the species characteristic of moorland pools without bog development. We sometimes collected larvae in pools with boggy *Sphagnum*, but it has to be proved whether an environment with bog development is less favourable for this species. See also Trophic conditions.

pH

Leuven et al. (1987) collected *P. platypus* only in very acid water at pH 3.46–5.22. Other workers also call the species characteristic of acid pools (Kreuzer, 1940; Buskens, 1989a; Duursema, 1996; Arts, 2000). The larvae are less common at pH > 7, but are found there occasionally, usually in low numbers (Reiss, 1984; Steenbergen, 1993; own data). Van Kleef & Esselink (2005) found *P. platypus* in restored moorland pools mainly when restoration had not been successful and acidification began again.

TROPHIC CONDITIONS AND SAPROBITY

P. platypus is usually found in acid water with low orthophosphate and chlorophyll-a concentrations, but it can also sometimes be collected in less acid and more eutrophic water (Steenbergen, 1993; Limnodata.nl; own data). It is a characteristic species of polyhumic (dystrophic) water (Brundin, 1949; Saether, 1979; own data). In peat cuttings the larvae tolerate low oxygen levels (probably even anoxia in the night) and very high ammonium-N concentrations (up to more than 3 mg/l) (Waajen, 1982; own data). Daytime oxygen concentrations lower than 5 mg/l have also been recorded (Steenbergen, 1993; Limnodata.nl).

SALINITY

Probably the species is absent from brackish water.

DISPERSAL

The very common occurrence of *P. platypus* in temporary moorland pools and its presence in ephemeral puddles indicates a rather high dispersal capacity of adult females. It is possible that the species does not fly in woodland, or only rarely (see under Water type).

Psectrocladius psilopterus group

Because of an mix-up of specimens during the rearing of larvae, the name gr. *psilopterus* has been used incorrectly for larvae of gr. *limbatellus* and gr. *sordidellus*. Especially in the East European literature, where the keys by Chernovskij (1949) and Pankratova

(1970) were used until the 1990s, this mistake has to be taken into account.
Two European species belong to this group: *psilopterus* and *bisetus*.

Psectrocladius psilopterus (Kieffer, 1906)

Psectrocladius psilopterus Cranston 1982: 122–125, fig. 50f (pro parte); Moller Pillot, 1984: 140–143, figs VI.42 f, k, l (pro parte)
Psectrocladius simulans Thienemann, 1944: 628 (pro parte); Pankratova, 1970: 218–220, fig. 135 (pro parte)
nec *Psectrocladius* gr. *psilopterus* Thienemann, 1944: 628, figs 163, 164, 166; Chernovskij, 1949: 112, fig. 97; Pankratova, 1970: 213–214, fig. 129
Psectrocladius sordidellus Brundin, 1949: 818, fig. 178, 182 (see Wülker, 1956: 25); ? p. 724; (nec aliis)

SYSTEMATICS AND IDENTIFICATION

Using Cranston (1982) or Moller Pillot (1984), larvae of *psilopterus* and *bisetus* are both identified as *P. psilopterus*. The larvae of *P. bisetus* are smaller (see under this species). Adult males of *P. bisetus* are also identified as *P. psilopterus* using Pinder (1978). Identification is possible using Langton & Pinder (2007). The exuviae of both species are distinguished both in Langton (1984) and in Langton (1991).
The following information about this species is based on the assumption that apart from bogs and moorland pools all records of gr. *psilopterus* in the Netherlands belong to the species *psilopterus*. This cannot be said for acid stagnant water bodies and for records from other regions.

DISTRIBUTION IN EUROPE AND THE NETHERLANDS

P. psilopterus has been recorded in many countries throughout Europe (Saether & Spies, 2010). In the Netherlands the species is common in Pleistocene regions and in the fen peat regions, but rare or absent from other Holocene regions and southern Limburg (Moller Pillot & Buskens, 1990; Steenbergen, 1993; Limnodata.nl).

LIFE CYCLE

Shilova (1976) stated only one generation in Russia, but Reiff (1994) collected exuviae from April to October in lakes in Bavaria and suggested three generations a year. We sometimes found larvae in third instar until late April. Probably the life cycle is the same as that described for most other species of the genus (see the introduction to the genus).

MICROHABITAT

The larvae live on the bottom and on submerged and emersed plants, more scarcely also on floating leaves (Koskenniemi & Paasivirta, 1987; Janecek, 2000; own data). The first authors stated that the species was most common when the bottom was covered with filamentous algae. Shilova (1976) reared the species from decaying organic bottom material in a pool.

SOIL

In the Dutch province of Noord-Holland the species has never been collected in water bodies on clay and has been collected significantly more often on fen peat than on sand (Steenbergen, 1993). In other parts of the Netherlands the species is also absent or almost absent from clay and chalk regions, but it is rather common on sandy soils (Limnodata.nl). It can be found more frequently in moorland pools on sand than in bog pits (own data).

WATER TYPE

Current

P. psilopterus can be found in running water more frequently than any other species of the genus. The species seems to be absent from brooks in areas with alkaline soil and from fast running brooks and streams (see table 7 in Chapter 9). In pools, ditches and lakes the species is more common and sometimes abundant; see, however, under soil and pH.

Dimensions

Langton (1989) supposed that in the southern part of its range *P. psilopterus* lives in small, shallow water bodies, and in the northern part more in large lakes. In Russia and in the Netherlands the species lives both in small and large stagnant water bodies, probably dependent on other factors (Shilova, 1976, as *P. simulans*; Limnodata.nl). In running water in the Netherlands its occurrence is confined to brooks rarely wider than 6 m; Shilova (1976) also reported its occurrence in wider streams.

Permanence

P. psilopterus is not rare in temporary moorland pools and upper courses of lowland brooks (own data). We also found the species in other temporary pools and even in a woodland track.

pH

In the Netherlands the species is common in very acid water (moorland pools and upper courses of brooks) and scarcer in non-acid lowland brooks and in lakes in the fen peat area with a pH between 7 and 8 (Steenbergen, 1993; van Kleef, unpublished; Limnodata.nl; own data). Elsewhere the species is rare. The data on occurrence at different pHs given in table 6 apply only to regions with fen peat or lowland brooks, where the species lives. In the literature from other countries we found no indications that the species is more common in acid or peaty environments, but we stated that the species is absent from calcareous lowland brooks in Germany.

TROPHIC CONDITIONS AND SAPROBITY

In oligohumic lakes, *P. psilopterus* appears to be characteristic of water with low conductivity, poor in phosphate and rarely eutrophic (Saether, 1979; Buskens, 1989; Reiff, 1994; Ruse, 2002). The species also lives in mesohumic and polyhumic lakes, pools and ditches where much decaying material can be present (e.g. Shilova, 1976), although van Kleef (2010) stated that the species became significantly more common after restoration (removal of organic matter, decreased ammonium content).

The larvae are resistant to a temporarily lower oxygen content and can be found sometimes in water with a BOD between 4 and 6 (Limnodata.nl). The relations between different aspects of water quality for this species are still obscure. The figures given in table 5 in chapter 9 concern different types of water and can be used only as a rough estimation.

SALINITY

Most probably the species is not able to live in brackish water.

DISPERSAL

According to investigations by van Kleef & Esselink (2005), settlement is highly dependent on the presence of populations within one or two kilometres. Probably the species has a low dispersal power.

Psectrocladius schlienzi Wülker, 1956

SYSTEMATICS AND IDENTIFICATION

P. schlienzi cannot be placed within one of the three species groups of the subgenus *Psectrocladius* (see Wülker, 1956: 17, 59). Only the pupa is very characteristic (see Langton, 1991). The larva has been described by de Beauvesère-Storm & Tempelman (2009) and resembles that of *P. psilopterus*, but the larvae from Germany, described by Wülker, have the dark mandible of the other species groups.

DISTRIBUTION IN EUROPE AND THE NETHERLANDS

P. schlienzi has been collected in only a few countries scattered across Europe. In the Netherlands there is only one record, from a brook in the southern Veluwe (de Beauvesère-Storm & Tempelman, 2009).

ECOLOGY

Most records in the literature are from lakes and pools, usually poor in nutrients or dystrophic (Wülker, 1956: 49; Reiss, 1984; Bitušík, 2000; Paasivirta, 2012). Its occurrence in the Netherlands in a lowland brook (with good water quality) is an exception.

Psectrocladius sordidellus (Zetterstedt, 1838)

nec *Psectrocladius sordidellus* Brundin, 1949: 724, 818 (see Wülker, 1956: 25); Pinder, 1978: 74, fig. 37I, 116B

SYSTEMATICS AND IDENTIFICATION

Three species can be distinguished in the *sordidellus* group (Wülker, 1956: 7–8): a large poly-form (*ventricosus*), a medium meso-form (*sordidellus*) and a small oligo-form (*brehmi*). It is doubtful whether *P. brehmi* is a separate species (see there).

For a long time there were big problems in identifying adult males of this species. Identifications before the publication of Langton & Pinder (2007) are often not reliable. Pupae and exuviae are correctly described and keyed in Langton (1980, 1984, 1991). Identification of the larvae may be possible using Zelentsov (1980); the key by Cranston (1982) is less reliable. Zelentsov described the eggs and the larva in all instars.

DISTRIBUTION IN EUROPE AND THE NETHERLANDS

P. sordidellus has been found in many European countries and will live almost everywhere on the continent (Saether & Spies, 2010). In the Netherlands there are records from most parts of the country, but the species is scarce in the province of Zeeland (Krebs, 1981, 1984).

LIFE CYCLE

There are at least two or three generations a year (Zelentsov, 1980; Reiff, 1994). Reiss (1968) and Reiff (1994) stated emergence from February until November, most numerously from the end of May to September. The data reported by Schleuter (1985) showing maximum numbers in early spring possibly apply to another species (see Identification above).

MICROHABITAT

The larvae can be found on silty and sandy bottoms, among filamentous algae and on plants and stones (Reiss, 1968; Zelentsov, 1980; Becker, 1995; Janecek, 2000; own data). It is not clear which microhabitat is preferred. According to Gouin (1936, as *P. stratiotis*), the larvae are free living, but construct a silken tube to pupate.

WATER TYPE

Current

P. sordidellus mainly inhabits stagnant water: lakes, pools, canals and ditches. Within the subgenus *Psectrocladius* only the species *P. psilopterus* and *P. sordidellus* are more or less regularly found in running water. However, *P. sordidellus* is scarce in (slowly running) brooks, streams and rivers and stable populations will probably rarely develop in running water. The species very rarely lives on stones in fast-flowing streams or in larger rivers (Becker, 1995; Orendt, 2002a; own data). Most probably *P. sordidellus* was the most common (but not numerous) Orthocladiinae species on a tidal sandy flat in the Haringvliet estuary in the Netherlands (see under Permanence). The suggestion by Langton (1980: 78) that females of this species do not oviposit in stream habitats is probably only true for fast-flowing streams.

Permanence

Probably this species was common although never numerous on a tidal sandy flat in the Haringvliet estuary in the Netherlands, even on sites which were exposed for half of the day (Smit et al., 1996). Elsewhere, *P. sordidellus* has been collected sometimes in temporary pools (Zelentsov, 1980; ? Schleuter, 1985; own data).

pH

The species has been collected scarcely in very acid pools in woods and heathland (? Schleuter, 1985; van Kleef, unpublished) and in acid lakes (Koskenniemi & Paasivirta, 1987). There are many records from alkaline water bodies and some from brackish water (Krebs, 1984; Wolfram, 1996; Schmale, 1999; Ruse, 2002). See table 6.

TROPHIC CONDITIONS AND SAPROBITY

P. sordidellus has often been collected in phosphate-poor water and appears to be common in less eutrophic conditions, consistent with what has been stated for the whole subgenus *Psectrocladius* (Buskens, 1989a; Steenbergen, 1993; Brodersen et al., 1998; Limnodata.nl; own data). However, Orendt (1993) stated a significant preference for eutrophic water and Bitušik (2000) called the species typical for very eutrophic conditions. The larvae are collected mainly from water rich in oxygen, but they tolerate very low oxygen contents (up to anoxia in the night) and have been found sometimes in ditches in α-mesosaprobic to polysaprobic condition (Grontmij | Aqua Sense, unpublished; own data). Anyhow *P. sordidellus* has been found more frequently than other species of the genus in hypertrophic or polluted lakes and ditches (see table 5 in Chapter 9).

SALINITY

In the Netherlands and England *P. sordidellus* has been collected sometimes in brackish water (Krebs, 1984; Ruse, 2002), but the species is very scarce in brackish pools and ditches in the Dutch province of Zeeland (Krebs, 1981, 1984, 1990). Also Steenbergen (1993) found larvae of the *sordidellus* group only scarcely in brackish water in the province of Noord-Holland.

Tourenq (1975) collected the species in a number of brackish water bodies in the Camargue (southern France). However, he wrote that he was not able to distinguish this species from *P. ventricosus*. Both species are present in the brackish Baltic Sea, but there *ventricosus* is the most common.

Psectrocladius ventricosus Kieffer, 1925

SYSTEMATICS AND IDENTIFICATION

P. ventricosus is the large poly-form of the *sordidellus* group (Wülker, 1956). The adult male is absent from Pinder (1978), but can be identified using Langton & Pinder (2007). The pupa and larva can hardly be identified; they have been described by Zelentsov (1983). The larva is larger than that of *sordidellus* and has very short anal tubules (Thienemann, 1937: 26; 1944: 628–629).

DISTRIBUTION IN EUROPE AND THE NETHERLANDS

According to Saether & Spies (2010) *P. ventricosus* has been found only in the northwestern and southeastern part of Europe. Wolfram (1996) also reported its occurrence in Neusiedler See in Austria. It is not clear whether the species is actually absent from the countries around the western part of the Mediterranean Sea: Tourenq (1975) was not able to distinguish this species from *P. sordidellus*.
In the Netherlands there is only one reliable record, from the province of Zeeland (Krebs, 1984: 92).

SWARMING

Zelentsov (1983) observed swarming in groups of 10 to 150 specimens, at a height of 1 to 4 m above small elevations or near shrubs and trees.

FEEDING

The larvae feed on algae and detritus (Zelentsov, 1983).

MICROHABITAT

Meuche (1938) collected the larvae among filamentous algae on reed stems, wood and stones. Zelentsov (1983) found the larvae living in tubes among plants and algae, on stones and on the bottom with plant remains.

WATER TYPE

The larvae live in lakes and pools (Thienemann, 1937; Meuche, 1938; Kreuzer, 1940; Krebs, 1984).

SALINITY

Meuche (1938) and Kreuzer (1940) found the species only in brackish water with up to 14‰ NaCl (=7 g Cl/l). Krebs (1984) reared the species from a wide ditch in the Dutch province of Zeeland with a chloride concentration of about 1 g/l. The record from Neusiedler See in Austria (Wolfram, 1996) indicates occurrence in lakes with high alkalinity in central Europe.
P. ventricosus is a common species in the brackish Baltic Sea (Paasivirta, 2000).

Psectrocladius zetterstedti Brundin, 1949

IDENTIFICATION

The adult male can be identified easily by the absence of an anal point in the male (Brundin, 1949: 816–818). The pupa has been keyed by Langton (1991), but can be hardly distinguished from that of *P. sordidellus*.

DISTRIBUTION AND ECOLOGY

P. zetterstedti has been found in northern and alpine regions (Saether & Spies, 2010).

The record from the Netherlands mentioned by these authors is doubtful.
The species lives in oligotrophic and polyhumic lakes on different types of bottom and also in isoetid vegetations (Brundin, 1949).

Pseudorthocladius Goetghebuer, 1932

Most species of *Pseudorthocladius* live mainly in terrestrial habitats. However, the larvae are often encountered in springs and also in moorland pools. Because this book treats only aquatic species, the terrestrial aspects will be discussed only briefly.

NOMENCLATURE AND IDENTIFICATION

The use of the name *Pseudorthocladius* for the genus under consideration is incorrect, but has been commonly used for more than fifty years.. As long as the case will be under consideration of the ICZN, current usage is to be maintained (Spies & Saether, 2004).
The male adults of most species can be identified using Saether & Sublette (1983). The pupae and larvae of most Palaearctic species are still unknown. For differences between *curtistylus* and *filiformis* adults, see under the former species.

LIFE CYCLE

In the Netherlands emergence has been stated from the end of April until September (own data). Larvae collected in autumn emerged only in this period. In one case we collected bottom material in early August and reared 42 adults the following May and June. Larvae collected in winter also emerged over a long period in May and June. In autumn and winter the larvae are in diapause in second, third and fourth instar.

Pseudorthocladius albiventris (Goetghebue r, 1938)

P. albiventris is only known from Austria. The adult male can be identified using Saether & Sublette (1983: 49).

Pseudorthocladius berthelemyi Moubayed, 1988

P. berthelemyi is only known from fast-flowing streams in southern Europe and the alpine region, and is not treated here.

Pseudorthocladius cristagus Stur & Saether, 2004

P. cristagus is a species with hairy wings and is separable from *P. pilosipennis* by its hypopygium (Stur & Saether, 2004). The species is only known from the locus typicus: a helocrene in Luxembourg (Stur et al., 2005). Its occurrence in the Netherlands is possible.

Pseudorthocladius curtistylus (Goetghebuer, 1921)

IDENTIFICATION

Based on more than a 100 investigated males and females from more than 20 localities in the Netherlands, we found small differences from the descriptions and key by Saether & Sublette (1983). The males of *filiformis* and *curtistylus* can be distinguished as follows:

- AR 0.85–1.2; squama with (4) 5–9 (12) setae; wing length 1.3–1.7 mm — *filiformis*
- AR 0.45–0.75; squama with 1–4 setae; wing length 1.0–1.3 mm — *curtistylus*

The females can be distinguished by:

- squama with (4) 5–8 (10) setae — *filiformis*
- squama with 1–4 setae — *curtistylus*

DISTRIBUTION IN EUROPE AND THE NETHERLANDS

P. curtistylus has been recorded throughout Europe (Saether & Spies, 2010). In the Netherlands it has been reared from almost all parts of the country.

LIFE CYCLE

See under the genus.

ECOLOGY

We did not find any difference in ecology between *curtistylus* and *filiformis* and so we treat both species together. Strenzke (1950: 230 et seq.) called all his larvae *curtistylus*, although he did not rear all specimens to adults (see e.g. his note on page 231). Most probably he collected both species.

Both species are very common on banks of brooks and streams, in marshland, alder carr and very wet grassland. The larvae live there also in dense vegetation, in contrast to some other terrestrial chironomids. They are rarely found in pioneer situations (Strenzke, ibid.; own data). Both species live in acid as well as more basic conditions, but they are possibly a little more common in acid environments, even among *Sphagnum*. The species has also been collected in the brackish Baltic Sea (Paasivirta, 2000).

The larvae are absent from drier bottoms and sometimes common in temporary flooded wetlands (Lundström et al., 2010). In acid moorland pools the larvae are often found up to 50 cm deep (Leuven et al., 1987; van Kleef, 2010). Such records are not confined to acid water. In aquatic environments the larvae live among plants or mosses or on the bottom. They do not swim and therefore they are rarely observed in aquatic samples.

Both species can be present in springs (e.g. Stur et al., 2005) and both are also found along the margins of brooks (own data; see also Caspers, 1980a). It is not clear whether adults caught in emergence traps in brooks and streams (e.g. Pinder, 1974; Ringe, 1974) apply to specimens which lived as larvae below the water level.

Pseudorthocladius filiformis (Kieffer, 1908)

As far as is known the distribution, life cycle and ecology are the same or very similar to those of *P. curtistylus*. See under this species.

Pseudorthocladius macrovirgatus Saether & Sublette, 1983

Pseudorthocladius cranstoni Saether & Sublette, 1983: 49, 91–93, fig. 54; Klink & Moller Pillot, 1996

NOMENCLATURE

Schnell (1991) stated that *P. cranstoni* and *P. macrovirgatus* show only minor variations and belong to the same species.

DISTRIBUTION IN EUROPE AND THE NETHERLANDS

P. macrovirgatus has been collected only in England, Ireland, France, Norway, the Netherlands and the Nearctic (Saether & Spies, 2010). In the Netherlands only three localities are known: on the Veluwe, in Twente and Noord-Brabant (leg. H. Siepel).

ECOLOGY

P. macrovirgatus has been collected in Europe in a peat pool in Ireland, a bog in England and moist, mesophilic heathland in Brittany (France) (Saether & Sublette, 1983; Delettre, 1994). The three Dutch localities are wet parts of heathlands.

Pseudorthocladius pilosipennis Brundin, 1956

P. pilosipennis has been recorded only in a few countries in north and northwest Europe (Saether & Spies, 2010). Brundin (1956) caught the adults in Sweden only on *Sphagnum* bottoms.

Pseudorthocladius rectangilobus Caspers & Siebert, 1980

Pseudorthocladius virgatus Saether & Sublette, 1983: 49, 81–84, figs 46–48

P. rectangilobus has been collected in Germany, Ireland and Norway (Schnell, 1991; Saether & Spies, 2010). Only the adult male has been described (Caspers & Siebert, 1980). The locus typicus is near a brook in Germany. In the Nearctic the species has been collected along a number of streams and rivers (Saether & Sublette, 1983, sub *P. virgatus*). For synonymy see Schnell (1991).

Pseudosmittia Edwards, 1932

SYSTEMATICS

The genus *Pseudosmittia* in its original sense is very heterogeneous. For this reason, Ferrington & Saether (2011) divided it into three genera: *Allocladius*, *Hydrosmittia* and *Pseudosmittia*. This classification is followed here. For the Netherlands the most important species placed in *Allocladius* are *A. arenarius*, *A. bothnicus* and *A. nanseni*. Placed in *Hydrosmittia* are (e.g.) *H. montana*, *H. oxoniana*, *H. ruttneri* and *H. virgo*. These two genera are treated separately.

ECOLOGY

All species of the genus *Pseudosmittia* s.s. are terrestrial or semiterrestrial. However, the larvae can live submerged for a long time without difficulty; after inundation they seem to remain on the bottom and do not swim (own data). Several species live in marshes, inundated floodplains or rain puddles and can emerge from water (Schleuter, 1985; Lundström et al., 2009; own data). Because of their small size and primarily terrestrial life, the genus is rarely collected in aquatic samples.

Rheocricotopus Thienemann & Harnisch, 1932

SYSTEMATICS AND IDENTIFICATION

Saether (1985) made a review of the genus with keys and descriptions for male imagines, pupae and larvae. Other keys written before 1985 are more or less incomplete. Mountain brooks are also inhabited by other species described after 1985 (see Saether & Schnell, 1988a; Scheibe, 2002).
The two subgenera *Psilocricotopus* and *Rheocricotopus* s.s. can be easily distinguished as adult and pupa. Saether divides the subgenus *Psilocricotopus* into an *atripes* group and a *chalybeatus* group. In Moller Pillot (1984) the name gr. *atripes* is used for the whole subgenus *Psilocricotopus*. The abdominal setae of the larvae of the *atripes* group are described in detail by Thienemann (1944: 629).

Rheocricotopus atripes (Kieffer, 1913)

Trichocladius foveatus Crisp & Lloyd, 1954: 273, 297
Rheocricotopus foveatus Lehmann, 1971: 489; Lindegaard et al., 1975: 116

DISTRIBUTION IN EUROPE AND THE NETHERLANDS

The species has been found in many countries in all parts of Europe (Saether & Spies, 2010). In the Netherlands *R. atripes* has been collected only in southern Limburg and in some springs on the Veluwe (Moller Pillot & Buskens, 1990).

LIFE CYCLE

Lehmann (1971) collected the adults only from May to October. We found fourth instar larvae all year round.

FEEDING

Crisp & Lloyd (1954: 297) suggested that the larvae in a mud flat fed on (quickly as well as slowly) decaying leaves and twigs.

WATER TYPE

Lehmann (1971: 528) called *R. atripes* a character species of the fast-flowing upper courses of the river Fulda. However, the exuviae have not been found in most of the pre-Alpine streams in Bavaria (Orendt, 2002a). In the Netherlands the larvae have been found almost exclusively in helocrene springs. Crisp & Lloyd (1954) caught the adults regularly from a mud flat with seepage in England. The species was scarce in the moss carpet of the Danish spring Ravnkilde (Lindegaard et al., 1975).

Rheocricotopus chalybeatus (Edwards, 1929)

DISTRIBUTION IN EUROPE AND THE NETHERLANDS

R. chalybeatus has been found in many countries in all parts of Europe (Saether & Spies, 2010). The species has been collected in the Netherlands in the large rivers and in brooks and streams scattered over the whole Pleistocene area and southern Limburg (Limnodata.nl; own data). Records from the fen peat region have to be verified.

LIFE CYCLE

Tokeshi (1986) stated two generations a year in a small stream in eastern England, emerging in spring and late summer. He suggested prolonged hatching (inferred from increasing population density) from the end of the summer until the following spring. However, a diapause of young larvae could not be excluded, because he did not investigate microhabitats other than plants. Other investigators stated emergence from the middle of April until October, rarely in March and November and most numerously in May/June and August/September (Wilson, 1977; Becker, 1995; own data). Tokeshi (1986a) found 85% of the total annual production in March to May, but elsewhere the highest numbers occurred in late summer.

FEEDING
In the guts of *R. chalybeatus* Tokeshi (1986a) found 0–60% diatoms, suggesting that the larvae have no specialised feeding behaviour.

MICROHABITAT
The larvae have been found on plants and stones (Lehmann, 1971; Millet et al., 1987); Wilson (1977) collected the exuviae only below riffles in the river Chew. Becker (1994) also reared some specimens from sand. We found most larvae on stones, possibly because plants in our larger rivers are almost absent.

WATER TYPE
Current
R. chalybeatus has been collected in fast-flowing (> 1m/s) streams and rivers, but can also be common in slow-flowing lowland brooks (Fahy, 1973; Orendt, 2002a; own data) and even in springs (Lehmann, 1971). Gendron & Laville (1997) called it a plain species (see, however, under Temperature). It is not clear why the species is absent from many (also rather fast-flowing) lowland brooks.
There are no records from lakes or other stagnant water bodies.

Dimensions
The species is often very common in large rivers, especially on stones (Hayes & Murray, 1989; Klink, 1991; Becker, 1995; Móra, 2008; Janecek, 2010), but larvae and exuviae have also been collected in large numbers in narrower streams and (in smaller numbers) even in brooks no more than 3 m wide (Lehmann, 1971; Orendt, 2002a; Millet et al., 1987; own data).

Permanence
There are no records from temporary water, possibly because the species lives more in the lower courses of streams. It cannot be excluded that larvae in diapause are able to survive drying up of brooks, because this has been stated for *R. fuscipes*.

Temperature
Gendron & Laville (1997) call *R. chalybeatus* a plain species that is only present in mountain streams when the temperature there is high enough. Punti et al. (2009) noted in Spanish streams an optimum temperature of 15° C. Rossaro (1991) stated a mean water temperature of 21.27 °C in Italian streams, which is very high for an orthoclad species. In this context it is interesting that emergence of this species never has been stated in winter, as in *R. fuscipes*. But the presence in springs as stated by different authors (e.g. Lehmann, 1971) is difficult to explain.
According to Klink (1989), the species was absent from the Lower Rhine in previous centuries and settlement in the 20th century could be caused by rising temperatures.

SAPROBITY

The species can be numerous in relatively unpolluted water, but has been found by most authors mainly in waters with some organic pollution. It appears to be absent in severely polluted streams. The larvae are rather tolerant of low oxygen concentrations, but most probably they tolerate more pollution at faster currents (Wilson & Wilson, 1984; Bazerque et al., 1989; Hayes & Murray, 1989; Klink, 1990; Moog, 1995; Bitušik, 2000: 112; own data). The data are summarised in table 5 on p. 283.

Rheocricotopus effusus (Walker, 1856)

SYSTEMATICS AND IDENTIFICATION

Ekrem et al. (2010) stated a high genetic divergence within *R. effusus*, possibly indicating that more species were involved. Larvae can be distinguished from those of *R. fuscipes* because the median teeth have no accessory tooth, but this character can lead to mistakes in specimens with worn teeth.

DISTRIBUTION IN EUROPE AND THE NETHERLANDS

The species has been found in many countries in all parts of Europe (Saether & Spies, 2010). In the Netherlands *R. effusus* is a rare species in Limburg, in the extreme eastern part of the country and on the Veluwe (Limnodata.nl; own data). Other records have to be verified.

LIFE CYCLE

In a Danish lowland stream the species probably had three generations a year, in spring and summer (Lindegaard-Petersen, 1972). Laville & Lavandier (1977) collected adults in the Pyrenees up to November.

MICROHABITAT

As mentioned for the other species of the genus, *R. effusus* larvae live on stones and water plants and only scarcely on the bottom of streams (Lehmann, 1971). Laville & Lavandier (1977) collected the species mainly in a mountain brook rich in mosses.

WATER TYPE

Current and dimensions

Lindegaard-Petersen (1972) found the larvae rather numerously in a Danish lowland stream with a mean current velocity of about 50 cm/s. In such brooks and streams in the Netherlands the species is very rare. The absence of *R. effusus* in most streams in the Netherlands cannot be ascribed only to current velocity, although elsewhere the species has been collected only in fast or very fast currents (Lehmann, 1971; Laville & Vinçon, 1991; Michiels, 1999). The species seems to be absent from large rivers.
Some specimens have been found in lakes in Bavaria (Reiff, 1994) and in a mountain lake in Slovakia (Bitušik, 2000: 112).

Temperature

According to Rossaro (1991), *R. effusus* is a typical cold water species. He collected it at a mean water temperature of 9.8 °C. Punti et al. (2009) stated an optimum temperature of 8.9 ° C. in Spanish streams, much colder than other species of the genus.

Rheocricotopus fuscipes (Kieffer, 1909)

Rheocricotopus dispar Wilson 1977: 12–16; Klink & Moller Pillot, 1982: 49, 52

DISTRIBUTION IN EUROPE AND THE NETHERLANDS

R. fuscipes has been found in many countries in all parts of Europe (Saether & Spies, 2010). In the Netherlands the species is common in the large rivers, in southern Limburg, on the Veluwe and in the most easterly part of the country (Klink & Moller Pillot, 1982; Moller Pillot & Buskens, 1990). All other data (also records in Limnodata.nl) have to be verified. In the lowland brooks of Noord-Brabant *R. fuscipes* seems to be rare; the main species here is *R. chalybeatus*.

LIFE CYCLE

Most probably *R. fuscipes* has three or four generations a year, emerging from the end of February to November, in England (and rarely in the Netherlands) also in winter (e.g. Wilson, 1977; Lindegaard & Mortensen, 1988; Michiels, 1999). Depending on local conditions, the highest numbers are found at different times of the year.
According to our data and those of Lindegaard & Mortensen (1988), in early winter most larvae are in second and third instar and only a few in fourth instar, suggesting a diapause. Moreover, their ability to survive drought in late summer (see under Water type: Permanence) suggests a dormancy. Lindegaard et al. (1975) collected very few larvae in winter (using a sieve with a mesh size of 200 μm) in the moss carpet of the Ravnkilde spring, where in spring and late summer densities of more than 5000 larvae/m² were found. Lindegaard & Mortensen suggest delayed hatching of eggs, consistent with the life cycle of *R. chalybeatus* assumed by Tokeshi (1986).

FEEDING

In winter we found mainly detritus in the guts of larvae. Probably diatoms are consumed more in spring, as has been stated for *R. chalybeatus*.

MICROHABITAT

In most investigations the largest numbers of larvae are found on stones, gravel, plants or coarse detritus. The larvae live in tubes attached to the substrate. Where such substrate is scarce, the larvae appear to live also on sandy bottoms (Tolkamp, 1980; Pinder, 1980; Kretzschmar & Böttger, 1994; Verdonschot & Lengkeek, 2006). Lindegaard et al. (1975) found the larvae very numerously in the moss carpet of the Ravnkilde spring in Denmark. The larvae lived mainly in the border of the moss carpet (where current velocities were highest) and were scarce in the interior part of the moss carpet.

DENSITIES

Lindegaard et al. (1975) found maximum densities of more than 8000 larvae/m² in the border of the moss carpet in the Ravnkilde spring in Denmark.

WATER TYPE

Current

R. fuscipes appears to be a common and often abundant inhabitant of fast-flowing brooks and streams (Lehmann, 1971; Ringe, 1974; Orendt, 2002a). In the Netherlands it is most common in the fast-flowing brooks and streams of southern Limburg. However, Laville & Vinçon (1991) collected *R. fuscipes* in relatively slowly flowing Pyrenean streams. In both lowland brooks and mountain streams the species lives more in the upper courses than in the lower courses (Lehmann, 1971; Moog, 1995; Gendron & Laville, 1997; own data). Its presence probably depends on various factors, such as

current velocity, temperature and food. It is necessary to take this into account when using table 7 in chapter 9. It is not clear why *R. fuscipes* is the most common species of the genus in the eastern part of the Netherlands, while in the province of Noord-Brabant *R. chalybeatus* is much more common.
Reiff (1994) collected the exuviae in a number of lakes in Bavaria.

Dimensions
R. fuscipes is a typical inhabitant of upper courses of streams (see also under Current and Temperature). We found large numbers in brooks 1 to 3 m wide, but larvae and exuviae have also been collected in goodish numbers in large rivers in the Netherlands, such as the Rhine and Meuse (Klink & Moller Pillot, 1982). Wilson & Wilson (1984) and Becker (1995) collected this species in the Rhine in Germany in much smaller numbers than *R. chalybeatus*.

Temperature
In contrast to *R. chalybeatus*, the species lives in streams with relatively cold water. Rossaro (1991) found a mean water temperature of 9.6 °C. In Dutch streams and in Bavarian lakes the mean temperature is very much higher (Reiff, 1994; own data). The real influence of temperature is still not clear, but it has to be seen in relation to other factors (see also section 2.10).

Shade
The species can be very numerous in shaded as well in unshaded brooks and streams.

Permanence
Within the appropriate stream type the species appears to be sometimes abundant in brooks that dry up in late summer. Possibly diminished concurrence plays a role here. It is not known whether larvae found in such brooks had been in (? egg) diapause from late summer or hatched after oviposition in November–December (see Life cycle).

TROPHIC CONDITIONS AND SAPROBITY

Reiff (1994) found the exuviae especially in highly productive lakes in Bavaria. However, the species is absent from very eutrophic pools and lakes in the Netherlands and northern Germany.
In streams *R. fuscipes* is often collected at organically polluted stations (Cuijpers & Damoiseaux, 1981; Röser & Neumann, 1985; Bazzanti & Bambacigno, 1987). Bazerque et al. (1989) call it more pollution resistant than *R. chalybeatus*, but the reverse is suggested by Milošević et al. (2012). Wilson (1987) collected the exuviae a little more numerously in less polluted water, but the species was still present in water with a BOD of about 10 and an ammonium-N content of about 5 mg/l. In heavily polluted conditions the larvae can live when the current velocity is high enough and therefore the oxygen content is never very low (e.g. Wilson, 1989). The species also lives in very clear woodland brooks; it has a wide amplitude regarding pollution, which is reflected in table 5 in Chapter 9.

OR

Rheocricotopus glabricollis (Meigen, 1830)

Trichocladius Gouini Gouin, 1936: 167–170

IDENTIFICATION OF THE LARVA

The larva can be distinguished from *R. atripes* by the S I with 6–7 apical teeth. However, in some specimens the setae anteriores appear to be bifid, with each branch split into 3–4 apical teeth (Saether, 1985: 92).

DISTRIBUTION IN EUROPE

The species appears to live in all parts of Europe, but there are only records from a relatively small number of countries (Saether & Spies, 2010). Until now it is not known from the Netherlands.

WATER TYPE

Gouin (1936: 152, 170) collected the larvae in running water between dead leaves. Most other records are from springs and small streams (Lehmann, 1971; Saether, 1985).

Rheosmittia Brundin, 1956

Tshernovskiella Pankratova, 1970: 319, fig. 207

SYSTEMATICS AND IDENTIFICATION

Rheosmittia is related to *Parakiefferiella* (Cranston et al., 1989). Only three European species are known, two of which have been described as **adult male** and can be distinguished using Cranston & Saether (1986): *R. spinicornis* and *R. languida*. The **exuviae** of three species are keyed in Langton (1991) and Langton & Visser (2003). Only the **larva** of *R. spinicornis* is known; for identification to genus level, see Cranston & Saether (1986). Two species are keyed in Schmid (1993). The figures in Pankratova (1970: p. 319) and Rossaro (1982) are of moderate quality; that of Cranston et al. (1983) possibly applies to an American species with aberrant mentum.

DISTRIBUTION IN EUROPE

Probably owing to their small size not very much is known about the distribution of the species. *R. spinicornis* can be found possibly throughout Europe, but has been stated at species level in only a few countries (Saether & Spies, 2010; cf. Pankratova, 1970: 319). *R. languida* is only known from northern and central Europe (Serra-Tosio & Laville, 1991; Saether & Spies, 2010).

ECOLOGY

The larvae live in the upper courses of streams, but can be found also in lower courses (Cranston & Saether, 1986). Rossaro (1991) found *R. spinicornis* especially at higher water temperatures (as a mean about 20 °C). Exuviae or larvae have been collected in fast-flowing streams in Scandinavia and Bavaria (Cranston & Saether, 1986; Michiels, 1999; Orendt, 2002a), but also in the river Po in Italy (Rossaro, 1982) and in lowland streams in Russia (Pankratova, 1970; Izvekova et al., 1996). The last authors collected the larvae in the cleanest, oligosaprobic part of the Babenka stream in Russia at a density of 2450 larvae/m^2 . Most probably the larvae live mainly in sandy bottoms (Cranston & Saether, 1986).

Smittia Holmgren, 1869

As far as is known, larvae of *Smittia* are only terrestrial. They avoid very wet places and when submerged by inundation they hardly creep at all (Moller Pillot, 2005: 117; 2008: 264). The suggestion that *S. aquatilis* is an aquatic species (e.g. in Coffman et al., 1986) has to be rejected. The identity of *S. aquatilis* is unknown (see Moller Pillot, 2008: 253).

The genus is not treated here; most ecological aspects have been described by Moller Pillot (2008).

Stackelbergina preclara Shilova & Zelentsov, 1978

Stackelbergina praeclara has been found only in temporary water bodies in central Russia (near Borok). The egg mass, larval instars I–IV and biology have been described by Zelentsov (1980a). The larvae have a diapause in second instar from June until spring. The species is not treated further in this book.

Symbiocladius rhithrogenae (Zavrel, 1924)

IDENTIFICATION

The adult male is not been keyed in Langton & Pinder (2007). The exuviae can be identified using Langton (1991). The larva has been keyed without description or figures by Cranston (1982). For figures and description see Schmid (1993). Beautiful photos can be found in Schiffels (2009).

DISTRIBUTION IN EUROPE

S. rhithrogenae has been found in the central and eastern part of Europe and in the Pyrenees (Serra-Tosio & Laville, 1991; Saether & Spies, 2010). The species is absent from lowlands, but occurrence in Dutch southern Limburg cannot be excluded.

ECOLOGY

The larvae live as parasites on nymphs of Heptageniidae, mainly *Rhithrogena*. They feed on the mayfly's hemolymph. The ecology has been described by Schiffels (2009), with many photos. Most probably its occurrence is confined to fast-flowing brooks and streams.

Synorthocladius semivirens (Kieffer, 1909)

DISTRIBUTION IN EUROPE AND THE NETHERLANDS

The species has been recorded throughout Europe (Saether & Spies, 2010). In the Netherlands the species has been found only in the eastern half of the country. It is common in the southeastern part and in the large rivers and is relatively scarce in the northeastern provinces (Limnodata.nl; own data).

LIFE CYCLE

Emergence is usually stated from March or April to October (Lehmann, 1971; Klink, 1985; Reiff, 1994; Becker, 1995; Michiels, 1999; own data), but there are also records from winter months (Drake, 1982, 1985). The highest numbers are stated in different times of the year, depending on environmental (possibly also biotic) factors. Mackey (1977) stated a development time of 8.7 days for artificially reared larvae, but Ladle et al. (1984) calculated the duration of larval life to be 93 days in summer in an experimental stream. Most probably there are about three generations in central Germany and the Netherlands, but only two in southern Sweden (see Reiff, 1994: 130). In Austria Schmid (1993a) found all larval instars during the whole year, but in winter we collected only larvae in second and third instar.

FEEDING

Much detritus can be found in the guts (Mackey, 1979), but also often mainly diatoms and sometimes filamentous algae (A. Klink, unpublished; own data). Gendron & Laville (2000) ascribed an increase of larvae in the lower course of the river Aude in the Pyrenees to silt enrichment after a flood.

MICROHABITAT

The larvae live in tubes, usually gelatinous, sometimes made of sand or detritus (Potthast, 1914; Lindegaard-Petersen, 1972). Large numbers can be found in lakes and streams on stones, but (possibly depending on current velocity) they can be numerous on plants or in the bottom (Brundin, 1949; Pinder, 1980; Becker, 1995). Pankratova (1970) noted their presence on the underside of *Nuphar* leaves. Schmid (1993a) found many larvae in the upper 10 cm of the gravel substrate in a stream and a low number still at 40 cm depth.

DENSITIES

Schmid (1993a) stated a maximal density of 102 larvae per dm^2 in the upper 10 cm of the bottom and a mean density of 36 larvae/dm^2 in a stream in Austria.

WATER TYPE

Current and dimensions

The species is scarce in typical lowland streams in the Netherlands and adjacent parts of Germany, and in the upper courses it is often absent. In such streams larger populations are found in and around fish ladders and in the lower courses of the streams. It is not known whether the sharp increase after 1990 (see Moller Pillot & Buskens, 1990) was caused by more accurate sampling or by the increase in the numbers of fish ladders in the last 20 years.

The larvae are often numerous in the lower parts of fast-flowing streams and in large rivers (Klink, 1985; Schmid, 1993a; Becker, 1995; Michiels, 1999; Garcia & Laville, 2001). Within fast-flowing streams the larvae can be more numerous in pools because these provide more food (sedimented particular organic matter), as stated by Syrovátka et al. (2009).

S. semivirens also lives abundantly in many lakes in northern regions (Brundin, 1949) and in some subalpine lakes (Reiff, 1995), and more scarcely in lakes in northern Germany (Meuche, 1938; Otto, 1991).

Most probably the species is absent from the undeep eutrophic lakes in the Netherlands. In England the species is scarcely found in storage reservoirs and gravel pits (Mundie, 1957; Ruse, 2002a), but there are no records from sand pits in the Netherlands.

Temperature

S. semivirens can be numerous in summer in lowland streams in and around fish ladders and therefore appears to tolerate relative high water temperatures. It is found in mountain streams too (e.g. Gendron & Laville, 1997).

Permanence

The species seems to be rare in temporary water, but has been found once in a city fountain (Hamerlik & Brodersen, 2010).

pH

S. semivirens has been found rarely in acid water, but Simpson (1983) collected several larvae of S. near *semivirens* at a pH around 6 and one larva at a pH around 5.

TROPHIC CONDITIONS AND SAPROBITY

Brundin (1949) collected *S. semivirens* in both oligotrophic and eutrophic lakes (but not in polyhumic lakes) in Sweden. In Dutch streams the species is most common at low pollution levels (BOD usually less than 2.2 mg O_2/l) and a good oxygen supply (Limnodata.nl). However, Laville & Viaud-Chauvet (1985) and Wilson (1989) found the exuviae also in water with relatively severe organic pollution and somewhat lowered oxygen content (in fast-flowing streams). In Dutch Limburg the numbers were sometimes high in the moderately polluted river Grensmaas (Meuse in southern Limburg) and very low in the more severely polluted Jeker (Klink, 1985; own data). Because its occurrence in polluted water seems to be confined to fast-flowing streams and in lowland streams it is more or less concentrated in faster flowing sites, oxygen supply seems to be important. See the tables in Chapter 9.

SALINITY

In northern countries the species occurs rarely in brackish water (Izvekova, 1996; Paasivirta, 2000: 596). In the Netherlands it will be confined to fresh water.

Tavastia Tuiskunen, 1985

Until now three European species of *Tavastia* have been described, of which only *T. yggdrasilia* is known from the West European lowland. *T. australis* has been collected only in Finland, in a spruce mire with springs and brooklets (Tuiskunen, 1985). *T. alticrista* has been caught in emergence traps from alpine calcareous springs in the Berchtesgaden National Park, Germany (Stur & Wiedenbrug, 2005).
Pupae and larvae of the whole genus are unknown.

Tavastia yggdrasilia Brodin, Lundström & Paasivirta, 2008

IDENTIFICATION

Only the adult male and female have been described (Brodin et al., 2008).

DISTRIBUTION IN EUROPE AND THE NETHERLANDS

T. yggdrasilia has been collected in Scandinavia, Russia, Belarus and the Netherlands and seems to be rather common in northern countries (Saether & Spies, 2010; Paasivirta, 2012). In the Netherlands the species has been found at four locations, in Utrecht and northwest Overijssel (own unpublished data).

LIFE CYCLE

In Sweden the adults were obtained from early May to late September, suggesting more than one generation per year (Brodin et al., 2008). All Dutch specimens emerged at the end of April or in May.

ECOLOGY

In Scandinavia the species is only known from wetlands, mainly from the vegetational zones of lakes (Brodin et al., 2008). It lives in humid soils, but emerges also during truly aquatic conditions from inundated wetlands, in eutrophic or mesotrophic to eutrophic conditions.
In the Netherlands its occurrence seems to be confined to less eutrophic marshes with groundwater seepage (e.g. Calthion) or a little acidified (trembling bogs). All

these localities were situated in nature reserves and have been investigated by Siepel (unpublished). The difference in trophic conditions between the Dutch and the Scandinavian populations is possibly not as large as the terminology suggests. Our records from Belarus are more or less from the same habitats as in the Netherlands.

Thalassosmittia thalassophila (Bequaert & Goetghebuer, 1913)

Pseudosmittia thalassophila Remmert, 1955: 36–37

SYSTEMATICS AND IDENTIFICATION

In Europe the genus is represented by one species, *T. thalassophila*. The larva has not been keyed and figured by Cranston (1982) and Moller Pillot (1984), but can be identified to genus level using Cranston et al. (1983) and Klink & Moller Pillot (2003), and to the species using Pankratova (1970). Pankratova and Klink & Moller Pillot also give figures of this species; the figures in Cranston et al. apply to a related American species with slightly different central mental tooth, setae submenti, labral lamellae and S I.

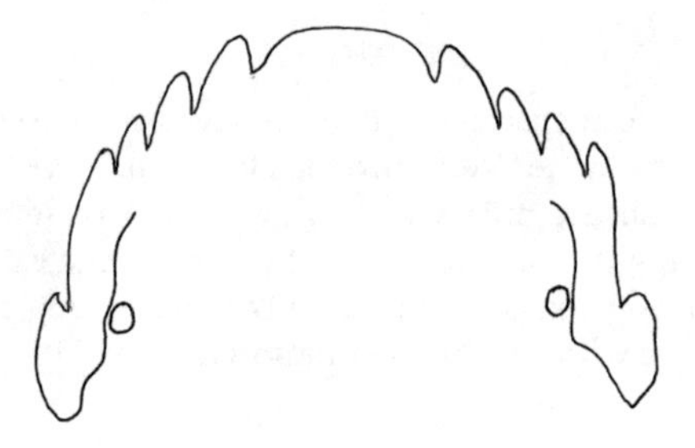

13 *Thalassosmittia thalassophila: mentum*

DISTRIBUTION IN EUROPE AND THE NETHERLANDS

T. thalassophila has been collected in many countries along the coasts of western and southern Europe (Saether & Spies, 2010). In the Netherlands there are no records from the North Sea coast, but it has been found in several tidal rivers, mainly in the province of Zuid-Holland, and along the IJsselmeer dam (Afsluitdijk) (Limnodata.nl; own data).

FEEDING

Ring (1989) stated that the American species feeds on diatoms and green algae. Larvae were able to feed selectively on certain diatoms. They were observed to feed on a ciliate-bacterial film in addition to the algae.

MICROHABITAT

Remmert (1955) found the larvae in thin layers of silt under algae (*Enteromorpha*) on stones and wood. The larvae are tube builders (Neumann, 1976). They can live above the water level for a long time.

WATER TYPE

The larvae live in the marine intertidal zone, but also in estuaries and tidal rivers and further inland in more or less fresh water, as in the Lek and Hollandsche IJssel rivers in the Netherlands.

SALINITY

Remmert (1955) stated that the larvae tolerate strong freshening and also high salinity to the levels found in sea water, and probably much higher, when the substrate lies above the water level.

Thienemannia Kieffer, 1909

Symmetriocnemus Thienemann, 1937: 30
Metriocnemus clavaticornis Potthast, 1914: 341–343, figs 125–128
Metriocnemus ex gr. *clavaticornis* Chernovskij, 1949: 142 [211 in ed. 1961]

SYSTEMATICS AND IDENTIFICATION

The genus is closely related to *Metriocnemus* (Saether, 1985a). The size variation in *T. gracilis* is quite large, in adults as well as in larvae (Saether, 1985a; own data). A key for **males** is given in Saether (1985a) and Langton & Pinder (2007); **pupae** can be identified using Saether (1985a) or Langton (1991), but the pupa of *T. gracei* is not known with certainty. The pupae studied by Potthast (1914) had thoracic horns, as stated by Thienemann (1937), and belongs therefore to *T. fulvofasciata* or *T. gracei*.
Because the head capsule length of the **larva** (of the genus) varies from 0.23 to 0.33 mm, the anal and supraanal setae can be shorter than given in the key by Moller Pillot (1984: 16). A better generic character is the relative length of the longest anal setae: in *Thienemannia* more than three-quarters of the head length, in *Metriocnemus* at most half as long as the head.
Most probably (based on associated material) the larvae of *T. gracilis* and *T. gracei* have the same characters as those given by Moller Pillot (1984) and Saether (1985a) for *T. gracilis*. An adult male of *T. gracilis* was collected from the same place (Bron Belletbeek 52039, 8.5.2012) as the larvae described by Saether (1985a: 118–120). However, because in the keys by Moller Pillot (1984) and Schmid (1993) all larvae of the genus are ascribed to *T. gracilis*, the use of this species name in the literature can be wrong.

LIFE CYCLE

Most adults have been collected in spring, but most probably there are two or more generations a year (Saether, 1985a; own data).

FEEDING

We found detritus, filaments of fungi and sand grains in the guts.

ECOLOGY

All species of the genus have been collected in springs and small foothill or montane streams (Potthast, 1914; Caspers, 1980; Cranston, 1982; Saether, 1985a; Stur et al., 2005). Sometimes two species were collected at the same location.

Thienemannia fulvofasciata (Kieffer, 1921)

? *Metriocnemus clavaticornis* Potthast, 1914: 341–343, figs 125–128 (possibly *T. gracei*)

DISTRIBUTION IN EUROPE

This species has been reported only from some countries in western and central Europe (Saether & Spies, 2010). Exuviae identified as *fulvofasciata* could belong to *gracei* (Langton, 1991: 148). Conversely, adult males identified using Pinder (1978) as *Metriocnemus gracei* could belong to *fulvofasciata*. There are no records from the Netherlands.

MICROHABITAT

Larvae have been found in mosses in rheocrenes and small streams (Saether, 1985a; Stur et al., 2005).

Thienemannia gracei (Edwards, 1929)

Metriocnemus gracei Pinder, 1978: 90, figs 44D, 131C; Caspers, 1980: 81
? *Metriocnemus clavaticornis* Potthast, 1914: 341–343, figs 125–128 (possibly *T. fulvofasciata*)

IDENTIFICATION

See under the genus. The wing length of the Dutch males is only 1.25–1.5 mm and not 1.7 mm as given by Saether (1985a). Larvae identified as *Thienemannia* were collected in the Hemelbeek near Elslo (the Netherlands), from which also adults of *T. gracei* have been reared (material in collections H. Cuppen, Waterschap Roer & Overmaas, H. Moller Pillot).

DISTRIBUTION IN EUROPE AND THE NETHERLANDS

T. gracei has been reported from a limited number of countries scattered across Europe, as yet not from Scandinavia (Saether & Spies, 2010). In the Netherlands the species is known from three localities in the western part of southern Limburg.

WATER TYPE AND MICROHABITAT

We reared the species from small spring brooks, mainly from the wet seepage zones near the brook up to 50 cm from water. Mosses were absent here and the bottom consisted of coarse or fine organic matter and sometimes much gravel. Caspers (1980) found *T. gracei* and *T. gracilis* in the same woodland brook.

Thienemannia gracilis Kieffer, 1909

Thienemannia gracilis Moller Pillot, 1984: 159–160, fig. VI.49 (pro parte)

DISTRIBUTION IN EUROPE AND THE NETHERLANDS

According to Saether & Spies (2010) this species has been found in all parts of Europe. See, however, the text under Identification above. In the Netherlands adults have been reared or collected at four locations in southern Limburg. Larvae collected in springs on the Veluwe (leg. H. Cuppen) probably belong to this species.

WATER TYPE AND MICROHABITAT

Lehmann (1971) mentioned records from different authors in fast-flowing upper courses and hygropetric environments, often in mosses. Caspers (1980) collected the species from a small woodland brook. The larvae also live in springs and spring brooks (Stur et al., 2005; own data). We also reared the species from stones without mosses at a large distance from the spring, but in lower courses of fast-flowing streams it seems to be rare (e.g. Michiels, 1999). Grzybkowska & Witczak (1990) found large numbers of larvae at fast currents in the lower reaches of the river Grabia in Poland; however, it is not clear whether they identified adults or only larvae.

Thienemannia libanica Laville & Moubayed, 1985

This species has been recorded in several countries in western Europe (Saether & Spies, 2010). Stur et al. (2005) found it in a spring brook in Luxembourg.

Thienemanniella Kieffer, 1911

SYSTEMATICS AND IDENTIFICATION

Thienemanniella is closely related to *Corynoneura* and can be included with this genus and *Corynoneurella* in a separate tribe, Corynoneurini, within the Orthocladiinae (Saether, 1977a, 1979a). In the older literature (Goetghebuer, 1939; Lenz, 1939a) this group of genera was considered to be a subfamily, Corynoneurinae. For further information see under *Corynoneura*.

A recent review of the genus has been made by Fu et al. (2010). However, the identification problems with the European species have not yet been solved. The **adult males** of some species have been described in detail by Schlee (1968) and the key to adult males by Langton & Pinder (2007) gives valuable additions to the key by Pinder (1978), but all these keys are incomplete and descriptions of most species are insufficient.
Identification of **pupal exuviae** using Langton (1991) or Langton & Visser (2003) often does not lead to a reliable species name (e.g. Orendt, 2002a). This applies also to the **larvae**, although Schmid (1993) described and figured several differences between larval types.
The result of this situation is that ecological information for many species is very incomplete and the names mentioned in the literature are often not reliable. Nevertheless, two species appear to be very common in many European streams: *T. majuscula* and *T. vittata*. These two species show some differences in their ecology, which is described below. Possibly in many cases some scarce species among the material remained undetected during the investigations.

LIFE CYCLE

Probably all species of the genus have more or less the same life cycle: a number of generations (two to five or six) from early spring until late autumn, in Atlantic or Mediterranean regions possibly emergence all year round. In colder regions a diapause in late autumn could be possible. There appear to be some differences between species (see under *majuscula* and *vittata*).
Gendron & Laville (1992) stated emergence mainly during daytime, in contrast to larger chironomids.

EGGS AND EGG DEPOSITION

Nolte (1993: 24, 25) described and figured the egg masses of *T. partita*. The egg masses had the shape of a coiled string with the (58–76) eggs in a single row, very different from the egg masses of *Corynoneura*. They had been deposited in a stream among moss just below the water's edge.

MICROHABITAT

Most species seem to choose their microhabitat according to oxygen supply and the presence of food. In fast-flowing streams the larvae are usually found in the bottom (e.g. Schmid, 1993), but in lowland brooks almost the whole population can be found near the water surface on plants and wood (e.g. Tokeshi, 1986; own data). The larvae creep freely around seeking food; only prepupae and pupae live in a tube of salivary secretion. This behaviour is probably the reason why the larvae are relatively often transported by drift, especially during flooding (Gendron & Laville, 2000; Moller Pillot, 2003).

DENSITIES

The larvae are often overlooked, partly because of their small size (Tokeshi, 1986). We think that the sampling method in the Netherlands also leads to underestimation, because in lowland streams most larvae live at the water surface.

WATER TYPE

Almost all species live in flowing water and current velocity plays an important role in species composition (figure 14). The whole genus is very rare in lakes (e.g. Reiff, 1994). The larvae are very scarce in large rivers, possibly because of the risk of being carried off. Some species (e.g. *T. acuticornis*) can also be found in hygropetric environments. The influence of shade is still completely unknown. In shaded brooks without vegetation we found most larvae on floating twigs and wood.

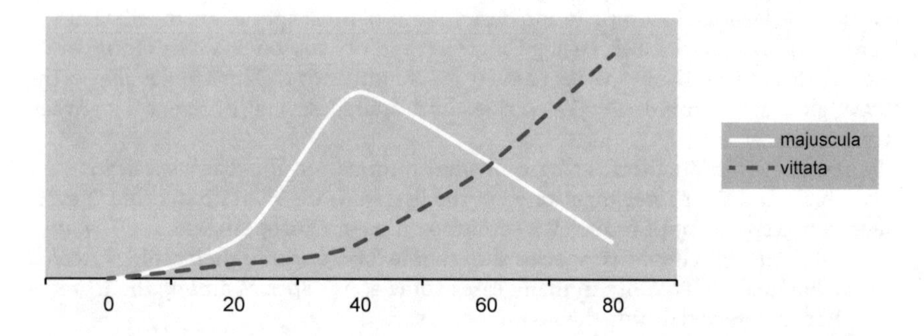

14 Schematic representation of the occurrence of Thienemanniella majuscula and Thienemanniella vittata in streams with different current velocity (cm/s)

SAPROBITY

The whole genus is often considered to be characteristic of water of low saprobity (see e.g. Moog, 1995). Ruse (2012) stated from material sampled in his investigations of the river Thames a negative correlation between presence of the exuviae of the genus and increasing nitrogen, ammonium-N and BOD. However, tolerance is more or less dependent on the oxygen supply, as discussed for each species below.

Thienemanniella acuticornis (Kieffer, 1912)

Thienemanniella (Microlenzia) fusca Lenz, 1939a: 17
Thienemanniella morosa Pinder, 1978: 98, fig. 48C, 140D

SYSTEMATICS AND IDENTIFICATION

The adult male does not exhibit aberrant characters from other species of the genus. However, the pupa and larva are very aberrant, suggesting another genus. According to Fu et al. (2010) there is no reason for separation. Identification (to all stages) presents no problems with the existing keys.

DISTRIBUTION IN EUROPE

T. acuticornis has been recorded from a number of countries scattered across Europe (Saether & Spies, 2010). In the Netherlands there is only one record, from a spring brook at the southern edge of the Veluwe (leg. A. Klink).

ECOLOGY

The larvae are usually collected among mosses in springs or below waterfalls and weirs (Lenz, 1939a; Cranston, 1982; Schmid, 1993). Most probably they also live further downstream in fast-flowing brooks and streams (Fittkau & Reiss, 1978; Casas & Vilchez-Quero, 1989; Paasivirta, 2012). According to Moog (1995), their occurrence is almost entirely confined to water without organic pollution, but their tolerance is unknown because the characteristic habitat is rarely polluted.

Thienemanniella clavicornis (Kieffer, 1911)

IDENTIFICATION

The adult males can be identified using Schlee (1968) and Langton & Pinder (2007) together. The species is keyed but not figured in Fu et al. (2010). Identification of pupae and larvae to species level is impossible. Schmid (1993) gives a short description and figures of the larva. The colour of the head is yellow according to Thienemann (1944: 652), dark brown according to Cranston (1982) and brown according to Schmid (1993).

DISTRIBUTION IN EUROPE

T. clavicornis has been found throughout Europe (Saether & Spies, 2010). Its occurrence in the Netherlands (Limburg) is very probable, but it has never been stated. The records given by Limnodata.nl apply without doubt to other species of the genus.

ECOLOGY

Schmid (1993) collected the larvae in the bed sediments of alpine brooks, up to 30 cm deep. Possibly they can be found elsewhere also on plants and wood near the water surface (see *T. vittata*). The species has been recorded from spring brooks, fast-flowing upper courses and foothill streams (Lehmann, 1971; Fahy, 1973) and seems to be relatively common and widespread (Edwards, 1950; Schlee, 1968).

Thienemanniella flavescens (Edwards, 1929)

IDENTIFICATION

T. flavescens can be identified only as adult male (Edwards, 1950; Langton & Pinder, 2007; Fu et al., 2010).

DISTRIBUTION IN EUROPE

The species has been recorded only in some West European countries (see Ecology) and in Russia (Saether & Spies, 2010). Its occurrence in the Netherlands is likely.

ECOLOGY

There are very few data about the ecology of this species. Moubayed-Breil (2008: 22, 30) found it in France only in lowland streams and some foothill streams. It seems to be absent from typical mountain streams. There are records from the lower parts of the Pyrenees (Serra-Tosio & Laville, 1991) and foothill streams in England and Ireland (Edwards, 1950; Fahy, 1973).

Thienemanniella lutea (Edwards, 1924)

IDENTIFICATION

It is not clear whether this species has been identified correctly in the past. According to Edwards (1950) the **male** antenna has only 11 distinct segments, whereas Fu et al. (2010) stated 12 or 13 segments. The most reliable character seems to be the absence of setae on the gonocoxite lobe (Pinder, 1978; Langton & Pinder, 2007: 143–144, fig. 74I). According to Edwards (1950), the **female** is recognisable because of its almost entirely yellow colour, with the scutal stripes orange or light brownish. Pupa and larva are unknown.

DISTRIBUTION IN EUROPE

T. lutea has been recorded only on the British Isles, in the Pyrenees and in Macedonia (Serra-Tosio & Laville, 1991; Saether & Spies, 2010).

ECOLOGY

Nothing is known about the ecology.

Thienemanniella majuscula (Edwards, 1924)

Thienemanniella flaviforceps agg. Moller Pillot, 1984
Thienemanniella majuscula Pinder, 1978: 98 pro parte (= *majuscula* + *obscura*)
nec *Thienemanniella majuscula* Langton, 1984: 80 (= *obscura*?)
nec *Thienemanniella majuscula* Langton, 1991: 84 (= *Corynoneurella paludosa*)
nec *Thienemanniella* cf. *majuscula* Lindegaard-Petersen, 1972: 493; Schmid, 1993: 200, fig. 140; Bitušik, 2000: 119, fig. 45 A, B; Klink, 2010 (= *obscura*?)
Thienemanniella species A Cranston, 1982: 138, fig. 56g; Tokeshi & Pinder, 1985: 314

IDENTIFICATION

Adult males can be identified using Langton & Pinder (2007). *T. majuscula* in Pinder (1978) applies to both *majuscula* and *obscura*. Most probably this has led to the incorrect suggestion that larvae with three central mental teeth (species C in Cranston, 1982: 138, cf. *majuscula* in Schmid, 1993: 200) belong to *majuscula*. In populations of *majuscula* (identified as adult males) Tokeshi (1986: 440), and we also, found only larvae with two central mental teeth. It is possible that a second species has such a mentum, but this seems to be very unlikely, both in the Netherlands and in the alpine region (see Water type: current).
Identification of the exuviae is as yet not possible; the species runs to *Thienemanniella* Pe2a in Langton (1991). There are many misidentifications (see above).

DISTRIBUTION IN EUROPE AND THE NETHERLANDS

T. majuscula has been recorded in a number of countries in northern, western and central Europe (Saether & Spies, 2010). These records have to be used with caution (see above). In the Netherlands the species is common in the whole Pleistocene area and scarce in southern Limburg and the fen peat regions, based on identification of adult males and larvae (Limnodata.nl; own data). There is one record from Flevoland (Klink & Mulder, 1992). Other records from Holocene regions have to be verified.

LIFE CYCLE

In the Netherlands emergence has been stated from March until November (own data). Tokeshi (1986) suggested five generations a year in England. Singh & Harrison (1984) found only two generations in a stream in Canada (with lower water temperatures). In winter Tokeshi found only young larvae, suggesting a diapause.

FEEDING

Tokeshi (1986a) stated that the gut contents of larvae in a brook in southern England contained a very high proportion of diatoms, at least in spring. In the Netherlands detritus seemed to be more important (as in *T. vittata*), but too few guts have been examined.

MICROHABITAT

The larvae are found in large numbers on aquatic plants near the water surface and in smaller numbers on wood and stones (Tokeshi, 1986; Verdonschot & Lengkeek, 2006; own data). It is not known whether they also occur in the bottom when the oxygen supply there is good (as has been stated in *T. vittata*).

WATER TYPE

Current

T. majuscula is only known from flowing water, mainly from lowland brooks. It seems to be absent from the fast-flowing streams in Austria and Slovakia, because larvae with two central mental teeth were not found by Schmid (1993) and Bitušik (2000). The name is absent from the literature on exuviae because the exuviae are undescribed. Laville & Vinçon (1991) collected the species scarcely in the lower parts of the Pyrenees. Singh & Harrison (1984) collected it in Canada (if it was indeed this species) at a current velocity of 0.3–1.3 m/s. We collected the species scarcely or not at all in fast-flowing brooks in southern Limburg (the Netherlands) and the adjacent part of Germany. However, in lowland brooks and streams where the species occurs it can be also found numerously in fish ladders at high current velocities. Possibly pH also plays a role here.

Dimensions

Generally the genus is not or rarely collected in rivers (Lehmann, 1971; Klink & Moller Pillot, 1982; Wilson & Wilson, 1984). Becker (1994) caught some adults of *T. majuscula* along the river Rhine only in light traps (possibly another species). Klink (2010) collected the larvae of *T. majuscula* in small rivers in the Seine basin. In the Netherlands the species is found in brooks and streams from 1 to 8 m wide.

Permanence

There are several records of larvae from brooks which rarely dried up; in one case (Moller Pillot, 2003) two larvae were collected in October in an upper course which had been completely dry for ten days in August. Considering the life cycle, repopulation depends mainly on the distance from brooks with a permanent population of the species.

pH

Most records from the Netherlands are from brooks and streams with a pH between 7 and 8. However, the species is not rare at pHs around 6 (Tokeshi, 1986; Moller Pillot, 2003). *T. majuscula* seems to be scarce in chalk streams in comparison with *T. vittata* (own data; cf. Tokeshi & Pinder, 1985, and other publications by Tokeshi).

SAPROBITY

T. majuscula has been collected in the Netherlands mainly in eutrophic water with little or no organic pollution. In some cases a higher BOD (up to 4 mg O_2/l), a somewhat lowered oxygen content or a high ammonium-N content (up to more than 1 mg N/l) has been stated (Peters et al., 1988; Duursema, 1997; Limnodata.nl). On one occasion we collected a low number of larvae in a severely polluted stream (the Jeker) with a relatively good oxygen supply. Their occurrence at lower current velocities indicates a lower oxygen demand in comparison with other species of the genus. See table 5 in Chapter 9.

Thienemanniella minuscula (Brundin, 1949)

See *Corynoneura minuscula*.

Thienemanniella obscura Brundin, 1947: 49, fig. 81

Thienemanniella majuscula Pinder, 1978: 98 pro parte

IDENTIFICATION

Identification of adult males and exuviae is problematic. Different descriptions and keys (e.g. Brundin, 1947: 49) lead to doubtful results and we think that a revision of the European species is necessary. Nevertheless, Langton (pers. comm.) wrote that he can distinguish the exuviae of *T. vittata* and *T. obscura* relatively easily. In older publications the species is often absent because in Pinder (1978) it is included in *T. majuscula*. Larvae named *T.* cf. *majuscula* in Schmid (1993) and Bitušik (2000) could apply to *obscura*. In view of all this, records of *T. obscura* have to be used with caution.

DISTRIBUTION IN EUROPE AND THE NETHERLANDS

T. obscura has been recorded in a number of countries in northern, western and central Europe (Saether & Spies, 2010). In Finland (Paasivirta, 2012) as well as in England (various authors) the species seems to be less common than *vittata* and *majuscula*. Its occurrence in the Netherlands is probable, but has never been proved.

ECOLOGY

Langton (pers. comm.) collected *obscura* exuviae and adults from macrophytes in springs and streams in England. Other sources in the literature create the impression that the ecology is not very different from that of *vittata*.

Thienemanniella partita Schlee, 1968

IDENTIFICATION

The adult male has not been keyed by Pinder (1978) and Langton & Pinder (2007). It has been described by Schlee (1968) and can be distinguished from related species by the absence of microtrichia on the eyes and by the shape of the endoskeletal structures. The pupa is absent from Langton (1991). The larva is briefly described and figured in Schmid (1993) and Bitušik (2000). When the central mental teeth have been worn off, identification may be possible by examining the lateral teeth of mentum and mandible.

DISTRIBUTION IN EUROPE

The species has been recorded in Germany, Austria, Spain and Italy, and possibly in southern France (e.g. Saether & Spies, 2010; Serra-Tosio & Laville, 1991). Without doubt it lives in many other countries too, but it has been overlooked because of the absence of the species from most keys.

EGGS

The egg masses have been described and figured by Nolte (1993: 24, 25).

MICROHABITAT

Nolte (1989) found the larvae on almost all types of substrata; within moss cushions on stones in submersed parts as well as in the spray zone above the water surface. Schmid (1993) collected larvae and pupae in the upper 20 cm of gravel sediments in an alpine gravel brook.

WATER TYPE

T. partita has been found in small brooks in foothill and mountain regions (Lehmann, 1971; Caspers, 1980a; Nolte, 1989; Schmid, 1993). According to Punti et al. (2009), in Spanish streams the species is characteristic of cold, siliceous upper courses.

Thienemanniella vittata (Edwards, 1924)

IDENTIFICATION

The name *T. vittata* is used here, although it may have been better to use *vittata* complex. As stated in the introduction to the genus, we (and many other workers) have based our identifications of **adult males** on Langton & Pinder (2007) and of exuviae on Langton (1991), neglecting *T. partita* and probably even more species. Adult males of *vittata* resemble *T. obscura* and identification based on the gonocoxite lobe, as given by Pinder (1978) and Langton & Pinder (2007), often leads to mistakes. The characters given by Brundin (1947: 49) do not lead to easy separation. In some cases identification of **exuviae** (Langton 1991: 84) also seems to lead to mistakes; the spinules on tergites VI–VIII are sometimes no longer than 7 µm.

The **larvae** have been described briefly by Schmid (1993: 201); if his cf. *majuscula* applies to *obscura*, most probably the combination of mental teeth, mandibular teeth and head colour are sufficient for identification of *T. vittata* to species level.

DISTRIBUTION IN EUROPE AND THE NETHERLANDS

T. vittata lives throughout Europe, although it has not been reliably recorded from many countries (see Saether & Spies, 2010). In Finland it seems to be the most common species of the genus (Paasivirta, 2012). In the Netherlands the species is common in southern Limburg and its occurrence in central Limburg has also been stated on the basis of reared specimens. Many larvae collected on the Veluwe, in Twente and Winterswijk most probably also belong to this species (H. Cuppen, pers. comm.; own data); we reared the species in goodish numbers from brooks in adjacent parts of Germany. The verified larvae collected in the Netherlands agree very well with the description by Schmid (1993; see above).

LIFE CYCLE

Emergence has been stated mainly from March to November, rarely at the end of February (Gendron & Laville, 1992; own data) and possibly sometimes as early as January (Drake, 1985). At the end of December we found only juvenile larvae, suggesting a diapause in late autumn (cf. *T. majuscula*). In contrast to what Tokeshi (1986: 437) stated for *T. majuscula*, in southern Limburg we found most *vittata* larvae in fourth instar by the end of January. In the Netherlands *vittata* seems to emerge earlier in spring than *majuscula*.

FEEDING

We found detritus and diatoms in the guts of the larvae in spring, and in summer almost only detritus. The central mental teeth of larvae living in sandy bottoms were often worn off.

MICROHABITAT

The larvae are most numerous near the water surface, on plants and wood. Tokeshi & Pinder (1985) stated that the larvae on firm and broad leaves increased in numbers towards the apex of the leaves. They live also on stones and on (and in) sandy bottoms (Pinder, 1980; Schmid, 1993; own data).

WATER TYPE

T. vittata larvae live almost only in flowing water; according to Langton (1991), they also live in montane lakes. In Austria, in the Pyrenees and other Spanish montane areas they are found in very fast to slow currents, but most of all in fast-flowing brooks (Laville & Vinçon, 1991; Schmid, 1993; Punti et al., 2009). They can be numerous there at current velocities of 1 m/s (Gendron & Laville, 1992). In the Netherlands the species lives mainly in foothill brooks and streams in the province of Limburg, usually at higher current velocities than *T. majuscula* (see figure 14 on p. 236). In lowland streams in the eastern part of the country and the adjacent part of Germany we collected it in relatively clean water or on fish ladders. There is no clear explanation for the absence of the species in lowland brooks in the province of Noord-Brabant (see however under pH and Saprobity). There are no records from large rivers.
There are no data from temporary streams. Considering the life cycle, such brooks and streams can be repopulated within a short time, as has been stated for *T. majuscula*.

pH

There are no records from acid water and relatively many from chalk streams. The species seems to prefer a higher pH than *T. majuscula*.

SAPROBITY AND OXYGEN

Most authors call the whole genus indicative of water with no or very little organic pollution (Moog, 1995; Bitušik, 2000: 119; Ruse, 2012), but Wilson & Ruse (2005) called it moderately tolerant. Most probably there are some differences between the species concerning saprobity (see table 5 on p. 283). Maasri et al. (2008) found *T. vittata* mainly in water without organic effluent. We collected the species sometimes numerously in unpolluted or very slightly polluted water and very scarcely in polluted brooks and streams. The larvae can live up to 20 cm deep in sandy bottoms (see above), but this has been stated only in very clean streams; possibly the choice of the microhabitat depends on the local oxygen supply (see under the genus).

Trissocladius Kieffer, 1908

This genus is related to *Hydrobaenus* and *Zalutschia*. In the older literature many species are described under this genus name, but the genus currently comprises only two species, which can be identified only as adult male using Saether (1976: 165). Kieffer & Thienemann (1908) found no differences between the larvae of both species. In most regions only *T. brevipalpis* seems to be present. We verified many males in the Netherlands and in Belarus and found no specimens of *T. heterocerus*. The latter species is absent from almost all keys to adults, pupae or larvae. There may be some doubt about whether *T. heterocerus* is a separate species on the basis of the aberrant antenna. The larva of both species is blood-red, in contrast to other genera of the subfamily Orthocladiinae.

Trissocladius brevipalpis Kieffer, 1908

IDENTIFICATION

As mentioned under the genus, only adult males can be identified using Saether (1976). For a description of the female imago, pupa and larva, see Saether (1980b).

DISTRIBUTION IN EUROPE AND THE NETHERLANDS

T. brevipalpis has been collected in large parts of Europe, but could be absent from some countries around the Mediterranean Sea (Saether & Spies, 2010). We found the species especially common in Belarus. In the Netherlands it is widely distributed in the Pleistocene part of the country, but it seems to be very scarce in the Holocene regions and possibly absent from southern Limburg (Moller Pillot & Buskens, 1990; Limnodata.nl).

LIFE CYCLE

We found the same life cycle in the Netherlands and Belarus. From early May until October almost all larvae are in second instar (rarely in third instar), usually in a cocoon. Very rarely a specimen can emerge in summer. Some larvae start further development in autumn, but most larvae do not grow further until winter. In the Netherlands the first emergences can be observed in January, but almost the whole population emerges in March and April. When second instar larvae were desiccated in May they did not make a cocoon and died. Without doubt the species has a period of dormancy in summer, as described in the genus *Hydrobaenus* (see there).
On 19 April 2004 we saw many adults copulating on the water surface of a pool in Belarus. We observed no swarming and rarely a little flying.

MICROHABITAT

The larvae are found on the bottom among dead leaves (often with algae) and among grasses and other plants.

WATER TYPE AND PH

The species is common in pools, marshes and inundated grassland on sandy or loamy soils which dry up in summer (also in woodland). In Belarus these water bodies are frozen or dry in winter. In such environments the larvae can also be collected in slow-flowing ditches and brooks, even further downstream in permanent stretches. It is not known whether the species survives all year round in permanent water. We never found it in acid water at pHs lower than 6, but Schleuter (1985: 88) caught some adults from temporary woodland pools with a pH of 5.0 (in winter).

TROPHIC CONDITIONS AND SAPROBITY

As far as the larvae lived in inundated parts or pools in grassland this applied always to not or moderately fertilised or dunged grassland. In marshland and woods there was usually much, but only gradually decaying, organic matter.

Trissocladius heterocerus Kieffer, 1908

This species has been collected only in Germany and Denmark (Saether & Spies, 2010). Kieffer & Thienemann (1908) and Potthast (1914: 328–329) described the egg masses. They found the larvae, pupae and adults numerously in temporary meadow pools in Germany. The ecology seems to be the same as for *T. brevipalpis*.

Tvetenia Kieffer, 1922

Eukiefferiella verralli gr. Lehmann, 1972: 399–401
Eukiefferiella gr. *discoloripes* Moller Pillot, 1984: 71, 73

SYSTEMATICS AND IDENTIFICATION

Saether & Halvorsen (1981) argued that within the original genus *Eukiefferiella* two groups can be clearly separated in all stages and merit the status of genus. They described the imago, pupa and larva of the new genus *Tvetenia*. Within this genus, Halvorsen does not use groups, because it is not sure that they are monophyletic (in litt. to U. Nolte, 1988). The two aggregates distinguished by Moller Pillot (1984: 71) can be separated most easily by examining the setae submenti, which are further posterior to the base of the ventromental plates in *calvescens* and *bavarica* (see fig. 15 and figs 144–147 in Schmid, 1993). For day-to-day identification of the larvae, it is advisable to note the colour of thorax and abdomen.

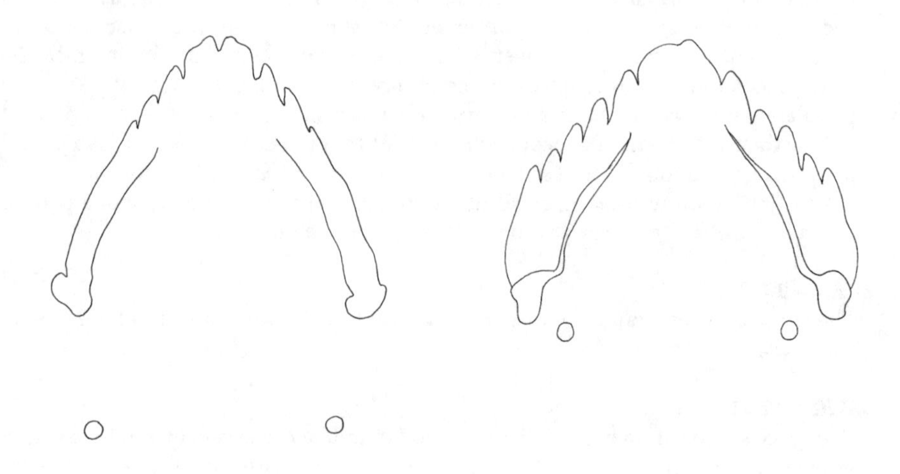

15 Tvetenia calvescens (l) and T. discoloripes (r): mentum with the insertions of the setae submenti

LIFE CYCLE

Most probably the life cycle of all species of the genus is more or less the same. However, the data required to give a broad survey are only available for *T. calvescens*. The number of generations can vary considerably and display a tendency to multivoltinism. A diapause in late autumn and winter seems to be possible, but may be facultative.

FEEDING

The whole genus appears to feed mainly on diatoms and detritus and seems to be not very selective (see under the species). Kawecka & Kownacki (1974) even found only detritus (mainly from rotting beech leaves) in the guts of the larvae of unidentified species of the genus in a very fast-flowing stream in Poland.

WATER TYPE

All species are inhabitants of running water.

pH

The whole genus seems to be absent from acid streams (Orendt, 1999; Janzen, 2003).

Tvetenia bavarica (Goetghebuer, 1934)

Eukiefferiella bavarica Lehmann, 1972: 387, 400, figs 65–68

SYSTEMATICS AND IDENTIFICATION

T. bavarica resembles *T. calvescens* in all stages. The adult male is absent from Pinder (1978), but can be identified using Langton & Pinder (2007). The exuviae are easiest to identify (Langton, 1991). The larva has been keyed by Schmid (1993: 204), but the length and the ratio of the antennal segments are more variable than given in his key. Therefore, it is not always possible to distinguish larvae of *calvescens* and *bavarica*. The adults exhibit much variation in size (Lehmann, 1972) and so the same can be true for larvae.

DISTRIBUTION IN EUROPE AND THE NETHERLANDS

T. bavarica has been recorded from many countries throughout Europe (Saether & Spies, 2010). In the Netherlands larvae have been identified from southern Limburg (see above). These records have to be verified by identification of exuviae.

LIFE CYCLE

Emergence has been stated in Germany from February or March until October (Lehmann, 1972; Michiels, 1999). Probably there are three (or more) generations a year.

MICROHABITAT

The larvae live among mosses on stones (Ertlová, 1970; Lehmann, 1971, 1972).

WATER TYPE

The larvae are collected in upper courses of streams with moderate to fast currents and seem to absent from lowland streams and large rivers (Lehmann, 1971, 1972; Braukmann, 1984: 212, 455; Caspers, 1991; Laville & Vinçon, 1991; Becker, 1995; Bitušik, 2000). This is consistent with the low mean water temperature of 7.7 °C. stated by Rossaro (1991). Ertlová (1970) stated the occurrence of the species in the river Danube (if correctly identified). The identified larvae from southern Limburg (the Netherlands) had been collected in small upper courses between the hills. Reiff (1994) collected the exuviae of one specimen in two lakes in Bavaria.

pH

A preference for calcareous springs is probable (Syrovátka et al., 2012).

SAPROBITY

The species is known almost exclusively from small upper courses in regions with very little pollution. Therefore nothing is known about tolerances, but it is usually considered as intolerant of pollution (e.g. Moog, 1995).

Tvetenia calvescens (Edwards, 1929)

Eukiefferiella lobulifera Thienemann, 1936: 46, 48, 54–55
Eukiefferiella calvescens Ringe, 1974: 236; Lehmann, 1972: 389, 400, figs 70–76

SYSTEMATICS AND IDENTIFICATION

The species resembles *T. bavarica*; see there and the introduction to the genus. Within this species Michiels (2004) distinguished three types with possible species status.

DISTRIBUTION IN EUROPE AND THE NETHERLANDS

The species has been collected throughout Europe (Saether & Spies, 2010). In the Netherlands it is common in southern Limburg and very scarce elsewhere in regions with flowing water (Limnodata.nl; own data).

LIFE CYCLE

Wilson (1977: 12), Tokeshi (1986: 434) and Schmid (1993a: 251) stated a tendency towards multivoltinism. In December and January almost all larvae are in second or third instar, suggesting a diapause (Tokeshi, 1986: fig. 1d; own data). However, many authors also collected low numbers of fourth instar larvae in winter.
The first emergences in western Europe are usually stated in February, but almost everywhere the largest numbers of exuviae and adults are collected in summer or even in autumn (Ringe, 1974; Michiels, 1999; own data). Wilson (1977) and Caspers (1980: 100) stated emergence also in December and January, in England and Germany respectively. These adults were much darker than the spring generation.

FEEDING

Tokeshi (1986: 501) stated fluctuating proportions of diatoms (10 to 80%) in the guts of *T. calvescens*. The larvae also eat much detritus and appear to ingest all the deposited materials present in their environment (see also Schmid, 1993a).

MICROHABITAT

The larvae are common and often numerous on plants and mosses (Thienemann, 1936; Tokeshi & Pinder, 1985; Tokeshi, 1986). They are equally abundant on gravel and stones (Pinder, 1980; own data) and can be found up to 40 cm deep in gravelly substrates (Schmid, 1993a). Schmid also stated a clumped dispersion pattern.
In rearing experiments the larvae initially creep freely around and gradually build more and more tubes, which are loosely constructed of salivary secretions to which algae and detritus are attached (Brennan & McLachlan, 1980; own data).

DENSITIES

Tokeshi (1986) often found 2 or 3 and sometimes up to more than 10 (young) larvae on 10 cm sections of *Myriophyllum spicatum*. Schmid (1993a: 244) stated a maximum of 1825 larvae/dm² in the top 10 cm of the sediment of a gravel stream. We found the highest densities (about 500 larvae/dm²) among mosses on a stone in a fast-flowing brooklet in Westphalia.

WATER TYPE

Current

In brooks and streams with a current velocity of 0.5 to 1 m/s *T. calvescens* is usually the most common species of the genus (Ringe, 1974; Gendron & Laville, 1997; Bitušik, 2000; Orendt, 2002a). In such streams it can be found also in stretches with slower currents. In typical lowland brooks with slower currents the species is absent or almost entirely confined to fish ladders, etc.
In subalpine regions the species also lives in lakes (Reiss, 1984; Reiff, 1995).

Dimensions

T. calvescens is especially numerous in small brooks and streams, but can be common in small rivers too (Klink, 2010). The species is scarce in the lower courses of large rivers (Klink & Moller Pillot, 1982; Wilson & Wilson, 1984; Becker, 1995).

Temperature

The presence of the species in lakes in Bavaria (Reiff, 1995) and in fish ladders in lowland streams indicates that it is not really cold-stenothermic, as suggested by Rossaro (1991).

Shade

The larvae live in open as well as heavily shaded streams (own data).

Permanence

The species seems to be absent from temporary brooks.

pH

A preference for calcareous streams is probable (Syrovátka et al., 2012).

TROPHIC CONDITIONS AND SAPROBITY

In lakes *T. calvescens* seems to be more or less restricted to oligotrophic and mesotrophic conditions (Reiss, 1984; Reiff, 1995; Bitušik, 2000). In streams the larvae can be found when there is no anthropogenic eutrophication at all, but they seem to be more abundant in slightly more eutrophic water (Learner et al., 1971; Wilson, 1987; own data). The numbers decrease with increasing pollution and the larvae are rare in streams with severe organic loading (Bazerque et al., 1989; Wilson, 1989). We collected the species very scarcely where there was severe pollution and only when the current was fast and the oxygen content rather high. Its presence in fish ladders in lowland streams with some organic pollution also suggests that oxygen is particularly important (see table 5 on p. 283).

Tvetenia discoloripes (Goetghebuer & Thienemann, 1936)

Eukiefferiella discoloripes Lehmann, 1972: 390–392, 400, fig. 77

SYSTEMATICS AND IDENTIFICATION

T. discoloripes is closely related to *T. verralli* (Lehmann, 1972: 400). Lehmann accidentally swapped the species names of the pupae of these two species (Cranston, 1982; Langton, 1991). Therefore, we are uncertain about all identifications of this species by Lehmann. Cranston (1982) distinguished the larvae of both species according to the length of the antennal blade; this character seems to be not always reliable (own data). See further under the genus.

DISTRIBUTION IN EUROPE AND THE NETHERLANDS

T. discoloripes has been recorded from many countries throughout Europe (Saether & Spies, 2010). In the Netherlands it is a common species in the provinces of Limburg and Gelderland; it has been collected scarcely in Noord-Brabant, Twente and Flevoland and in the large rivers (Nijboer & Verdonschot, unpublished; Limnodata.nl; Klink & Mulder, 1992; own data).

LIFE CYCLE

Emergence has been stated from February to November (Becker, 1995; own data). Tolkamp (1980: 150) found small larvae most abundantly from October to December. There are too few data to describe hibernation, but probably the whole life cycle resembles that of *T. calvescens*, only with more larvae in fourth instar in winter, and possibly a diapause in early winter in some of the larvae. See also the introduction to the genus.

FEEDING

We found diatoms and detritus in the guts. Most probably the larvae are not very selective feeders, as has been stated in *T. calvescens*.

MICROHABITAT

The larvae are often numerous on plants as well as on stones and wood; they also live on sandy bottoms, but probably in lower numbers (Becker, 1995; own data). In the small stream Snijdersveerbeek, where plants and stones were scarce, Tolkamp (1980) found most larvae between leaves and coarse detritus (as *discoloripes* group). The larvae usually construct tubes, but sometimes they creep freely around (Thienemann, 1936; own observations).

WATER TYPE

Current

As far as is known, *T. discoloripes* lives exclusively in flowing water. It appears to be most common in lowland brooks and streams, where it is found mainly on fish-ladders and other places where the current is relatively fast.. In large parts of the Netherlands it is the only species of the genus. In the fast-flowing brooks and streams (without organic pollution) in southern Limburg and in Bavaria, *T. verralli* is much more abundant (see under the latter species). The tables for current, saprobity and oxygen in Chapter 9 have to be used in combination.

Dimensions

The larvae live in small brooks and streams as well as in rivers (Becker, 1995; Klink, 2010; own data).

SAPROBITY AND OXYGEN

The larvae seem to be more common following pollution with sewage effluent (Wilson, 1989; Maasri et al., 2008; own data). We found them in small numbers in a severely polluted fast-flowing stream (Jeker) in southern Limburg with a BOD of sometimes 10 or more, but always rather rich in oxygen. The preference for riffles or fish ladders in lowland streams is also an indication of a need for much oxygen. We found the larvae sometimes in large numbers in brooks without organic pollution; in such cases much natural decomposing material was usually present (see also Tolkamp, 1980: 150).

Tvetenia tshernovskii (Pankratova, 1968)

Eukiefferiella vitracies Saether, 1969: 49–53
Eukiefferiella tshernovskii Pankratova, 1970: 165–166, fig. 96
Tvetenia sp. A Cranston, 1982: 74–75, fig. 28a; Schmid, 1993: 204, 208, 490–491, fig. 147

IDENTIFICATION

Male, female and pupa are absent from most keys; a description can be found in Saether (1969: 49–53). The larvae can be identified using Cranston (1982) or Schmid (1993). According to Przhiboro & Saether (2010: 61), in most larvae the median hump of the central mental tooth as figured in Pankratova (1970) is lacking. Przhiboro & Saether (ibid.) give additions to the description by Schmid (1993). An important character of *T. tshernovskii* is that the third antennal segment is subequal to or longer than the fourth segment.

DISTRIBUTION IN EUROPE

The species has been found in a number of countries in western and eastern Europe (Przhiboro & Saether, 2010). There are some unverified records from large rivers in the Netherlands.

FEEDING

In the guts of fourth instar larvae Przhiboro & Saether (2010) found mainly detritus, but also sand grains, diatoms and filamentous algae. In two specimens diatoms predominated.

MICROHABITAT

Pankratova (1968) collected the larvae in large numbers from stones and scarcely from sand on the bottom of the river Neva in Russia. Schmid (1993) noted the occurrence of larvae in the surface layer (0–10 cm) of bed sediments in the Danube in Austria.

WATER TYPE

The larvae have been collected in streams and rivers, in Russia also in lakes (Przhiboro & Saether, 2010).

Tvetenia verralli (Edwards, 1929)

Eukiefferiella verralli Lehmann, 1972: 392–394, 400, figs 78–84; Lindegaard et al., 1975: 116, 118, 130, 132

SYSTEMATICS AND IDENTIFICATION

The species is related to *T. dicoloripes*. See there and under the genus for identification and possible mistakes in the past.

DISTRIBUTION IN EUROPE AND THE NETHERLANDS

T. verralli has been recorded from many countries throughout Europe (Saether & Spies, 2010). In the Netherlands it has been collected in southern Limburg and in the large rivers (Limnodata.nl; own data).

LIFE CYCLE

Emergence has been stated mainly from March to October, with the largest numbers in late spring or in summer (Ringe, 1974; Lindegaard & Mortensen, 1988; Michiels, 1999). Wilson (1977) collected exuviae in England throughout the year. Probably there are at least three generations. In winter in southern Limburg (the Netherlands) we found about the same numbers of larvae in second, third and fourth instar (cf. Lindegaard & Mortensen, 1988, who probably missed most second instar larvae). See also the introduction to the genus.

MICROHABITAT

Schmid (1993) found the larvae in all lotic habitats. We collected the larvae from plants and stones, but not from open bottoms. Lindegaard et al. (1975) collected the larvae in a Danish spring among moss covered with water, but mainly on stones in the rills and in the borders of the moss carpet. The larvae live in loosely constructed tubes of salivary secretions to which small quantities of detritus are attached (Brennan & McLachlan, 1980).

DENSITY

Lindegaard & Mortensen (1988) measured a mean density of 5822 larvae/m^2 and estimated 7800 larvae/m^2 in a spring brook in Denmark.

WATER TYPE

Current

Apart from some exuviae collected in lakes in Bavaria (Reiff, 1995), all records are from flowing water. We never found the species in slow-flowing lowland brooks in the Netherlands and adjacent parts of Germany, but current velocity seems to play a secondary role in determining whether *verralli* or *discoloripes* will be more numerous (see table 7 on p. 287). *T. verralli* is the most common species (of these two) in the fast-flowing streams in Bavaria and the Pyrenees (Laville & Vinçon, 1991; Orendt, 2002a, with a mistake in column 4) and we stated the same in southern Limburg in the Netherlands.
Because of mistakes in identifying the exuviae of both species in the past (see under *T. discoloripes*), it is not always possible to use the literature from before 1984.

Dimensions

The species lives in springs, brooks and streams and can sometimes also be found in rivers. Becker (1995) collected *T. verralli* only once along the river Rhine, and Klink (2010) found the species in the upper courses of the Seine only scarcely, considerably less than *T. dicoloripes*.

Temperature
Rossaro (1991) stated a mean water temperature of 17.5 °C at locations in Italian streams inhabited by this species; this is much higher than in other rheophilous species.

SAPROBITY AND OXYGEN
T. verralli has hardly ever been collected in polluted streams, and more frequently than *T. discoloripes* in natural streams (see table 5 in Chapter 9). Its preference for faster currents also indicates a high oxygen demand. Rossaro (1991) stated a much higher mean water temperature for *verralli* than for *calvescens*, which increases its need for oxygen.

Zalutschia Lipina, 1939

SYSTEMATICS AND IDENTIFICATION
The genus *Zalutschia* is closely related to *Hydrobaenus* and *Trissocladius* and most species have been described and keyed as *Trissocladius* (e.g. in *Pankratova*, 1970: 134–245). Most species can be identified as male, female, pupa and larva using Saether (1976: 186–193). However, the only Dutch species, *Z. humphriesiae*, has been described some years later (Dowling & Murray, 1980).

DISTRIBUTION IN EUROPE
The genus is rather rich in species in the northern part of the Holarctic. Some species can be found further south in eastern Europe (mainly only known from Russia) or even to the Czech Republic (*Z. tatrica*, *Z. potamophilus*, see Pankratova, 1970 and Saether, 1976) or Serbia (*Z. mucronata*). *Z. zalutschicola* lives further south into central Russia, Denmark and Germany. One unidentified species has been collected in France (coll. British Museum). *Z. humphriesiae* is only known from western Europe. (Saether & Spies, 2010).

WATER TYPE
Almost all species of *Zalutschia* are confined to oligotrophic or mesotrophic to polyhumic, often acid lakes and pools (Saether, 1976, 1979).

OR

Zalutschia humphriesiae Dowling & Murray 1980

Zalutschia sp. Ireland Saether, 1976: 187

IDENTIFICATION
The adult male, pupa and larva have been described by Dowling & Murray (1980) and can be identified using keys published after that year. For comparison with related species, see the introduction of the genus.

DISTRIBUTION IN EUROPE AND THE NETHERLANDS
Z. humphriesiae has been collected in Spain, Portugal, Britain, Ireland and the Netherlands,
and possibly in Finland (Saether & Spies, 2010; Paasivirta, 2012). In the Netherlands the species is not rare on acid soils in the whole Pleistocene part of the country (cf. Moller Pillot & Buskens, 1990).

LIFE CYCLE

In Ireland Dowling & Murray (1980) collected adults from February to May. We also found exuviae and adults mainly from March to May, and fourth instar larvae from October to May (mainly in spring); exuviae were collected scarcely in autumn and we found the exuviae of one specimen in July. Most probably the species has a dormancy period in summer in second and/or third instar, as described in *Hydrobaenus* (see there). Possibly there is usually only one spring generation and sometimes also an autumn generation. Two generations in spring cannot be excluded (cf. *Hydrobaenus*).

MICROHABITAT

Dowling & Murray (1980) found the larvae among *Sphagnum* and filamentous algae. We collected them mainly on organic bottoms.

WATER TYPE AND pH

Almost all data apply to acid stagnant water, usually bog or moorland pools (Dowling & Murray, 1980; Klink, 1986; Duursema, 1996; Ruse, 2002; own data). In many cases the pools dried up in summer (sometimes partly). As in *Hydrobaenus* (see there), desiccation in summer is probably not a precondition for survival, although the species seems to be rare in permanent water bodies. Moller Pillot (2003) also found the larvae and exuviae numerously in an upper course flowing from such pools, in wet years even more than one kilometre downstream. The species seemed to survive all year round in this upper course, probably only in temporary stretches without a strong current.

DISPERSAL

The larvae have been collected relatively often in drift samples (see Moller Pillot, 2003).

Zalutschia mucronata (Brundin, 1949)

Zalutschia mucronata lives in Scandinavia and eastern Europe, westwards to Poland, Hungary and Serbia (Saether & Spies, 2010). The mention of a record from the Netherlands in Beuk (2002) was based on a mistake.
The larvae live in acid, polyhumic lakes (Brundin, 1949; Saether, 1976, 1979; Mossberg & Nyberg, 1980: 85). Other records have to be verified.

Zalutschia zalutschicola Lipina 1939

The species is distributed in northern Europe, extending southwards to central Russia and Germany. The larvae live in the littoral and profundal zones of oligotrophic or mesotrophic to polyhumic lakes (Saether, 1976, 1979; Mossberg & Nyberg, 1980). Eriksson & Johnson (2000) stated the presence of larvae only after the pH decreased from 5.9 to 4.7 between 1800 and 1975, but in Finnish lakes they live at a pH from 6.4 to 7.4 (Särkkä, 1983; Palomäki, 1989).

4 SYSTEMATICS, BIOLOGY AND ECOLOGY OF THE PRODIAMESINAE

The subfamily Prodiamesinae is closely related to the Diamesinae and Orthocladiinae (Saether, 2000). In the older literature the subfamily was included in the Orthocladiinae (e.g. Brundin, 1949; Lehmann, 1971). This practice was continued in Moller Pillot (1984), because it made identification of larvae a little easier. In this book we treat the subfamilies in different chapters, but we place the species of all subfamilies alphabetically together in the tables in Chapter 9.

In Europe the subfamily contains only four genera: *Monodiamesa*, *Odontomesa*, *Prodiamesa* and *Propsilocerus*. The last genus was included in the Orthocladiinae until 2012.

Monodiamesa Kieffer, 1922

The five European species have been listed by Langton & Visser (2003) and Ashe & O'Connor (2009). Three species are known as pupa (Langton & Visser, 2003). Only one species, *M. bathyphila*, can be expected in the lowland of western Europe. The larvae of this species can be identified by the relatively large head capsule: 0.69–0.75 mm; in *M. alpicola*, *M. ekmani* and *M. nitida* the head length is less than 0.65 mm (Schmid, 1993; Langton & McLarnon, 1998). The larva of *M. nigra* is unknown. Only *M. bathypila* and *M. ekmani* are treated here.

Monodiamesa bathyphila (Kieffer, 1918)

Prodiamesa bathyphila Pankratova, 1970: 109–111, fig. 56

DISTRIBUTION IN EUROPE AND THE NETHERLANDS

M. bathyphila has been recorded throughout most of Europe, but is possibly absent from Spain and Portugal (Saether & Spies, 2010). In the Netherlands the species has been found fossil in river sediments (Klink, 1989) and seems to have become almost extinct. In recent times the larvae have been found only in a small stream near Diepenheim and Goor (Klink & Moller Pillot, 1996; Limnodata.nl).

FEEDING

According to Brundin (1942: 90), the larvae feed on planktonic detritus and not on remains of plants. Paasivirta (1974) suggested that the larvae are predators, also feeding on bacteria. This seems more probable for larvae living in brooks and streams.

MICROHABITAT

The larvae are bottom dwellers. In the profundal zone of lakes they live on organic silt, often with sand (Brundin, 1949, cf. Särkkä, 1983). In streams they live mainly in sandy bottoms, sometimes with stones or gravel or on clayish substrate (Pankratova, 1964; Zinchenko, 2002; Janzen, 2003).

WATER TYPE

Current and dimensions

In Russia the larvae are common bottom dwellers in lowland brooks, streams and rivers with a current speed of 7 to 50 cm/sec (Pankratova, 1964; Zinchenko, 2002). They have also been scarcely collected in the mid-river bed of the river Warta in Poland (Grzybkowska & Dukowska, 2002) and in the river Danube (Schmid, 1993; Bitušik, 2000). Janzen (2003) found the species in several sandy lowland streams in Germany and there are some records from a lowland stream in the Netherlands (see Distribution).

Stagnant water

M. bathyphila is an inhabitant of the profundal zone of cold oligotrophic and mesotrophic lakes (Brundin, 1949; Saether, 1979), in some cases living in the sublittoral zone (e.g. Särkkä, 1983). In many cases the larvae are most common in the epiprofundal and scarce in deeper parts of the lake (e.g. Paasivirta, 1974: fig. 2).

TROPHIC CONDITIONS AND SAPROBITY

According to Brundin (1942: 90), the larvae can be rather common in the profundal of moderately eutrophic lakes, but they are absent from most polyhumic lakes, because they require a high oxygen supply and feed on planktonic detritus. Their occurrence is positively correlated with oxygen content. It is not clear why the larvae are confined to oligotrophic environments in the Nearctic. They can occur abundantly in mesotrophic lakes in Europe (Moore, 1979a: 319).
In western European streams the species seems to be almost extinct because of its high demands on oxygen content.

Monodiamesa ekmani Brundin, 1949

This species has been collected in montane lakes and streams in northern Europe, Scotland, Ireland, the Alps and the Pyrenees. The adult male is keyed in Langton & Pinder (2007); the pupa keys out as *Monodiamesa* Pe 1a in Langton (1991). The female, pupa and larva are described in Langton & McLarnon (1998). These authors suggest that the larvae require an undisturbed sandy substratum and year-round high oxygenation.

Odontomesa fulva (Kieffer, 1919)

SYSTEMATICS AND IDENTIFICATION

O. fulva is the only species of the genus in Europe. Identification presents no problems in all stages. The female has been described by Saether (1985c). The larval head capsule is unique in nearly all features.

DISTRIBUTION IN EUROPE AND THE NETHERLANDS

The species is distributed throughout Europe, but is absent from Ireland (Saether & Spies, 2010). In the Netherlands it is widely distributed in Pleistocene areas and in southern Limburg, but very scarce in the Holocene areas (Limnodata.nl), for example in Flevoland (Klink & Mulder, 1992).

LIFE CYCLE

Adults emerge in the Netherlands from March to October, in Germany from April to October and in Russia only from the end of May until the end of August (Reiss, 1968; Lehmann, 1971; Shilova, 1976; Reiff, 1994; own data). Shilova (1966, 1976) and Pankratova (1970: 28) stated two generations in Russia, but Pinder (1983) suggested five generations in England. In November and December Pinder found only larvae in second instar, but we collected the fourth instar in early December and large numbers of third and fourth instar larvae in early January. Nevertheless, a short diapause in autumn cannot be excluded.

FEEDING

Shilova (1966, 1976) described in detail how the larvae feed. Their whole behaviour is unique among chironomids. They are passive filter-feeders, taking in much water and pushing its out by peristaltic movements. The food (mainly unicellular algae) is retained by the specialised mouthparts.

MICROHABITAT

The larvae live in sandy bottoms, often with organic silt (e.g. Lehmann, 1971; Shilova, 1976; Srokosz, 1980; Pinder, 1980; Verdonschot & Lengkeek, 2006). Young larvae were scarcely found at a depth of 10–50 cm in the sediment (Schmid, 1993). As a rule they are free-living, without a tube (Shilova, 1976). Given the manner of feeding, variation in the stream bottom substrate may be important in periods of low as well as high discharge (see under Dispersal).

WATER TYPE

Current

The larvae live in sandy bottoms and are therefore most common in lowland brooks and streams. However, they are also quite common in less fast-flowing stretches of foothill streams, and even sometimes at such stretches in montane brooks (Lehmann, 1971; Gendron & Laville, 1997; Bitušik, 2000; Orendt, 2002a; Janzen, 2003; Limnodata.nl). See table 7 in Chapter 9.

The species is rare in stagnant water. Most records from stagnant water apply to lakes or canals (e.g. Reiff, 1994; Limnodata.nl).

Dimensions

The species is mainly common in small brooks and streams and scarcely found in small rivers; it seems to be absent from large rivers.

Shade

The larvae are scarcely collected in shaded stretches of brooks and streams.

Permanence

There are very few records from temporary water. One reason for this may be because most temporary brooks and ditches in the Netherlands have hardly any substrate variation and therefore may not provide the required microhabitat.

pH

O. fulva was not rare in a small Dutch brook (Roodloop) at pH values around 6 (Moller Pillot, 2003). Most other data apply to water with pH values of 7 or more. It is probable that the food conditions are most important and not the acidity itself.

TROPHIC CONDITIONS AND SAPROBITY

The species is absent or hardly known from oligotrophic water. In brooks and streams it can be found in oligosaprobic water, but it appears to be more common in water with some organic loading (BOD1–3) and has been collected sometimes at a BOD of around 5. A fall in oxygen content at night below 5 mg/l appears to be no problem, but very low oxygen values have not been stated (Bazerque et al., 1989; Wilson & Ruse, 2005; Limnodata.nl; own data). It is probable that the special manner of feeding and microhabitat availability are determining factors for the presence of the larvae, and both can be disturbed by pollution. The species will be more susceptible to pollution in stagnant water. See further under Microhabitat and Shade.

DISPERSAL

Shilova (1976) observed that the larvae often crept around. Moller Pillot (2003) found many more *O. fulva* larvae in drift samples than larvae of many other species. The manner of feeding seems to require an appropriate combination of microhabitat and current velocity.

Prodiamesa Kieffer, 1906

Four species have been described in Europe, two of which occur in the Netherlands and adjacent lowlands. *P. bureschi* Michailova 1977 is only known from Hungary and Bulgaria and *P. delphinensis* from the alpine region in central Europe; both are not treated here. Identification of all stages of *P. olivacea* does not present any problems. The female has been described by Saether (1985b). The characteristic intestinal canal has been described by Shilova & Kurazhskovskaya (1980). For the other species see under *P. rufovittata*.

Prodiamesa olivacea (Meigen, 1818)

IDENTIFICATION AND MORPHOLOGY

The species can be easily identified in all stages (see above). More than in other chironomid species, the size of the head capsule is very variable, probably under the influence of environmental factors. Schmid (1993) measured a head length of 338–586 μm in third instar larvae and 652–866 μm in fourth instar larvae.

DISTRIBUTION IN EUROPE AND THE NETHERLANDS

P. olivacea is distributed throughout Europe (Saether & Spies, 2010). In the Netherlands it is widely distributed in the Pleistocene area, in southern Limburg and in the dune region; elsewhere it is a rare species (Moller Pillot & Buskens, 1990; Steenbergen, 1993; Limnodata.nl).

LIFE CYCLE

Emergence has been stated from March until October or November, very rarely in winter. As a rule there are two generations, emerging in spring and late summer (e.g. Wilson, 1977; Pinder, 1983; Janssens de Bisthoven et al., 1992). However, relatively often some adults emerge between these two emergence periods. A small third generation in this period may be possible (see *Pseudodiamesa branickii*). Schmid (1992a) stated that some larvae of the late summer generation do not pupate in autumn and emerge the following spring.
In many cases one generation is absent or very small. This may be caused by poor oxygen supply (no current in summer) or the absence of larvae and their food as a result of spates in winter. Often it is not clear why there is only one generation (Pankratova, 1970; Moller Pillot, 1971, 2003; Pinder, 1983).

OVIPOSITION

The eggs are deposited on macrophytes at the water's edge. The egg masses contain approximately 800–2000 eggs (Nolte, 1993).

FEEDING

The most important food appears to be decomposing organic material rich in bacteria. The larvae feed on coarse organic material, but also on fine detritus or on the sheathed bacterium *Sphaerotilus natans* (Crisp & Lloyd, 1954; Janković, 1974; Mackey, 1976b). According to Janković, they are not able to feed by filtration. To a lesser extent they also consume algae and small animals (Konstantinov, 1958; Williams & Hynes, 1974; Balushkina, 1987). Williams & Hynes stated more carnivorous feeding possibly because they investigated bottoms without much decomposing material (see Microhabitat).

MICROHABITAT

The larvae are almost entirely bottom inhabitants and are rarely found on stones or among vegetation (e.g. Srokosz, 1980; Verdonschot & Lengkeek, 2006). They creep freely around and make no tube. They can be numerous in organic silt, but Tolkamp (1980: 142–144) stated a significant over-representation of the larvae in sand with coarse and fine detritus. Williams & Hynes (1974) found fairly large specimens for most of the year to depths of 20 and 30 cm in a gravelly bottoms, but no deeper, probably because aerated surface water regularly penetrated to this depth, but no further than 30 cm.

DENSITIES

High densities can be found only where much decomposing organic material is available (coarse detritus and silt). Balushkina (1987) and Izvekova et al. (1996) stated at such localities in Russian streams mean densities of about 4000 larvae/ m^2 . Janssens de Bisthoven et al. (1992) found up to more than 30,000 larvae/m^2 in October, more than 20,000 of which in third instar. However, such densities are very rare.

WATER TYPE

Current

The larvae are common in flowing water and scarce in stagnant water. They can be found in lowland as well in montane streams, but thrive well only in stretches with a slow flow, where enough decomposing material and silt is available. Because the oxygen supply is important (see Saprobity), survival is determined by a combination of food availability, current velocity and temperature. Table 7 in Chapter 9 gives the mean current velocity of the stream, not at the place where the larvae live.
In stagnant water the larvae are mainly found in lakes with more or less stable oxygen content, but sometimes also in unstable water (e.g. Waginger See: Reiff, 1994). Kreuzer (1940) and Schleuter (1985) even collected a larva in a woodland pool. In the Netherlands there are very few records from stagnant water.

Dimensions

The larvae live in narrow brooks as well as in streams, but are scarce in large rivers (see under Current). In springs and small brooks they can be numerous when there is only a very shallow layer of water so that diffusion of oxygen is no problem.

TEMPERATURE

Janković (1974) found the highest feeding rate, assimilation efficiency and weight increase at 10 °C. At higher temperatures the feeding intensity decreased. These larvae were reared in Petri dishes at a population density of 3000 larvae/m^2 and therefore the oxygen supply will have played a role: the oxygen demand of the larvae rises with temperature (see below and in sections 2.10 and 2.15). Konstantinov (1971) found the highest oxygen consumption at 20 °C.

Shade
P. olivacea lives in shaded and unshaded brooks and streams, most probably dependent only on the presence of quickly decomposing organic material.

Permanence
The species easily settles in temporary habitats by oviposition (see Dispersal below). Because of the distance from permanent populations or the absence of water during the flying period in late summer, the larvae are usually absent from suitable temporary books and streams, unless larvae can survive during dry periods in the bottom (Moller Pillot, 2003).

pH
P. olivacea is most common and numerous in circumneutral or slightly alkaline water (Raddum & Saether, 1981; Steenbergen, 1993; Orendt, 1999; Janzen, 2003; Limnodata.nl). However, the larvae have been collected in large numbers at a pH of 6 and sometimes even at a pH of 4.2 to 5 (Hall, 1951; Hawczak et al., 2009; Janzen, 2003; Moller Pillot, 2003). It is very probable that the larvae are tolerant of acid conditions, but that they are more or less dependent on bacterial activity for feeding. It is not known if they can complete their life cycle in very acid water.

SAPROBITY AND OXYGEN
P. olivacea is often considered to be a species very tolerant of severe organic pollution or even characteristic of polluted water (Peters et al., 1988; Moog, 1995; Wilson & Ruse, 2005). However, the species is usually most numerous in slightly to moderately polluted water and often absent from the most polluted stretches of a stream (Moller Pillot, 1971; Wilson, 1987; Bazerque et al., 1989). In more than 3500 samples in Dutch brooks and streams the mean BOD was 2.33 and the mean ammonium-N content was 0.36 mg N/l (Limnodata.nl). In dune brooks in the province of Noord-Holland the larvae are most common in less eutrophic water (Steenbergen, 1993).
The occurrence of the species can only be understood when the saprobity in its proper sense (intensity of decomposition) and oxygen supply are looked at together. The larvae prefer places with much decomposing organic material, but also require a good oxygen supply. Walshe (1948) stated that the larvae cannot survive anaerobiosis, in contrast to some other chironomids such as *Chironomus*, *Psectrotanypus* and *Tanypus*. Konstantinov (1971) observed that fourth instar larvae have to make undulating movements all the time at an oxygen content of 1.4 mg/l and that they stop these movements at a concentration of 1 mg/l. As a consequence, the larvae can live in decomposing organic mud with a thin layer of flowing water (Crisp & Lloyd, 1954), but not at a depth of 30 cm in a slow-flowing stream with hardly or no oxygen at night (Moller Pillot, 1971). In winter and spring this is usually no problem, because the oxygen content is higher and the oxygen demand of the larvae is lower (see under Water type). Another consequence is that the larvae prefer higher levels of pollution when the current velocity is higher. In table 5 in Chapter 9 saprobity and oxygen must be looked at together, which clearly shows that high saprobity in combination with a low oxygen content is unsuitable.

SALINITY

The larvae have been found rarely in slightly brackish water (Hall, 1951; Steenbergen, 1993), most probably because flowing water is rarely brackish. Stagnant brackish water appears to be usually unsuitable.

DISPERSAL

Moller Pillot (2003) observed that brooks are often repopulated via the air; apparently females easily bridge distances of 500 m. The larvae enter drift during the whole year, in winter probably under the influence of high current velocities or shortage of food, in summer mainly at low discharge, probably under the influence of oxygen shortage.

Prodiamesa rufovittata Goetghebuer, 1932

IDENTIFICATION

The adult male is keyed in Pinder (1978) and Langton & Pinder (2007). The pupa cannot be distinguished from that of *P. delphinensis* and both are named *Prodiamesa* Pe 1 in Langton (1991) and Langton & Visser (2003). The larva is absent from Moller Pillot (1984), but can be identified using Ferrarese & Rossaro (1981), Schmid (1993), Bitušik (2000) or Janecek (2007). Only Schmid and Janecek also key the larva of (cf.) *delphinensis*.

DISTRIBUTION IN EUROPE AND THE NETHERLANDS

The species lives in large parts of Europe, but is absent from northern Europe, Ireland and probably some parts of the Mediterranean area (Saether & Spies, 2010). In the Netherlands it is known from only three localities: in Limburg (Hemelbeek) and Twente (Springendal, Ruenbergerbeek); other records have to be verified (Limnodata.nl; own data).

MICROHABITAT

Most larvae have been collected from sandy bottoms with some organic silt (Lindegaard-Petersen, 1972; Ferrarese & Rossaro, 1981). A larva in Springendal (the Netherlands) was found in coarse organic material.

WATER TYPE

The larvae live in rather fast-flowing water, in narrow lowland and foothill brooks (Springendal and Hemelbeek in the Netherlands) as well as in subalpine streams (Inn and Salzach in Austria) and large rivers (Po and Brenta in Italy: see Ferrarese & Rossaro, 1981: 91).

SAPROBITY

The three localities in the Netherlands are very clean brooks with very little or no organic pollution. Balushkina (1987) collected many larvae in the unpolluted upper course of the Izhor stream near St Petersburg, but also some in the severely polluted lower course. According to Moog (1995), the species can live also in α-mesosaprobic water. It is possible that the larvae only require a very good oxygen supply or that they have been transported to polluted water by drift. In any case, it is not clear why *P. rufovittata* is a rare species everywhere.

Propsilocerus lacustris Kieffer, 1923

Orthocladiinae gen? *orielica* Chernovskij, 1949: 139–140
Propsilocerus orielicus Pankratova, 1970: 116–117, fig. 61; Shcherbina, 1989: 298–299
Propsilocerus lusatiensis Cranston et al., 1983: fig. 9.60; Langton, 1991: 96

SYSTEMATICS AND IDENTIFICATION

The genus *Propsilocerus* was previously placed in the subfamily Orthocladiinae. Cranston et al. (2012) stated on the basis of molecular gene investigations that it belongs to the Prodiamesinae. In Europe three species have been described, of which *P. jacuticus* and *P. paradoxus* have been found only in northern and eastern Europe; these two species are not treated here.
Identification of all stages to species level is possible using Saether & Wang (1996); for the larva of *P. paradoxus* see Wang & Saether (2001), for *P. jacuticus* see Saether (1997). The larva of *P. lacustris* is figured in Pankratova (1970) and Cranston et al. (1983).

DISTRIBUTION IN EUROPE

P. lacustris has been recorded in eastern and central Europe, westward to northern Germany (Saether & Spies, 2010). There are no recent records from the Netherlands.

LIFE CYCLE AND MATING

There is only one generation a year, with emergence in spring. The maximal growth of the larvae takes place in autumn and winter. The adults do not swarm and mate on or near the water body (Shcherbina, 1989; Saether & Wang, 1996).

ECOLOGY

The ecology has been described by Shcherbina (1989). The larvae live in bottom sediments in the littoral zone of lakes and in pools, in eastern Europe also in streams. They have been collected also in brackish water (Pankratova, 1970; Cranston et al., 1983; Janecek, 2007).

5 SYSTEMATICS, BIOLOGY AND ECOLOGY OF THE DIAMESINAE

The subfamily Diamesinae is closely related to the Prodiamesinae and Orthocladiinae (Saether, 2000). For practical reasons these three subfamilies have been treated in one key by Pankratova (1970) and Moller Pillot (1984), because it is not easy to separate the larvae of these subfamilies, especially under low magnification. For a key to subfamilies see Cranston & Reiss (1983). The head of most Diamesinae larvae has a strongly contrasting black occipital margin; the third antennal segment is usually annulated.
Although we treat the subfamilies in different chapters, we place the species of all subfamilies alphabetically together in the tables in Chapter 9.
Within the Diamesinae, three tribes are represented in Europe: Boreoheptagyiini, Diamesini and Protanypodini (Oliver, 1983; Ashe & O'Connor, 2009). In this chapter all species are treated in alphabetical order.
The life cycle and feeding of Diamesinae larvae do not show joint peculiarities and are treated per species. Most Diamesinae inhabit flowing water. The number of species is greatest in glacial brooks and alpine streams and decreases downstream (Lindegaard & Brodersen, 1995). Only a few species live in lakes and lowland streams (e.g. *Protanypus*, *Potthastia*).

Boreoheptagyia Brundin, 1966

SYSTEMATICS AND IDENTIFICATION

Boreoheptagyia is the only genus in the tribe Boreoheptagyiini. The eight European species are listed in Langton & Visser (2003). Keys to most adult males, pupae and larvae are given in Serra-Tosio (1989a). Exuviae are also keyed in Langton (1991). A number of larvae have been keyed by Bitušik (2000).

DISTRIBUTION IN EUROPE

The genus has been recorded throughout the central and southern parts of Europe as far north as Germany and Poland. It is absent from most lowland regions, such as Denmark and the Netherlands (Saether & Spies, 2010). It is probable that *B. legeri* is the only species of the genus that lives outside montane areas (see e.g. Serra-Tosio & Laville, 1991).

ECOLOGY

The larvae live in rheomadicolous habitats: places where a constant rain of water falls onto stones, especially in the lowest parts of waterfalls, more rarely on such places along streams (Oliver, 1983; Serra-Tosio, 1989). The genus has been recorded almost exclusively in mountain regions, but it can also be found rarely in foothill streams (Serra-Tosio & Laville, 1991; Janzen, 2003).

Diamesa Meigen, 1835

The genus *Diamesa* is very rich in species in Europe; for a complete list see Langton

& Visser (2003) or Ashe & O'Connor (2009). Identification of larvae to species level is usually not possible; only *D. insignipes* can be more or less reliably identified. The most useful keys to larvae are by Schmid (1993), Bitušik (2000) and Janecek (2007). Identification of exuviae is often difficult; Langton (1991) has keyed a large number of species.

Most species live only in springs and brooks in high mountains and are more or less cold-stenothermic. In the West European lowland few species can be expected; only three of them are treated here. Other species may have been present in the past (e.g. Klink, 1989). Several other species have been found in the Russian lowland (Zinchenko, 2002).

The guts contents of the larvae usually contain mainly diatoms, but in many cases much detritus has also been consumed (Kawecka & Kownacki, 1974; Kawecka et al., 1978; see also Moller Pillot, 1984: 67; Moog, 1995).

Diamesa insignipes Kieffer, 1908

DISTRIBUTION IN EUROPE AND THE NETHERLANDS

D. insignipes lives throughout Europe, except for Scandinavia and the southwestern part of Finland (Saether & Spies, 2010; Paasivirta, 2012). It can be found in larger areas of the European lowland than any other species of the genus (Thienemann, 1952; Braukmann, 1984: 211; Serra-Tosio & Laville, 1991). In the Netherlands it is almost confined to the foothill brooks and streams in southern Limburg, but there are controlled records from lowland brooks near Barneveld and in Twente (Limnodata.nl; Moller Pillot & Buskens, 1990: 26, 59; R. Heusinkveld, unpublished).

LIFE CYCLE

Many authors stated two emergence periods: in early spring and in summer or autumn. However, fourth instar larvae are usually collected most numerously in May. It is very probable that larvae have been missed often in winter; moreover, one generation can be small in number (Pankratova, 1970: 28; Lehmann, 1971; Lindegaard-Petersen, 1972; Ringe, 1974; Pinder, 1980; Nolte, 1989, 1991; own data).

OVIPOSITION

Nolte (1989) found ovipositing females and strings of eggs frequently near the source of the Breitenbach in Germany. Artificial substrata some 500 m downstream were not accepted. The eggs were laid on substrata at the water surface, such as roots, blades of grass, stones or mosses.

MICROHABITAT

The larvae are found mainly on stones or gravel, but in many cases also on plants (e.g. Lindegaard-Petersen, 1972; Pinder, 1980; own data). Nolte (1989) found the larvae in large numbers on the submersed part of stones in moss cushions, especially on the anterior side, where less detritus was deposited. Small larvae were found mainly within the moss cushions, where they lived immobile between leaf and stem of the mosses. Larger larvae of the same species lived more on stone surfaces in more lotic environments.

WATER TYPE

The larvae live in springs and fast-flowing brooks and streams; they are absent from large rivers (Lehmann, 1971; Kownacka & Kownacki, 1972; Braukmann, 1984; Brabec, 2000). They can be found at current velocities lower than 50 cm/sec only in the neighbourhood of cold springs. There are very few records from lakes (Langton, 1991; Reiff, 1994). The species is not pronouncedly cold-stenothermic and tolerates some organic pollution (Ferrarese & Rossaro, 1981; Moog, 1995).

Diamesa permacra (Walker, 1856)

Diamesa permacer Lehmann, 1971: 476; Schmid, 1993: 40, fig. 13 (incorrect spelling)
? *Syndiamesa hygropetrica* Moller Pillot, 1984: 17, 157, fig. VI.47; Moller Pillot & Buskens, 1990: 15, 31, 76

SYSTEMATICS AND IDENTIFICATION

D. permacra belongs to the *dampfi* group. The larvae of this group have a well developed procercus with usually 6 anal setae. Sometimes it can be difficult to distinguish these larvae from those of *Syndiamesa* (see there). The larva of *D. permacra* has been described and keyed by Schmid (1993), Bitušik (2000) and Janecek (2007).

DISTRIBUTION IN EUROPE AND THE NETHERLANDS

The species has been collected in many European countries, but there are no records from eastern Europe. In contrast to most other *Diamesa* species, it has been found also in lowland regions, such as Denmark (Saether & Spies, 2010). The records from the Netherlands (southern Limburg, Veluwe) published as *Syndiamesa hygropetrica* (see above; identified as larvae) belong to the *Diamesa dampfi* group and probably to *D. permacra*. We collected at least one other species of the *dampfi* group in adjacent parts of Germany.

ECOLOGY

D. permacra has been collected mainly in springs and brooks in montane and upland regions, often lower than 500 m asl (e.g. Lehmann, 1971; Serra-Tosio & Laville, 1991; Langton, 1991). See also Distribution. According to Rossaro (1991), the larvae are cold-stenothermic, but the place in the DCA ordination diagram by Lindegaard & Brodersen (1995: fig. 5) indicates no direct relation with low temperatures.

Diamesa tonsa (Haliday, 1856)

Diamesa thienemanni Pankratova, 1970: 79–80; Lehmann, 1971: 476–477

SYSTEMATICS AND IDENTIFICATION

D. tonsa belongs to the *cinerella* group. The larvae of this group have been described and figured by Schmid (1993), but identification to species is only possible for exuviae or adult males.

DISTRIBUTION IN EUROPE

D. tonsa has been collected throughout Europe, even in lowland regions such as Denmark and lower-lying parts of Germany and France (Serra-Tosio & Laville, 1991; Saether & Spies, 2010). There are no records from the Netherlands. The species appears to have a good dispersal capacity, because it is present on several islands, even on the Faeroes (as *D. camptoneura*: Ashe & O'Connor, 2009: 287).

ECOLOGY

The larvae live on stones in springs, brooks and streams. They are found in montane as well as in foothill streams (e.g. Braukmann, 1984: 211, cf. 455; Laville & Vinçon, 1991; Michiels, 2004) and more rarely in upland lakes (Lehmann, 1971; Reiff, 1994). In contrast to many other species of the genus, *D. tonsa* occurs mainly in streams with some organic pollution (Ferrarese & Rossaro, 1981; Moog, 1995).

Potthastia Kieffer, 1922

SYSTEMATICS AND IDENTIFICATION

Four *Potthastia* species have been described in Europe. However, Langton (1991) found at least five different exuviae. Within the genus two groups can be distinguished: the *gaedii* group and the *longimanus* group; these groups are only clearly different in the larvae. The larvae of the *gaedii* group resemble those of *Sympotthastia*: see there.

ECOLOGY

All species of *Potthastia* live mainly in flowing water. In contrast to most other Diamesinae they are not cold-stenothermic (Ferrarese & Rossaro, 1981). Emergence starts rarely in early spring and the larvae develop more in summer than in winter months.

Potthastia gaedii (Meigen, 1838)

IDENTIFICATION

Adult males and pupae are keyed in all existing keys. Because the larvae of related species are still unknown, identification of larvae can be a problem in regions where other species can occur.

DISTRIBUTION IN EUROPE AND THE NETHERLANDS

P. gaedii has been recorded throughout Europe (Saether & Spies, 2010). In the Netherlands the larvae have been collected scarcely in the large rivers and in some brooks and streams in Limburg and the southeastern part of Noord-Brabant (Limnodata.nl; own data).

LIFE CYCLE

Emergence has been stated from May to October. There are two, possibly up to four generations a year. The autumn generation can be smaller or larger in number than the spring generation (Kawecka & Kownacki, 1974; Pinder, 1983; own data). Nothing is known about hibernation.

FEEDING

Kawecka & Kownacki (1974) found the guts to contain mainly diatoms and detritus, and few animal remains.

MICROHABITAT

The larvae are bottom dwellers, most abundant on sand and gravel and in low numbers on stones (Pinder, 1980, 1983; Becker, 1995; own data).

WATER TYPE

Current

The species is most common in fast-flowing streams (e.g. Rossaro, 1991, Orendt, 2002a). Janzen (2003) collected the larvae only in foothill streams in Germany. In other regions (England, Russia) *P. gaedii* is a relatively common inhabitant of fast-flowing lowland streams with gravelly or stony bottoms (Pinder, 1980, 1983; Pankratova, 1970: 33). In the Netherlands, records from lowland streams with sandy bottoms and relatively low current velocities are rare. In Finland *P. gaedii* is widely distributed in lakes (Paasivirta, 2012). Mundie (1957) collected the adults near an English storage reservoir. In central Europe the species is almost absent from lakes and other stagnant water bodies.

Dimensions

The larvae live mainly in larger streams and small rivers, scarcely in large rivers and in brooks less than 6 m wide. In stagnant water the species is almost confined to northern lakes (see above).

SAPROBITY

Most streams inhabited by this species are unpolluted or only slightly polluted. Bazerque et al. (1989) found *P. gaedii* in the river Somme in France, but not in stretches with severe pollution; it was present again further downstream where the water quality had improved, but the substrate was still somewhat organically polluted. Klink (1985) did not find the species in the polluted Grensmaas (river Meuse in southern Limburg) in the Netherlands. Moog (1995) and Wilson & Ruse (2005) call the species tolerant of slightly to moderately polluted water. The relative influence of the oxygen supply cannot be deduced from the available data.

SALINITY

Paasivirta (2000) collected one specimen in brackish water in the Gulf of Bothnia in Finland. In other parts of Europe the species lives only in fresh water.

Potthastia longimanus Kieffer, 1922

Potthastia campestris Pankratova, 1970: 103–105, fig. 53

IDENTIFICATION

See under the genus. The larval type is very characteristic. Because the larvae of related species are unknown, identification of larvae is only possible to group level. The occurrence of such related species in the Netherlands has never been stated, but is not impossible.

DISTRIBUTION IN EUROPE AND THE NETHERLANDS

The species lives in many countries throughout Europe (Saether & Spies, 2010). In the Netherlands it is widely distributed in the Pleistocene part of the country and in southern Limburg. In the Holocene area it is mainly collected in lakes in the fen peat region and more rarely in the large rivers (Limnodata.nl; own data).

LIFE CYCLE

Emergence has been stated from March to October. In the Netherlands, most pupae and exuviae have been collected in spring and autumn, suggesting two generations. However, many authors in Europe stated emergence also in summer. Possibly this is an intermediate generation, as in *Pseudodiamesa branickii*. The time of year the species is most numerous seems to depend on local factors. In early winter we found many larvae in third and fourth instar.

FEEDING

The guts have often been found to contain many algae and detritus (e.g. Moog, 1995). However, A. Klink (unpublished) saw very many chaetae of Naididae in two larvae from different streams. Because the mouthparts are a strong indication of predatory feeding, algae and detritus could originate from the prey of the larva. The presence of enough prey animals could be an important factor in determining the presence or absence of the species in lakes and streams (see below).

MICROHABITAT

Some authors call *P. longimanus* psammophilous (e.g. Oliver, 1983; Orendt, 2002). However, the larvae are often collected on plants, among mosses and on stones (Meuche, 1938; Lehmann, 1971; Becker, 1995; Brodersen et al., 1998; Bitušik, 2000; own data). They do not make tubes.

WATER TYPE

Current

P. longimanus is mainly an inhabitant of flowing water. It is almost absent in very slow-flowing lowland streams, but Lehmann (1971) collected the species much more in the potamal than in the rhithral zone. The optimum environment seems to be fast-flowing lowland streams and foothill streams (see table 7 in Chapter 9), but the larvae are sometimes also found in montane brooks (Braukmann, 1984; Peters et al., 1988; Laville & Vinçon, 1991; Janzen, 2003; Syrovátka et al., 2009; Verberk et al., 2012). It is very probable that the optimum depends on a combination of current, oxygen content and food.

In lakes the species is often absent or scarce, locally (but not in the Netherlands) more common (Meuche, 1938; Reiss, 1968; Orendt, 1993; Reiff, 1994; Brodersen et al., 1998; Buskens, unpublished).

Dimensions

The larvae are scarce in narrow brooks less than 3 m wide, but they can probably be more common if oxygen supply, permanence and trophic conditions are optimal. The larvae are less common in large rivers than in streams. In stagnant water the species is almost confined to lakes, probably because of the better oxygen supply in larger water bodies. The larvae live mainly in the littoral zone, but can be found sometimes at a depth of 10 m (Brundin, 1949; Bitušik, 2000).

Temperature

The species tolerates relative high temperatures in summer, up to more than 20 °C.

Shade

We never found the species in shaded brooks or streams. However, this has never been specifically investigated.

Permanence

The species has been rarely found in temporary water.

pH

The larvae rarely live in acid water (Limnodata.nl). We once collected a larva in an upper course at pH values lower than 6, but it is questionable whether the life cycle could be completed there.

TROPHIC CONDITIONS AND SAPROBITY

P. longimanus is a typical inhabitant of eutrophic water and is rarely found in oligotrophic or mesotrophic lakes (Brundin, 1949). In Danish lakes Brodersen et al. (1998) found an optimum chlorophyll-a concentration of 29 µg/l. In streams the species is almost confined to more or less polluted water; however, it is probable that the larvae prefer less polluted conditions when the current is slow. Peters et al. (1988) found the larvae to be mainly abundant in moderately polluted lowland brooks and streams, but still present where there was rather severe organic pollution. Limnodata.nl gives a minimum oxygen saturation of 50% during daytime and a BOD between 1 and 5 mg O/l, rarely more.

SALINITY

Paasivirta (2000) collected one specimen in brackish water in the Gulf of Bothnia in Finland. In other parts of Europe the species lives only in fresh water.

Potthastia montium (Edwards, 1929)

DISTRIBUTION IN EUROPE

P. montium has been recorded only in western and central Europe. However, because the species lives also in the East Palaearctic and the Near East, it can be expected also in other parts of Europe (Saether & Spies, 2010). It has not been collected in the Netherlands.

IDENTIFICATION

The adult male has been keyed by Langton & Pinder (2007), the exuviae by Langton (1991). The larva is unknown.

ECOLOGY

The larvae live in montane and foothill streams (e.g. Fahy & Murray, 1973; Casas & Vilchez-Quero, 1989; Michiels, 2004).

Potthastia pastoris (Edwards, 1933)

Only the adult male of *P. pastoris* has been described and has been keyed by Langton & Pinder (2007). It seems to be related to *P. longimanus*. The species has been reported from the British Isles, France and Finland (Saether & Spies, 2010). According to Paasivirta (2012), it lives in flowing water.

Protanypus Kieffer, 1906

Protanypus is the only genus in the tribe Protanypodini. The larvae are conspicuous because of the numerous setae on the head capsule (Oliver, 1983). Keys to adult males, pupae and larvae can be found in Saether (1975a). The characteristic intestinal canal has been described by Shilova & Kurazhskovskaya (1980).
The two European species, *P. morio* (Zetterstedt, 1938) and *P. caudatus* Edwards, 1924, live in oligotrophic to mesotrophic lakes in northern Europe and the Alps

(mainly in the profundal zone); *P. morio* is also found in northeastern Poland and in the British Isles (Brundin, 1949; Saether, 1975; Saether & Spies, 2010). There are no recent records from the Netherlands. *P. morio* has been collected in acid as well as in less acid lakes (Raddum & Saether, 1981). This species has two generations in Fennoscandia and in the Chiemsee in Bavaria, but only one in the Bodensee (Reiss, 1968; Saether, 1975a; Reiff, 1994). The larvae are carnivores (Brundin, 1949: 721).

Pseudodiamesa branickii Nowicki, 1873

SYTEMATICS AND IDENTIFICATION

The genus *Pseudodiamesa* contains four European species, of which *P. nivosa* lives only in mountain regions and two other species are only known from Russia (Ashe & O'Connor, 2009). The larvae of *P. (Pseudodiamesa) branickii* and *P. (Pachydiamesa) nivosa* can be identified using Ferrarese & Rossaro (1981), Oliver (1983), Schmid (1993), Bitušík (2000) and Janecek (2007).

DISTRIBUTION IN EUROPE

P. branickii lives throughout Europe, but is absent from most lowland regions (Saether & Spies, 2010). See, however, Water type.

LIFE CYCLE AND OVIPOSITION

Emergence has been stated from March to October. There are two or three generations a year. Eggs laid in early March lead to a generation time of five months (two generations a year); eggs laid in late March or April lead to a (small) summer generation, emerging in June and July, and three generations a year (Lehmann, 1971; Nolte & Hoffmann, 1992). The egg masses contain 450–1700 eggs and are attached to firm substrata just above the water's edge (Nolte, 1993).

FEEDING

The larvae are predators, feeding on chironomids and sometimes larger prey; to a lesser extent also algae and detritus are consumed (Ferrarrese & Rossaro, 1981). Nolte & Hofmann (1992) reared them from egg to adult on algae and detritus.

MICROHABITAT

The larvae live in sandy bottoms with much detritus, sometimes also on stones encrusted with sand or among mosses or sedges (Lehmann, 1971; Ferrarese & Rossaro, 1981; Nolte & Hofmann, 1992; Zinchenko, 2002).

WATER TYPE

In western and central Europe, *P. branickii* is mainly an inhabitant of montane springs and brooks, more scarcely living in foothill streams. The species is often found in montane oligotrophic lakes (e.g. Lehmann, 1971; Laville & Vinçon, 1991; Bitušík, 2000). We collected the larvae in four different springs in the lowland of Belarus (Moller Pillot & Moroz, 2002). In Russia the larvae have been collected also at the edges of lowland streams (Zinchenko, 2002: 37, fig. 16; see also Pankratova, 1970).

Pseudokiefferiella Zavřel, 1941

SYSTEMATICS AND IDENTIFICATION

Pseudokiefferiella is related to *Diamesa* and it is not always easy to distinguish these genera in all stages. Only one species, *P. parva*, has been named and described, but there are still several undescribed species. Thienemann (1952) keyed a larva from the High Tatra mountains as sp. 2. Langton (1991: 59) keyed the exuviae of a species from Czechoslovakia as Pe 1 (see also Janecek, 2007: 24).

The larvae of *P. parva* resemble those of the *Diamesa dampfi* group and *Syndiamesa*, but can be distinguished by the long dark body setae and the labral lamellae with two pectinate lobes (Ferrarese & Rossaro, 1981; Oliver, 1983; Schmid, 1993; Bitušík, 2000).

DISTRIBUTION AND ECOLOGY

P. parva has been recorded in many montane areas throughout Europe, but seems to be absent from lowland areas (Thienemann, 1952; Saether & Spies, 2010). The larvae inhabit bed sediments and mosses in small streams and hygropetric places (Schmid, 1993).

Sympotthastia Pagast, 1947

SYSTEMATICS AND IDENTIFICATION

The genus *Sympotthastia* is quite different from *Potthastia* in the adult and pupal stages, but the larvae resemble those of the *Potthastia gaedii* group. Probably the only reliable differences are that in *Potthastia gaedii* the galea of the maxilla bears 5 to 7 peg-like lamellae and the premandible lacks inner teeth. In *Sympotthastia* (almost) only setae-like lamellae are present (Epler, 2001: 5.3) and the premandible has small inner teeth. The larva of *S. spinifera* has been described and figured by Ferrarese & Rossaro (1981).

DISTRIBUTION IN EUROPE

Three species have been recorded in Europe. *S. spinifera* is known from central and southern Europe, *S. macrocera* only from France, Germany and Austria. *S. zavreli* has been collected in central and southern Europe and also in England and Ireland (Ashe & O'Connor, 2009; Saether & Spies, 2010). Possibly the mentum of a specimen of this genus has been found subfossil in the river Rhine in the Netherlands (cf. Klink & Moller Pillot, 1996).

ECOLOGY

All three species of the genus have been found in montane as well as foothill streams. Drake (1985) collected adult males of *S. zavreli* in the river Kennet in southern England in April and May. The bottom existed here of gravelly flints. The larvae build tubes of detritus and feed on diatoms (Bitušík, 2000).

Syndiamesa Kieffer, 1918

Diamesa hygropetrica Potthast, 1914: 353–355, fig. 137–142; Thienemann, 1952: 251 nec *Syndiamesa hygropetrica* Moller Pillot, 1984: 17, 157, fig. VI.47; Moller Pillot & Buskens, 1990: 15, 31, 76 (misidentification)

SYSTEMATICS AND IDENTIFICATION

Syndiamesa is closely related to *Diamesa*. It is probable that only two species occur in western Europe outside mountain regions: *S. edwardsi* and *S. hygropetrica*. Possibly the larvae cannot always be distinguished from those of the *Diamesa dampfi* group. In contrast to what is stated by Oliver (1983: 117 sub 6), both have a procercus with 5–7 anal setae. Usually the larvae of *Syndiamesa* can be recognised because the central mental tooth is less than two times as wide as a first lateral tooth; the deep incisions between the teeth are usually still visible after wear. Further, the third antennal segment in *Syndiamesa* is longer than the second segment and the AR is 2.2–3 (in *Diamesa dampfi* gr. 1.7–2.3) (Potthast, 1914; Oliver, 1983; Schmid, 1993; Bitušik, 2000; Janecek, 2007). See also under *Diamesa permacra*.

DISTRIBUTION IN EUROPE

S. edwardsi is only known from the British Isles, France, Spain and Italy. *S. hygropetrica* has been recorded in many countries in central and southern Europe (Saether & Spies, 2010). The records from the Netherlands are based on misidentifications, but the species may occur in southern Limburg or on the Veluwe in view of the record from Sauerland, Germany (Potthast, 1914). The genus has been collected also in Poland and Finland.

ECOLOGY

The larvae live mainly in hygropetric places, in springs as well as along brooks and streams. They are found in mountain regions, but also in foothill streams (e.g. Potthast, 1914). It is not clear if the larvae collected in foothill streams in Germany by Janzen (2003) are correctly identified.

6 SYSTEMATICS, BIOLOGY AND ECOLOGY OF THE BUCHONOMYIINAE

Murray & Ashe (1985) placed the Buchonomyiinae near the Podonominae and Tanypodinae. Saether (2000) preferred to place this subfamily near the other subfamilies, most likely in a semifamily Chironominae (against Telmatogetoninae and Tanypodinae). Cranston et al. (2012) stated on the basis of molecular gene investigations that Buchonomyiinae is the sister group of all other subfamilies, conforming more to Murray and Ashe's argumentation. The only European species is *Buchonomyia thienemanni*.

Buchonomyia thienemanni Fittkau, 1955

IDENTIFICATION

The adult male and the pupa can be identified easily with all existing keys. The female has been described and figured by Murray & Ashe (1985). The fourth instar larva has not been described; the third instar has been described by Ashe (1995). The larva is easily distinguished from most Orthocladiinae larvae because of the short antennae, the absence of a procercus and the more or less triangular shape of the head. Janecek (2007: 5), who saw also the fourth instar larva, emphasises the striking contrast between the pale head capsule and the brown mouthparts.

DISTRIBUTION IN EUROPE

B. thienemanni has been recorded in many countries in western and central Europe (Saether & Spies, 2010). The species seems to be scarce everywhere and very scarce in typical lowland regions (see Ecology). It has not been collected in the Netherlands, but the exuviae have been collected in the river Roer just before the Dutch border (B. van Maanen, pers. comm.).

LIFE CYCLE

Reiss (1984a) stated emergence in Bavaria mainly in summer, but in smaller numbers also in October, suggesting a second generation.

SWARMING AND OVIPOSITION

The adults form small swarms near a stream and mate in the end to end position. In a Petri dish the egg mass was laid under water and attached to the bottom of the dish. The egg mass and eggs have been described by Ashe & Murray (1983). See also Nolte (1993: 39).

ECOLOGY

Ashe (1995) argued that the case in which the third instar larva was found had probably been made by another insect, possibly a trichopteran of the family Hydropsychidae. Also, the build of the larva and pupa itself suggests a commensal, parasitic or predatory association with another invertebrate animal. The fact that so far hardly any larvae have been collected reinforces this argument. Almost all data are based on the collection of exuviae.

B. thienemanni has been found only in flowing water. It seems to be most common in small, rather fast-flowing, summer-warm streams (Reiss, 1984a). Many of these streams are foothill streams or flow through lowlands with accidented relief. There are also records from small, sometimes summer-cold rivers (Michiels, 1999; Garcia & Laville, 2001; Klink, 2010).

7 SYSTEMATICS, BIOLOGY AND ECOLOGY OF THE PODONOMINAE

The Podonominae are related to the Tanypodinae (Saether, 2000; Cranston et al., 2012). In Europe it is a very small subfamily and just a few species occur in the West European lowland. The larvae are easily distinguished from other subfamilies, because all European species have a very high procercus in combination with a well developed mentum (Brundin, 1983). Shilova & Kurazhskovskaya (1980) describe the characteristic intestinal canal. The larvae are free living and inhabit mainly brooks and springs or bogs. The pupae swim and resemble certain pupae of Tanypodinae; they often hang at the water surface like Tanypodinae and Culicidae pupae (Lenz, 1939).

Lasiodiamesa Kieffer, 1924

Studies on the ecology of *Lasiodiamesa* are limited mainly to the Netherlands, where several relic populations have been found. All these populations live in degraded bog remnants in nature reserves. The ecology of species in the genus may differ in other localities from that observed in the Netherlands, such as in more natural bogs and especially in more northern regions where climatic conditions are different.

SYSTEMATICS AND IDENTIFICATION

Lasiodiamesa is the only genus of the subfamily that has been found in the European lowland. The adult males can be identified using Brundin (1966). The characteristic gonostylus of *L. gracilis* and *L. sphagnicola* have been figured also in Brundin (1989). For *L. bipectinata*, see Saether (1967). Most pupae can be identified using Langton (1991). There is no key to the larvae.

DISTRIBUTION IN EUROPE AND THE NETHERLANDS

In Europe the genus *Lasiodiamesa* is confined to thenorthern part, rarely southward to the Alps. *L. sphagnicola* is the most widespread species, recorded from Scandinavia to Belarus, the British Isles, Ireland, the Netherlands and Bavaria. These southern populations are without doubt relics from the last glacial period (Brundin, 1966). *L. gracilis* lives southward to Poland and the Netherlands. *L. bipectinata* is only known from Norway and northern Germany, *L. armata* from Scandinavia and Finland (Saether & Spies, 2010).

In the Netherlands the genus has been collected in four different regions (Bargerveen, Engbertsdijkvenen, Winterswijk, Peel). *L. gracilis* has been identified from Winterswijk (Korenburgerveen, Wooldse Veen) and the Peel region, *L. sphagnicola* only from the Peel region (Verberk et al., 2003; van der Loo, 2005; van Duinen, 2008).

LIFE CYCLE

The adults fly from April to September; different populations emerge during different periods (Lenz, 1939; Brundin, 1966). In all investigated regions in the Netherlands the larvae have been found only in spring. In Estonia larvae were collected in spring as well in September–October (van Duinen, pers. comm.). See further under Water type: Permanence.

FEEDING

According to Lenz (1939), the larvae feed mainly on detritus and to a lesser extent on epiphytic algae such as diatoms. Verberk et al. (2003) found only algae in the guts of five larvae from Dutch bog remnants.

MICROHABITAT

The larvae are usually found in small, shallow puddles among rather dense vegetation of *Sphagnum* or submerged tussocks of *Molinea* using their procercus for gripping the structures (Verberk et al., 2003; van der Loo, 2005). Paasivirta (2012) even calls all species of *Lasiodiamesa* semiterrestrial. Older larvae seem to move from dense *Sphagnum* vegetation to more open water (Brundin, 1966; van der Loo, 2005).

WATER TYPE

In western and central Europe, *Lasiodiamesa* has been found only in *Sphagnum* bogs. In the Netherlands there are only degraded bog remnants, where the larvae live in small *Sphagnum* puddles, peat cutting pits and trenches (Lenz, 1939; Brundin, 1983; Verberk et al., 2003; van Duinen, 2008). In Swedish Lapland the larvae also occur in oligohumic tarns and ponds, springs and spring brooks (Brundin, 1966). In Estonia, G.-J. van Duinen found high numbers of larvae only in places where the influence of buffered and mineral rich groundwater could be observed, often at the edge of the bogs or in moving surface water. In exceptional cases, where the larvae lived in the bog centre, a higher degree of mineralisation seemed to be observable (van Duinen et al., 2008, pers. comm.). The same has been reported by Ashe (1987), who found the larvae in an Irish bog in pools 'associated with a slight flush (i.e. surface water movement or trickles which may be minerotrophic)'. This also fits with the observation by Verberk et al. (2010) that the species *L. gracilis* occupied fewer localities following rewetting efforts, which reduced minerotrophic influence. Additionally, van Duinen et al. (2003), who compared rewetted sites with non-rewetted sites only observed this species in the latter.This preference for water bodies with a higher degree of minerotrophic influence or water movement is likely related to enhanced food quality and not to altered oxygen supply (see below). Van Orsouw & Kolenbrander (2004) found no relation between conductivity and the occurrence of *Lasiodiamesa* in Korenburgerveen (the Netherlands). In the Peel region the genus appeared to be absent from pools influenced by the influx of river water (van Duinen, 2008) and the species was restricted to a site that had not undergone any cultivation or rewetting and constituted a remnant that was situated at the edge of a what was a much larger bog landscape in the past.

Permanence

In the Peel region the larvae were found almost only in peat cutting pits prone to desiccation (van Duinen, 2008). This may be the reason why there was no summer generation. However van der Loo (2005) found that larval occurrence in Wooldse Veen (the Netherlands) was restricted to places, where the ground water level fluctuations were small (<30 cm). This could suggest that larger fluctuations caused drying-up of the pools in summer, causing the disappearance of the species. It is also possible that temporary pools are preferred in spring (e.g. because of the influence of predation) and that the presence of temporary and permanent pools in the area is necessary for the species to survive all year round. Van der Loo noted that larvae tended to avoid waters with many predators. At the moment the importance of permanent and temporary water bodies throughout a season is yet to be studied.

OXYGEN

Nothing is known about the influence of oxygen supply on the occurrence of the larvae. Because other species living in acid peat cutting pits (*Psectrocladius platypus*, *Paralimnophyes longiseta*) tolerate very low oxygen concentrations, up to anaerobiosis in the night, *Lasiodiamesa* is probably also quite tolerant to hypoxia.

Parochlus kiefferi (Garrett, 1925)

Parochlus is the most plesiomorphic genus of the Podonominae, with only one species in Europe. The adult male has been described by Saether (1967), with many figures. The larva has been keyed and figured by Brundin (1983).
P. kiefferi is distributed in northern and central Europe, Ireland and the British Isles (Saether & Spies, 2010). In Finland it is very widespread (Paasivirta, 2012), but it is absent from the Netherlands and adjacent lowlands. The larvae live in springs and fast-flowing brooks, but have been recorded also in larger fast-flowing streams and rivers (Brundin, 1983; Langton, 1991).

8 SYSTEMATICS, BIOLOGY AND ECOLOGY OF THE TELMATOGETONINAE

The subfamily Telmatogetoninae occupies a special place within the family Chironomidae. The subfamily is very old (pre-Jurassic). For a discussion about the phylogeny see Saether (2000) and Cranston et al. (2012). The genera resemble one another in the larval stage (Cranston, 1983). They are easily distinguished from most Orthocladiinae larvae because of the short antennae, the absence of a procercus and the rather characteristic mentum (fig. 16), in combination with their habitat. See where necessary the key to subfamilies by Cranston & Reiss (1983).

All Holarctic species of the subfamily live in the intertidal marine zone. In the Netherlands, larvae have also been found further inland in brackish water. Records from fresh water are very rare (see below).

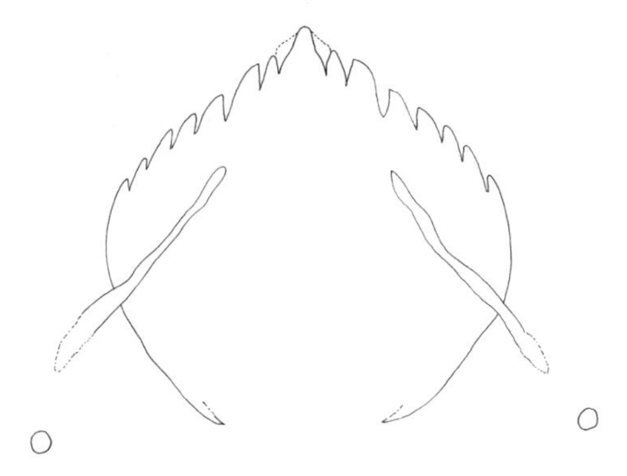

16 Telmatogeton: mentum

Telmatogeton Schiner, 1866

DISTRIBUTION IN EUROPE AND THE NETHERLANDS

T. japonicus Tokunaga, 1933 has been recorded from many countries along the Atlantic Ocean, the North Sea and the Baltic (Saether & Spies, 2010). As far as it has been verified, the Dutch material belongs to this species (det. pupae O. Duijts). Contrary to the description by Cranston (1983), the mentum of Dutch larvae has 8 pairs of lateral teeth. *T. pectinatus* (Deby, 1889) is only known from the British Isles and France (Saether & Spies, 2010).
In the Netherlands, larvae and pupae have been collected at several locations along the North Sea coast, in Oostvoornse Meer (lake) and in the North Sea Canal (Noordzeekanaal) as far as Amsterdam.

LIFE CYCLE

Tokunaga (1935) stated two emergence periods in Japan: in spring and summer. In the Netherlands pupae and adults have been found in August and September.

TE

SWARMING AND OVIPOSITION

The adults can fly, but do not swarm in the air. Mating takes place on rocks exposed to waves (Neumann, 1976; Cranston, 1989). The eggs are laid separately or stuck together in pairs on stones, often in crevices or h oles, submerged or in the spray zone (Nolte, 1993). The last author has described and figured the eggs. According to Lenz (1950, based on Tokunaga, 1935), the total number of eggs laid by one female is 150–190.

FEEDING

The larvae feed on living and decomposing algae (Lenz, 1950, based on Tokunaga, 1935).

MICROHABITAT

The larvae have been found in tubes on stones, buoys and in rock pools (Lenz, 1950; R. Munts, pers. comm.; own data). In the North Sea Canal (Noordzeekanaal) the larvae lived on large stones without any silt, just below the water level between filamentous algae (J. van Dijk, pers. comm.).

WATER TYPE AND SALINITY

The larvae are usually found in the marine intertidal zone, sometimes in places with organic pollution (Cranston, 1989; R. Munts, pers. comm.), but also in rock pools (own data). In the North Sea Canal they are exposed to fast currents by ships. The larvae often also live in brackish water (e.g. near Amsterdam) and can survive fresh water (Lenz. 1950). However, records of *T. japonicus* from fresh water in nature are very rare (Otto, 1991: 19).

DISPERSAL

Possibly the larvae are dispersed by intercontinental shipping (Cranston, 1989).

Thalassomya frauenfeldi Schiner, 1856

Thalassomya frauenfeldi Pankratova, 1970: 30, 120–121, fig. 63

SYSTEMATICS AND IDENTIFICATION

T. frauenfeldi is the only representative of the genus in Europe. Adult males can be distinguished from those of *Telmatogeton japonicus* using Langton & Pinder (2007: vol. 2: 154). The exuviae have been keyed and figured by Langton (1991), the larvae by Cranston (1983).

DISTRIBUTION IN EUROPE

T. frauenfeldi has been collected in a number of countries along the Atlantic Ocean, the North Sea, the Mediterranean and the Black Sea (Saether & Spies, 2010). It has not been recorded in the Netherlands.

FEEDING

The larvae prefer to feed on blue-green algae (Pankratova, 1970: 30–31).

ECOLOGY

The larvae live in similar places as *Telmatogeton*: in tubes on stones among algae in the intertidal marine zone, usually above the normal water level. They survive dryness for up to ten days. They have been studied by Valkanov (1949) in the Black Sea (partly translated by Lenz, 1950: 20–21 and Pankratova, 1970: 121). Possibly the species prefers waters with reduced salinity around harbours and river mouths (Cranston, 1983).

9 TABLES OF BIOLOGICAL AND ECOLOGICAL PROPERTIES OF SPECIES

(in principle for the Dutch situation)

The values in the tables are based on data from the literature and the numerous unpublished studies in the Netherlands referred to in Part II (Moller Pillot, 2009: 6).

blank	unknown
?	not sure

Explanation of tables 5–7
For each factor a species is awarded a score out of ten indicating the relative chance of encountering the species in a one square metre quadrant. This method was introduced by Zelinka & Marvan (1961) and used, among others, by Sládeček (1973) and Moog (1995). The values are based on the Dutch climate and Dutch circumstances. For a discussion of the usefulness of the tables outside the Netherlands, see the relevant section in Part II (Moller Pillot, 2009: 6).

The figures in our tables can be very different from those given by Moog (1995) or Verberk et al. (2012). These differences are partly caused by the fact that our tables apply to a different range of water bodies: no montane streams and not only water types usually investigated by water authorities.

In contrast to Part II, rare occurrence is indicated by a +. As a rule, values are not given for a species when no data were available for the whole range in question. In such cases, please consult the species text in Chapters 3 to 8.

The figures in the tables apply only to water which meets the other requirements of the species. The figures for species that live almost only in brooks and small streams, for example, do not apply to pools or rivers. Occurrence in polysaprobic water applies to water bodies that meet the oxygen requirements, etc. Bear in mind that as a rule a species can endure a negative influence better when all other factors are optimal.

TABLE 4 GENERAL ECOLOGY

gener = number of generations/year (see the comments in section 2.4)
adult = flying period of adults (and egg deposition) in months
developm = duration of larval development in months (if more generations: in summer)**
hibern = hibernation: larval instar (or eggs) in early December
eggs = number of eggs in one egg mass; bold figures are from two or more data (see the comments in section 2.5)
food = only the most important food for third and fourth instar is given; younger larvae eat more detritus and other fine materials. AN = animals; AL = algae; diat = diatoms; FP = fine particulate organic material; FD = fine detritus; CD = coarse detritus; D = detritus; PL = plant tissue; d- = dead
habitat = preferred microhabitat of the larvae: bo = near or in the bottom (D, CD see under food); pl = on plants; hs = hard substrate; terr = terrestrial

Values between brackets are found more rarely

* see text on this species
** The given duration is only an indication: it is dependent on temperature and food conditions, see section 2.4. It is also possible that species grow faster in spring because food (especially diatoms) is more available in spring.

	gener	adult	developm	hibern	eggs	food	habitat
Brillia bifida	2-4	(1-)4-12	2-4	1-4	**340**	CD	CD
Brillia longifurca	2-3	(1-)4-12	3	1-4	171	CD	wood
Chaetocladius piger	2	10-4	4, 8	1-4		D	edge
Corynoneura	3-7	(1-)4-12	1	1-4	**50**	al, D	pl, hs
Cricotopus bicinctus	2-5?	3-11	2?	1-3	**300**	al, D	pl, hs
Cricotopus sylvestris	3-5	4-10	1-2	2-3	**300**	d-al, D	pl, hs
Cricotopus triannulatus	3-5?	(2-)4-10	1½-2	2 (?3)	**175**	al,FP ?	hs
Cricotopus trifascia	2-3	(3-)4-10	2 (4)	2-3	**526**		(pl), hs
Cricotopus trifasciatus	4-5	4-10	1-2	eggs	**250**	PL	pl
Epoicocladius ephemerae	1-2	5-9	(3), 12	2-4		FP	*
Eukiefferiella claripennis	(?1-)3	(1-)3-10	1½	2-3(-4)	85	diat, FP	pl, hs
Eukiefferiella coerulescens	3	3-10	2	2-4			(pl), hs
Eukiefferiella ilkleyensis	3(-4?)	(2-)4-10	2½	1-2		diat, FP	pl, hs
Heleniella ornaticollis	2	3-9	3	2-3(>4)		D	bo
Heterotrissocladius marcidus	2(-3?)	4-10	(2?-) 4	3, 4	231	D	D
Hydrobaenus lugubris	1	3-5	12	*	**166**	al	
Limnophyes asquamatus	3-5?	(2-)3-11	2	1-4	**75**	D	edge
Nanocladius dichromus	3	4-10	3				pl, hs
Nanocladius rectinervis	3	3-11	3	2-3(-4?)			pl, hs
Odontomesa fulva	2-5?	3-10	1½ - 2	2-4 *		al	bo

	gener	adult	developm	hibern	eggs	food	habitat
Orthocladius oblidens *	1-3	4-5(-12)	1-3?	eggs		diat, D	pl, hs
Orthocladius rhyacobius *	1-2	3-5,8-10	1, 4	eggs		diat	pl, hs
Orthocladius rubicundus	(1-)2(-3)	3-10	2-3				(pl), hs
Orthocladius thienemanni	1-2 (3?)	3-5(-10)	2?	*		diat, D	
Paracladius conversus	(?1-)3	3-10	2-(?12)				bo
Paracricotopus niger	2-3?	4-10	?	2-3			bo, hs
Paralimnophyes longiseta	3-5	3-10	1-2	3-4	**37**	D	bo, pl
Parametriocnemus stylatus	(?1-)2-3	3-10	3-4(?-12)	2-4		D	bo
Paraphaenocladius impensus	2-3	4-10	3-4 ?	2-3		D	banks
Paratrichocladius rufiventris	2-4	(2-)4-11	2?	2-3(-4)		al, D	pl, hs
Potthastia longimanus	2-3	(3-)4-10	2-4?	3-4		AN	bo,pl,hs
Prodiamesa olivacea	2 (3?)	3-10	(?2-)4	2-4	**1300**	FP ?CP	bo
Psectrocladius limbatellus	3-7	3-10	1-2	2-4		al, FP	bo, CD
Psectrocladius obvius	2-3(-4?)	4-10			500	al	bo, pl
Psectrocladius oxyura	3 (4?)	4-10	2-3				bo
Psectrocladius platypus	4-5	4-10	1-2	2-3		al	pl
Psectrocladius sordidellus	3	3-9	(2?-)3	2-4	**180**		bo,pl,hs
Pseudodiamesa branickii	2-3	3-10	3, 5	4	**1000**	AN, D	bo
Pseudorthocladius	2-3	5-9		2-4			terr
Rheocricotopus fuscipes	3-4	(12-) 3-11	2	2-3(-4) eggs?		D (?)	pl, hs
Synorthocladius semivirens	2-3	(3-)4-10	2-3	2-3(-4)		diat, D	pl,st,bo
Thienemanniella vittata	4-6?	3-11	1	2-3		D (diat)	pl, hs
Trissocladius brevipalpis	1	(1-)3-4	12	*			bo
Tvetenia calvescens	(2-)3-5	(2-)3-11	1½	2-3(-4)		diat, D	pl, bo

T

TABLE 5: SAPROBITY AND OXYGEN

(see sections 2.14 and 2.15)

We adopt the definition of saprobity by Sládeček (1973: 28): the amount and intensity of decomposition of organic matter. This author (ibid., p. 41) mainly follows the system of Caspers (1966), which is derived largely from conditions in stagnant water. However, many other workers use this system mainly for flowing water, sometimes even only for flowing water (e.g. Moog, 1995: I-22). The main difference between stagnant and flowing waters in this context lies in the availability of oxygen. In this book we try to give the demands of each species with respect to the most important factors found in (undeep) stagnant as well in flowing water. For this a reason we keep saprobity and oxygen content separate in an attempt to show whether a species tolerates (or even prefers) a given amount of organic pollution and in what way the oxygen content determines the occurrence of the larvae.
We use oligosaprobity only when hardly decomposition takes place.

The oxygen contents were measured in the water column, but estimation of saprobity has to take more than just the water column into account. An important part of the decomposition takes place on the bottom. Our polysaprobic level is the beta-polysaprobic level of Caspers (1966), Sládeček (1973) and Moog (1995: I-26), in which anoxybiosis is the rule in the bottom sediments. Ice cover in winter has not been taken into account because it is rarely significant in the Netherlands. When ice and snow cover continues for long periods in winter, a lower saprobity can result in anoxia, especially in undeep water bodies.

ol = oligosaprobic
B = β-mesosaprobic
A = α-mesosaprobic
p = polysaprobic
stab = stable oxygen regime: day and night above 50% saturation
unst = unstable: minimum between 10% and 50% saturation
low = sometimes (but not longer than a few hours) less than 5% saturation
rott = rotting: in summer almost daily less than 5% saturation for hours

\+ = rarely
(+) = supplied by current from upper course
* = see text on this species

For species hardly living in stagnant water in the Netherlands (+ or 0 in the first column in table 7) the figures for saprobity apply only to flowing water. These species can live in stagnant water with a good oxygen supply and/or low temperature.

	saprobity							oxygen			
	ol	ol/B	B	B/A	A	A/p	p	stab	unst	low	r
Acricotopus lucens *	0	2	3	3	2	+	+	4	5	1	
Brillia bifida	2	4	3	1	+	+	(+)	9	1	0	
Brillia longifurca	+	1	3	3	2	1	+	8	2	(+)	
Chaetocladius piger *	3	3	3	0.5	0.5	+	+	8	2	+	
Corynoneura coronata	0	3	6	1	0	0	0				
Corynoneura lobata	2	5	3	0	0	0	0	9	1	0	
Corynoneura scutellata	0	1	3	3	2	1	+	4	4	2	
Cricotopus annulator	+	2	4	3	1	+	0	7	3	0	

	saprobity							oxygen			
	ol	ol/B	B	B/A	A	A/p	p	stab	unst	low	rott
Cricotopus bicinctus	0	1	2	3	3	1	+	5	4.5	0.5	0
Crcotopus festivellus	0	2	3	3	2?	0	0				
Cricotopus intersectus	0	+	3	4	2	1	+	5	4.5	0.5	+
Cricotopus sylvestris	0	0.5	2	3	3	1	0.5	2	5	2.5	0.5
Cricotopus triannulatus	0	1	2	3	3	1	0	7	3	0	0
Cricotopus trifascia	0	2	3	3	2	+	0	8	2	0	0
Epoicocladius ephemerae	1	3	5	1	0	0	0	9	1?	0	0
Eukiefferiella brevicalcar	1	3	4	2	0	0	0	9	1	0	0
Eukiefferiella claripennis	0	1	1.5	2	2	2.5	1	7	3	0	0
Eukiefferiella coerulescens	3	3	2	2	0	0	0	9	1	0	0
Heleniella ornaticollis	2	4	4	0	0	0	0	10?	0?	0	0
Heterotanytarsus apicalis	2	4	4	+	0	0	0	8	2	0	0
Heterotrissocladius marcidus	1	4	4	1	0	0	0	8	2	0	0
Metriocnemus 'hirtellus type'	1	2	3	1	1	1	1	6	2	1	1
Nanocladius dichromus	0	1	2	2.5	2.5	2	0	7	3	0	0
Nanocladius rectinervis	0	1	2	3	2	2	0	8	2	0	0
Odontomesa fulva	+	2	4	3	1	0	0	8	2	0	0
Orthocladius frigidus	3	4	3	0	0	0	0	10	0	0	0
Orthocladius oblidens	1	2	5	2	(+)	0	0	8	2	0	0
Orthocladius rhyacobius	1	3	4	2	(+)	0	0	9	1	0	0
Orthocladius rivulorum	1	4	4	1	0	0	0	10?	?	0	0
Orthocladius rubicundus	1	3	4	2	+	0	0	9	1	0	0
Orthocladius thienemanni	1	3	4	2	(+)	0	0	9	1	0	0
Paracladius conversus	0	2	4	3	1	+	0	8	2	0	0
Paracricotopus niger	1	4	3	2	0?	0	0	8?	2?	0	0
Parakiefferiella smolandica	2	4	4	+	0	0	0				
Paralimnophyes longiseta	0	0	1	3	3	2	1	2	4	3	1
Paratrichocladius rufiventris	0	0.5	1.5	2	2.5	3	0.5	6	4	0	0
Paratrissocladius excerptus	0?	3	5	2	0	0	0	10	0	0	0
Pothastia gaedii	1	3	4	2	0	0	0				
Potthastia longimanus	0	1	4	3	2	+	0	9	1	0	0
Prodiamesa olivacea	+	1	3	3	2	1	+	5	4	1	0
Psectrocladius bisetus								8	2	0	0
Psectrocladius obvius	0	1	5	3	1	0	0	7	2	1	0
Psectrocladius oxyura	0	2	6	2	0	0	0	9	1	0	0
Psectrocladius platypus								5	4	1	0
Psectrocladius psilopterus	1	2	5	1.5	0.5	0	0	8	2	+	0
Psectrocladius sordidellus	0	1	5	3	1	+	0	6	3	1	0
Rheocricotopus chalybeatus	0	1	4	3	1	1	0	7	3	0	0
Rheocricotopus fuscipes	0	2	3	2	2	1	0	8	2	0	0
Synorthocladius semivirens	+	2	3	3	1	1	0	9	1	0	0
Thienemanniella majuscula	0.5	1	4	3	1	0.5	0	8	2	0	0
Thienemanniella vittata	1	4	4	1	+	0	0	9	1	0	0
Tvetenia calvescens	1.5	3	3	2	0.5	+	0	8	2	0	0
Tvetenia discoloripes	0	1	3	3	2	1	0	8	2	0	0
Tvetenia verralli	1	2	3	3	1	0	0	9	1	0	0

TABLE 6: pH AND CHLORINITY

(see sections 2.12 and 2.19)

The pH and chlorinity mentioned are for the water column. Both can be much higher in the bottom. In the tables we use the chlorinity, not the total salt content (1 ‰ salinity = 0.54 g Cl/l).
Usually Limnodata.nl gives aberrant pH ranges because water authorities do not investigate all types of (small) water bodies.

+ = rarely
(+) = supplied by current from upper course
[+] = chlorinity only based on data from northern or southern regions
* = see text on this species

	pH					chlorinity (g Cl/l)				
	< 4.5	5	6	7	>7.5	<0.3		1–3		>10
Acricotopus lucens	0	+	2	5	3	7	3	+	0	0
Allocladius arenarius						2	4	3	1	0
Brillia bifida*	+	1	2	3.5	3.5	10	0	0	0	0
Brillia longifurca	0	0	0?	5	5	10	0	0	0	0
Chaetocladius piger*	1	1	2	3	3	7	3	+	0	0
Cricotopus bicinctus	0	0	1	4	5	9	1	[+]	0	0
Cricotopus ornatus	0	0	0	0	10	+	3	5	2	0
Cricotopus sylvestris	0	+	1	4	5	4	4	2	[+]	0
Diplocladius cultriger	0	0	1	7	2	9	1	0	0	0
Halocladius varians	0	0	0	0	10	0	+	0.5	4.5	5
Heterotanytarsus apicalis*	+	3	4	3	+	10	0	0	0	0
Heterotrissocladius marcidus*	0	1	4	5	+	10	0	0	0	0
Metriocnemus 'hirtellus type'	0	+	1	3	6	9	1	+	0	0
Nanocladius balticus	0	0	1	4	5	10	[+]	[+]	0	0
Nanocladius dichromus	0	0	1	4	5	9	1	[+]	[+]	0
Odontomesa fulva	0	0	1	4.5	4.5	10	0	0	0	0
Orthocladius	0	0	+	4	6	10	+	[+]	[+]	0
Paracladius conversus	0	0	0	1	9	10	+	0	0	0
Parakiefferiella bathophila	0	1	2	4	3	10	[+]	[+]	0	0
Paralimnophyes longiseta	3	4	2	1	+					
Parametriocnemus stylatus	0	0	+	4	6	10	[+]	[+]	0	0
Paratrichocladius rufiventris	0	0	+	4	6	10	[+]	[+]	0	0
Potthastia longimanus	0	0	+	5	5	10	[+}	{+]	0	0
Prodiamesa olivacea	+	+	1	5	4	10	+	0	0	0
Psectrocladius barbimanus	0	0	0	1	9	8	1.5	0.5	0	0
Psectrocladius bisetus	4	4	2	0	0	10	0	0	0	0
Psectrocladius limbatellus	0	1	1	4	4					

	pH					chlorinity (g Cl/l)				
	‹ 4.5	5	6	7	›7.5	‹0.3		1–3		›10
Psectrocladius obvius	0.5	0.5	2	5	2	10	+	0	0	0
Psectrocladius oxyura	0	0	1	4	5					
Psectrocladius platypus	4	4	2	+	+	10	0	0	0	0
Psectrocladius psilopterus *	2.5	2.5	2	1.5	1.5	10	0	0	0	0
Psectrocladius sordidellus	+	+	1	4	5	9	1	+	0	0
Thalassosmittia thalassophila						+	3	3	2	2
Thienemanniella majuscula	0	0	2	6	2					
Tvetenia	0	0	0?	4	6					
Zalutschia humphriesiae	4	4	2	(+)	0	10	0	0	0	0

TABLE 7: CURRENT AND PERMANENCE
(see section 2.10)

Current
The current velocities apply only to linear water bodies, because the current velocities in the littoral zone of lakes are not comparable with current velocities in streams. They are thought to have been measured 10 cm below the surface (if not near the bottom) in the middle of the stream. The figures in the tables do not reflect the situation in large rivers; the current (as defined above) in rivers is much faster and most larvae live at other sites, such as the underside of stones. We take into account the period during which the larvae are present (often more in spring).

Permanence
The figures given for permanence apply mainly to water bodies that dry up in summer, but in principle not to pools or ditches, which are sometimes in contact with nearby permanent water. In the latter case, species characteristic of permanent water may be carried by streams to places where no water was previously present. Neither do the figures apply to temporary exposure of parts of an estuary or river during low tide or a low stage of the river. A water body is considered dry only when the bottom is dry.

+ = rarely
(+) = supplied by current from upper course

	current (cm/s)					permanence: dry (in weeks/y.)				
	< 10	10-25	25-50	50-75	75-100	> 12	6–12	< 6	rarely	not
Acricotopus lucens	8	2	+	(+)	0	0.5	1	1.5	3	4
Brillia bifida	+	1	2	4	3	+	1	1	2	6
Brillia longifurca	0	1	3	3	3					
Cardiocladius fuscus	0	0	(+)	(+)	10	0	0	0	0	10
Chaetocladius piger	1	6	2	1	+	1	3	3	2.5	0.5
Corynoneura coronata	1	2	4	2	1	0.5	0.5	2	3	4
Corynoneura edwardsi	5	4	1	0	0					
Corynoneura lobata	0.5	1.5	3	3	2					
Corynoneura scutellata	7	2	1	+	0					
Cricotopus albiforceps	2	3	2.5	1.5	1					
Cricotopus annulator	0	+	1	4	5					
Cricotopus bicinctus	+	2	4	3	1	0	+	+	1	9
Cricotopus intersectus	8	2	+	+	0	0	0	+	+	10
Cricotopus similis	0	0	(+)	2	8					
Cricotopus sylvestris	4	3	2	1	+	1	1.5	1.5	3	3
Cricotopus tremulus	0	0	0	3	7					
Cricotopus triannulatus	0	1	2	3	4					
Cricotopus trifascia	0	0	(+)	3	7					
Cricotopus trifasciatus	2	4	3	1	+	0	+	1	2	7
Diamesa insignipes	0	0	+	2	8					
Diplocladius cultriger	2	4	3	1	+	1	2	2.5	2.5	2
Epoicocladius ephemerae	0	+	2	5	3	0	0	0	0	10
Eukiefferiella brevicalcar	0	0.5	1.5	5	3	0	0.5	0.5	1	8
Eukiefferiella claripennis	+	0.5	1.5	5	3	0	0.5	0.5	1	8

	current (cm/s)					permanence: dry (in weeks/y.)				
	< 10	10-25	25-50	50-75	75-100	> 12	6–12	< 6	rarely	not
Eukiefferiella clypeata	0	0	1	3	6					
Eukiefferiella coerulescens	0	0	1	4	5					
Eukiefferiella gracei	0	0	1	4	5					
Eukiefferiella ilkleyensis	0	0	0.5	3.5	6					
Heterotanytarsus apicalis	1	3	4	2	+					
Heterotrissocladius marcidus	1	3	4	2	+					
Hydrobaenus pilipes	2	5	3	+	+	2	2	3	2	1
Nanocladius dichromus	1	2	4	2	1					
Nanocladius rectinervis	0	1	2	3	4					
Odontomesa fulva	+	3	4	2	1	0	+	+	1	9
Orthocladius excavatus	1	1	3	3	2					
Orthocladius holsatus	7	2	1	0	0					
Orthocladius oblidens	+	1	3	3	3	+	+	+	2	8
Orthocladius rhyacobius	0	0	1	5	4	1	1	1	2	5
Orthocladius rivulorum	0	0	+	3	7					
Orthocladius rubicundus	0	0	+	3	7					
Orthocladius ruffoi	0	0	0	1	9					
Orthocladius thienemanni	0	+	1	5	4	1	1	1	2	5
Paracladius conversus	5	5	+	(+)	0	0	0	0	0	10
Paralimnophyes longiseta	9	1	(+)	0	0	2	3	3	1	1
Paracricotopus niger	0	0	0	2	8					
Paratrichocladius rufiventris	0	+	1	3	6	+	+	+	1	9
Paratrissocladius excerptus	0	0	0	1	9					
Potthastia gaedii	0	0	+	4	6					
Potthastia longimanus	0	+	2	4	4	0	+	+	1	9
Prodiamesa olivacea	+	3	4	2	1	0	0.5	0.5	1	8
Psectrocladius limbatellus	10	+	+	(+)	0	2	2	2	2	2
Psectrocladius obvius	8	2	0	0	0					
Psectrocladius platypus	8	2	+	0	0	1.5	1.5	2	2	3
Psectrocladius psilopterus	7	2	1	0	0	1	1	1	3	4
Psectrocladius sordidellus	9	1	0	0	0	+	0.5	0.5	1	8
Rheocricotopus chalybeatus	0	1	2	4	3					
Rheocricotopus effusus	0	0	+	2	8					
Rheocricotopus fuscipes	0	1	1	5	3	1	2	2	2	3
Synorthocladius semivirens	0	0	+	4	6					
Thienemanniella majuscula	0	1	5	3	1	?	?	0.5	1.5	8
Thienemanniella vittata	0	(+)	1	3	6					
Tvetenia calvescens	0	+	+	4	6					
Tvetenia discoloripes	0	+	1	5	4	+	+	+	+	10
Tvetenia verralli	0	0	+	4	6					
Zalutschia humphriesiae	9	1	(+)	0	0	3	3	2	2	+

REFERENCES

Ali, A. & M.S. Mulla, 1976. Substrate type as a factor influencing spatial distribution of chironomid midges in an urban flood control channel system. – Environmental Entomology 5: 631-636.

Anderson, N.H., 1989. Xylophagous Chironomidae from Oregon streams. – Aq. Insects 11: 33-45.

Anderson, N.H. & J.R. Sedell, 1979. Detritus processing by macroinvertebrates in stream ecosystems. – Ann. Rev. Entomol. 24: 351-377.

Armitage, P.D., P.S. Cranston & L.C.V. Pinder, 1995. The Chironomidae – Biology and ecology of non-biting midges. – Chapman & Hall (London).

Armitage, P.D., K. Lattmann, N. Kneebone & I. Harris, 2001. Bank profile and structure as determinants of macroinvertebrate assemblages – seasonal changes and management. – Regulated rivers: Research and Management 17: 543-556.

Arts, G.H.P., 2000. Natuurlijke levensgemeenschappen van de Nederlandse binnenwateren. 13, Vennen. – Rapport EC-LNV AS-13. Wageningen. 1-80.

Ashe, P., 1983. A catalogue of chironomid genera and subgenera of the world including synonyms (Diptera: Chironomidae). – Ent. scand. Suppl. 17: 1-68.

Ashe, P., 1987. *Lasiodiamesa sphagnicola*, *Cardiocladius capucinus* and *Orthocladius (Eudactylocladius) fuscimanus* (Diptera: Chironomidae) new to Ireland. – Ir. Nat. J. 22: 194-195.

Ashe, P., 1992. Corrections to the Chironomidae part of the catalogue of Palaearctic Diptera. – Neth. J. Aq. Ecol. 26: 215-221.

Ashe, P., 1995. Description of a late-instar larva of *Buchonomyia thienemanni* Fittkau and further data on its ecology with diagnoses for the subfamily Buchonomyiinae and the genus *Buchonomyia* (Diptera: Chironomidae). - In: Cranston, P. (ed.): Chironomids: From genes to ecosystems. CSIRO Publ., Melbourne: 425-430.

Ashe, P. & D.A. Murray, 1980. *Nostococladius*, a new subgenus of *Cricotopus* (Diptera: Chironomidae). – In: Murray, D.A. (ed.): Chironomidae. Ecology, Systematics, Cytology and Physiology. Oxford, Pergamon Press: 105-111.

Ashe, P. & D.A. Murray, 1983. Observations on and descriptions of the egg-mass and eggs of *Buchonomyia thienemanni* Fitt. (Diptera: Chironomidae). – Mem. Amer. ent. Soc. 34: 3-13.

Ashe, P. & J.P. O'Connor, 2009. A world catalogue of Chironomidae (Diptera). Part 1. – Irish Biogeogr. Soc. & Nat. Museum Ireland, Dublin. 445 pp.

Ashe, P., J.P. O'Connor & D.A. Murray, 2000. Larvae of *Eurycnemus crassipes* (Panzer) (Diptera: Chironomidae) ectoparasitic on prepupae/pupae of *Hydropsyche siltalai* Döhler (Trichoptera: Hydropsychidae), with a summary of known chironomid / trichopteran associations. – Spixiana 23: 267-274.

Balushkina, E.V., 1987. Functional importance of the larvae of chironomids in continental water bodies. – Trudy zool. inst. Akad. Nauk SSSR 142: 1-179. (In Russian).

Baranov, V. A., 2011. New and rare species of Orthocladiinae (Diptera, Chironomidae) from Crimea (Ukraine). – Vestnik zoologii 45: 79-84.

Bazerque, M.F., H. Laville & Y. Brouquet, 1989. Biological quality assessment in two rivers in the northern plain of France (Picardie) with special reference to chironomid and diatom indices. – Acta Biol. Debr. Oecol. Hung. 3: 29-39.

Bazzanti, M. & F. Bambacigno, 1987. Chironomids as water quality indicators in the River Mignone (Central Italy). – Hydrobiol. Bull. 21: 213-222.

Beauvesère-Storm, A. de & D. Tempelman, 2009. De dansmug Psectrocladius schlienzi nieuw voor Nederland, met een beschrijving van de larve (Diptera: Chironomidae). – Ned. Faunistische Meded. 30: 17-22.

Becker, C., 1995. Ein Beitrag zur Zuckmückenfauna des Rheins (Diptera: Chironomidae). – Thesis Bonn. Aachen: Shaker Verlag. 265 pp.

Berczik, A., 1967. Vorkommen einiger Chironomiden aus zwei Natrongewässern. – Opusc. Zool. Budapest 7: 75-82.

Berg, C.O., 1950. Biology of certain Chironomidae reared from *Potamogeton*. – Ecol. Monographs 20: 83-101.

Berg, M.B. & R.A. Hellenthal, 1992. Life histories and growth of lotic chironomids (Diptera: Chironomidae). – Ann. Entomol. Soc. Am. 85: 578-589.

Beuk, P.L.T. (ed.), 2002. Checklist of the Diptera of the Netherlands. - KNNV Publ. Zeist. 1-447.

Bitušik, P., 2000. Priručka na určovanie lariev pakomárov (Diptera: Chironomidae) Slovenska. Čast' I. Buchonomyinae, Diamesinae, Prodiamesinae a Orthocladiinae. – Techn. Univ. vo Zvolene, Fak. ekol. environm. Katedra Biológie. 1-133.

Botts, P.S. & B.C. Cowell, 1992. Feeding electivity of two epiphytic chironomids in a subtropical lake. – Oecologia 89: 331-337.

Brabec, K., 1997. Distribution of chironomid

larvae (Diptera, Chironomidae) in the river section influenced by a reservoir. – Folia Fac. Sci. Nat. Univ. Masarykianae Brunensis, Biologia 95: 27-36.

Brabec, K., 2000. Effects of changed thermal regime on chironomid community in a dammed river (A preliminary study). – In: Hoffrichter, E.O. (ed.): Late 20th century research on Chironomidae. Shaker Verlag, Aachen: 471-476.

Braukmann, U., 1984. Biologischer Beitrag zu einer allgemeinen regionalen Bachtypologie. – Thesis Giessen Univ. 473 pp.

Brennan, A. & A.J. Mclachlan, 1979. Tubes and tube-building in a lotic chironomid community. – Hydrobiologia 67: 173-178.

Brennan, A. & A.J. Mclachlan, 1980 (1981). Species of *Eukiefferiella* Thienemann (Dipt., Chironomidae) from a Northern river with notes on larval dwellings. – Entomologist's mon. Mag. 116: 109-111.

Brennan, A., A.J. Mclachlan & R.S. Wotton, 1978. Particulate material and midge larvae (Chironomidae: Diptera) in an upland river. – Hydrobiologia 59: 67-73.

Brodersen, K.P., P.C. Dall & C. Lindegaard, 1998. The fauna in the upper stony littoral of Danish lakes: macroinvertebrates as trophic indicators. – Freshw. Biol. 39: 577-592.

Brodin, Y.-W. & M. Gransberg, 1993. Responses of insects, especially Chironomidae (Diptera), and mites to 130 years of acidification in a Scottish lake. – Hydrobiologia 250: 201-212.

Brodin, Y., J.O. Lundström & L. Paasivirta, 2008. *Tavastia yggdrasilia*, a new orthoclad midge (Diptera: Chironomidae) from Europe. – Aquatic Insects 30: 261-267.

Brooks, S.J., P.G. Langdon & O. Heiri, 2007. The identification and use of Palaearctic Chironomidae larvae in palaeoecology. – Quaternary Research Association Technical Guide 10. 1-276.

Brundin, L., 1942. Zur Limnologie Jämtländischer Seen. – Meddn St. Unders. o. FörsAnst. SöttvattFisk. 20: 1-104.

Brundin, L., 1947. Zur Kenntnis der schwedischen Chironomiden. – Ark. f. Zool. 39A: 1-95.

Brundin, L., 1949. Chironomiden und andere Bodentiere der südschwedischen Urgebirgsseen. – Rep. Inst. Freshw. Res. Drottningholm 30: 1-914.

Brundin, L., 1956. Zur Systematik der Orthocladiinae (Dipt. Chironomidae). – Rep. Inst. Freshw. Res. Drottningholm 37: 5-185.

Brundin, L., 1956a. Die bodenfaunistischen Seetypen und ihre Anwendbarkeit auf die Südhalbkugel. Zugleich eine Theorie der produktionsbiologischen Bedeutung der glazialen Erosion. – Rep. Inst. Freshw. Res. Drottningholm 37: 186-191.

Brundin, L., 1966. Transantarctic relationships and their significance, as evidenced by chironomid midges. With a monograph of the subfamilies Podonominae and Aphroteniinae and the austral Heptagyiae. – K. Svenska Vetensk. Akad. Handl. 11: 1-472.

Brundin, L., 1983. The larvae of Podonominae (Diptera: Chironomidae) of the Holarctic region – Keys and diagnoses. – Ent. scand. Suppl. 19: 23-31.

Brundin, L. Z., 1989. The adult males of Podonominae (Diptera: Chironomidae) of the Holarctic region – Keys and diagnoses. – Ent. scand. Suppl. 34: 23-36.

Buskens, R.F.M., 1989. Monitoring of chironomid larvae and exuviae in the Beuven, a soft water pool in the Netherlands, and comparisons with palaeolimnological data. – Acta Biol. Debr. Suppl. Oecol. Hung. 3: 41-50.

Buskens, R.F.M., 1989a. Beuven: Herstel van een ecosysteem. – Vakgroep Aq. Oecol. en Biogeol. Kath. Univ. Nijmegen. 154 pp.

Buskens, R.F.M. & G.A.M. Verwijmeren, 1989. The chironomid communities of deep sand pits in the Netherlands. – Acta Biol. Debrecina Suppl. Oecologica Hungarica 1989 (3): 51-60.

Cannings, R.A. & G.G.E. Scudder, 1978. The littoral Chironomidae (Diptera) of saline lakes in central British Columbia. – Can. J. Zool. 56: 1144-1155.

Carter, C.E. & R.W.G. Carter, 1983. Factors influencing the chironomid community of a nearshore sand area. – Mem. Amer. Entomol. Soc. 34: 47-59.

Casas, J.J. & A. Vilchez-Quero, 1989. A faunistic study of the lotic chironomids (Diptera) of the Sierra Nevada (S.E. of Spain): changes in the structure and composition of the populations between spring and summer. – Acta Biol. Debr. Oecol. Hung. 3: 83-93.

Caspers, H., 1966. Stoffwechseldynamische Gesichtspunkte zur Definition der Saprobitätsstufen. – Verh. Internat. Verein. Limnol. 16: 801-808.

Caspers, N. 1980. Die Makrozoobenthos-Gesellschaften des Rheins bei Bonn. – Decheniana (Bonn) 133: 93-106.

Caspers , N., 1980a. Die Emergenz eines kleinen Waldbaches bei Bonn. – Decheniana beih. 23: 1-175.

Caspers, N., 1983. Chironomiden-Emergenz zweier Lunzer Bäche, 1972. – Arch. Hydrobiol. Suppl. 65: 484-549.

Caspers, N., 1991. The actual biocoenotic zonation of the river Rhine exemplified by the chironomid midges (Insecta, Diptera). – Verh. Internat. Verein. Limnol. 24: 1829-1834.

Caspers, N. & F. Reiss, 1987. Chironomidae des Lunzer Seengebietes in Niederösterreich (Insecta, Diptera, Nematocera). – Spixiana 10: 13-35.

Caspers, N. & M. Siebert, 1980. *Pseudorthocladius rectangilobus* sp. n., eine

R

neue Chironomide aus dem Hunsrück (Deutschland) (Diptera: Chironomidae). – Mitt. Schweiz. ent. Ges. 53: 181-183.
Chernovskij, A.A., 1949. Opredelitelj lichinok komarov semejstva Tendipedidae. – Opredel. po faune SSSR 31: 1-186.
Coffman, W.P., P.S. Cranston, D.R. Oliver & O.A. Saether, 1986. The pupae of Orthocladiinae (Diptera: Chironomidae) of the Holarctic region – Keys and diagnoses. – Ent. scand. Suppl. 28: 147-296.
Cranston, P.S., 1974. Corrections and additions to the list of British Chironomidae (Diptera). - Entomologist's mon. Mag. 110: 87-95.
Cranston, P.S., 1979. The development stages of *Limnophyes globifer* (Lundström) (Diptera: Chironomidae). – Hydrobiologia 67: 19-25.
Cranston, P.S., 1982. A key to the larvae of the British Orthocladiinae (Chironomidae). – Freshw. Biol. Ass. Sci. Publ. 45: 1-152.
Cranston, P.S., 1982a. The metamorphosis of *Symposiocladius lignicola* (Kieffer), n.gen., n.comb., a wood mining Chironomidae (Diptera). – Ent. scand. 13: 419-429.
Cranston, P.S., 1983. The larvae of Telmatogetoninae (Diptera: Chironomidae) of the Holarctic region - Keys and diagnoses. – Ent. scand. Suppl. 19: 17-22.
Cranston, P.S., 1984. The taxonomy and ecology of *Orthocladius (Eudactylocladius) fuscimanus* (Kieffer), a hygropetric chironomid (Diptera). – J. Nat. History 18: 873-895.
Cranston, P.S. , 1989. The adult males of Telmatogetoninae (Diptera: Chironomidae) of the Holarctic region – Keys and diagnoses. – Ent. scand. Suppl. 34: 17-21.
Cranston, P.S., 1999. Nearctic *Orthocladius* subgenus *Eudactylocladius* revised (Diptera: Chironomidae). – J. Kansas ent. Soc. 71 (1998): 272-295.
Cranston, P.S., N.B. Hardy & G.E. Morse, 2012. A dated molecular phylogeny for the Chironomidae (Diptera). – Systematic Ent. 37: 172-188.
Cranston, P.S. & D.R. Oliver, 1988. Aquatic xylophagus Orthocladiinae – systematics and ecology (Diptera, Chironomidae). – Spixiana Suppl. 14: 143-154.
Cranston, P.S., D.R. Oliver & O.A. Saether, 1983. The larvae of Orthocladiinae (Diptera: Chironomidae) of the Holarctic region – Keys and diagnoses. – Ent. scand. Suppl. 19: 149-291.
Cranston, P.S., D.R. Oliver & O.A. Saether, 1989. The adult males of Orthocladiinae (Diptera: Chironomidae) of the Holarctic region – Keys and diagnoses. – Ent. scand. Suppl. 34: 165-352.
Cranston, P.S. & F. Reiss, 1983. The larvae of Chironomidae (Diptera) of the Holarctic region – Key to subfamilies. Ent. scand. Suppl. 19: 11-15.
Cranston, P.S. & O.A. Saether, 1982. A redefinition of *Acamptocladius* Brundin, 1956 (syn. *Phycoidella* Saether, 1971, n. syn.) (Diptera: Chironomidae), with the description of *A. reissi* n. sp. – Ent. scand. 13: 25-32.
Cranston, P.S. & O.A. Saether, 1986. *Rheosmittia* (Diptera: Chironomidae): a generic revalidation and revision of the western Palaearctic species. – J. Nat. History 20: 31-51.
Crawford, P.J. & D.M. Rosenberg, 1984. Breakdown of conifer needle debris in a new northern reservoir, Southern Indian Lake, Manitoba. – Can. J. Fish. Aquat. Sci. 41: 649-658.
Crisp, G. & L. Lloyd, 1954. The community of insects in a patch of woodland mud. – Trans. R. ent. Soc. London 105: 269-313.
Cummins, K.W., 1973. Trophic relations of aquatic insects. – Ann. Rev. Entomol. 18: 183-206. *zie copie 2 pp. in St*
Cuppen, H., 2006. Monitoring onderzoek Hierdense Beek 2005. – Rapport Landschapsecol. en Hydrobiol. Adv. bur. Cuppen. 1-35.
Cuppen, H. & A. van Nieuwenhuuijzen, 2005. Larve *Metriocnemus scirpi* met zekerheid in Nederland gevonden. – Macrofauna Nieuwsmail 54, april 2005.
Cuijpers, P. & M. Damoiseaux, 1981. De Geul. Biologische beoordeling van de waterkwaliteit, met behulp van diverse systemen. – Rep. Natuurh. Genootsch. Limburg, Maastricht: 1-120 + app. (unpubl.).
Czeczuga, B., 1962. Ecological-physiological aspects of the distribution of some species of Tendipedidae (Diptera) larvae in a water reservoir. – Roczn. Acad. Med. Białymstoku, Suppl. 8: 1-99. (Polish with english summary)
Darby, R.E., 1962. Midges associated with California rice fields, with special reference to their ecology (Diptera: Chironomidae). – Hilgardia 32: 1-206.
Davies, L.J. &H.A. Hawkes, 1981. Some effects of organic pollution on the distribution and seasonal incidence of Chironomidae in riffles in the River Cole. – Freshw. Biol. 11: 549-559.
Dees, A. & H. Moller Pillot, 2010. *Chaetocladius* spec. 'Herkenbosch' opgehelderd (Chironomidae, Orthocladiinae). – Macrofaunanieuwsmail 94, 26 okt. 2010.
Delettre, Y.R., 1984. Recherches sur les Chironomides (Diptera) à larves édaphiques: biologie, écologie, mécanismes adaptifs. – Thesis Univ. Rennes: 1-310.
Delettre, Y.R., 1986. La colonisation de biotopes multiples: une alternative à la résistance in situ aux conditions mésologiques défavorables. Cas de *Limnophyes minimus* (Mg.), Diptère Chironomidae à larves édaphiques des landes armoricaines. – Rev. Écol. Biol. Sol 23: 29-38.

Delettre, Y.R., 1989. Influence de la durée et de l' assèchement sur l' abondance et la phénologie des Chironomides (Diptera) d'une mare semi-permanente peu profonde. – Arch. Hydrobiol. 114: 383-399.

Delettre, Y.R., 1994. Fire disturbance of a chironomid (Diptera) community on heathlands. – J. Appl. Ecol. 31: 560-570.

Delettre, Y., P. Tréhen & P. Grootaert, 1992. Space heterogeneity, space use and short-range dispersal in Diptera: a case study. – Landscape Ecol. 6: 175-181.

Dettinger-Klemm, P.-M. A., 2000. Influence of temperature and photoperiod on development in three species of Chironomidae (Diptera) – Chironomus dorsalis Meigen, 1818, Polypedilum uncinatum (Goetghebuer, 1921) and Paralimnophyes hydrophilus (Goetghebuer, 1921) – living in temporary ponds. – In: Hoffrichter, O. (ed.): Late 20th Century Research on Chironomidae: 295-308. Aachen, Shaker Verlag.

Dettinger-Klemm, P.-M. A., 2001. The metamorphosis of *Orthocladius (Symposiocladius) holsatus* Goetghebuer, 1937, with the description of *Orthocladius (Symposiocladius) lunzensis* sp.n. (Diptera: Chironomidae). – Aquatic Insects 23: 45-62.

Dettinger-Klemm, P.-M. A., 2001a. Drought tolerance and parthenogenesis in the semiaquatic/terrestrial chironomid *Limnohpyes asquamatus* Andersen, 1937 (Diptera: Chironomidae). – Tagungsbericht 2000 (Magdeburg) Deutsche Ges. Limnologie, Tutzing: 355-359.

Dettinger-Klemm, P.-M. A., 2003. Chironomids (Diptera, Nematocera) of temporary pools – an ecological case-study. – Thesis Marburg Univ. 371 pp.

Dettinger-Klemm, P.-M. A. & H.W. Bohle, 1996. Überlebensstrategien und Faunistik von Chironomiden (Chironomidae, Diptera) temporärer Tümpel. – Limnologica 28: 403-421.

Dowling, C. & D.A. Murray, 1980. Zalutschia humphriesiae sp. n., a new species of Orthocladiinae (Diptera, Chironomidae) from Ireland. – Acta Universitatis Carolinae – Biologica 1978: 49-58.

Drake, C.M., 1982. Seasonal dynamics of Chironomidae (Diptera) on the Bulrush *Schoenoplectus lacustris* in a chalk stream. – Freshw. Biol. 12: 225-240.

Drake, C.M., 1985. Emergence patterns of Diptera in a chalk stream. – Aquatic Insects 7: 97-110.

Driver, E.A., 1977. Chironomid communities in small prairie ponds: some characteristics and controls. – Freshw. Biol. 7: 121-133.

Duinen, G.-J. van, 2008. Evaluatie hoogveenherstel in Mariapeel en Deurnese Peel. Watermacrofauna. – Stichting Bargerveen, Nijmegen. 64 pp. + app.

Duinen G.-J, A., A.M.T. Brock, J.T. Kuper, R.S.E.W. Leuven, T.M.J. Peeters, J.G.M. Roelofs, G. van der Velde, W.C.E.P. Verberk & H. Esselink, 2003. Do restoration measures rehabilitate fauna diversity in raised bogs? A comparative study on aquatic macroinvertebrates. - Wetlands Ecology and Management 11: 447-459.

Duinen, G.-J. van, A. Dees en H. Esselink, 2008. Hoogveen-karakteristieke en zeldzame water-macrofaunasoorten in het Bargerveen. 2006-2007. Eindrapportage. – Stichting Bargerveen, Nijmegen. 44 pp.

Duinen, G.A. van, T.Timm, A.J.P. Smolders, A.M.T. Brock, W.C.E.P. Verberk & H. Esselink, 2006. Differential response of aquatic oligochaete species to increased nutrient availability – a comparative study between Estonian and Dutch raised bogs. – Hydrobiologia 564: 143-155.

Duursema, G., 1996. Vennen in Drenthe. Een onderzoek naar ecologie en natuur op basis van macrofauna. – Assen, Zuiveringsschap Drenthe. 140 pp.

Duursema, G., 1997. Hydrobiologisch onderzoek in het Rolderdiep. – Zuiveringsschap Drenthe, Assen. 1-47.

Edwards, F.W., 1950. Family Chironomidae. – In: Coe, R.L.: Handb. Identific. Brit. Insects 9: 121-206.

Edwards, F.W., 1929. British non-biting midges (Diptera, Chironomidae). – Trans. R. ent. Soc. Lond. 77: 279-430.

Ekrem, T., E. Stur & P.D.N. Hebert, 2010. Females do count: Documenting Chironomidae (Diptera) species diversity using DNA barcoding. – Organisms, Diversity and Evolution 10: 397-408.

Epler, J.H., 2001. Identification manual for the larval Chironomidae (Diptera) of North and South Carolina. – J.H. Epler, Crawfordville (private publ.). 534 pp.

Eriksson, L. & R.K. Johnson. 2000. Palaeolimnological studies of selected Swedish reference lakes. – In: Hoffrichter, E.O. (ed.): Late 20th century research on Chironomidae. Shaker Verlag, Aachen: 485-489.

Ertlová, E., 1970. Chironomidae (Diptera) aus Donauaufwuchs. - Biológia (Bratislava) 25: 291-300.

Ertlová, E., 1971. Vorkommen der Art Eukiefferiella lutethorax Goetghebuer 1949 (Chironomidae, Diptera) in der Donau. – Biológia, Bratislava 26: 139-142.

Ertlová, E., 1974. Einige Erkenntnisse über Chironomiden (Diptera, Chironomidae) aus Bryozoen. – Biológia (Bratislava) 29: 869-876.

Fahy, E., 1973. Observations on lotic chironomid life cycles in Western Ireland. – Ent. Tidskr. 94: 83-88.

Fahy, E. & D.A. Murray, 1972. Chironomidae from a small stream in Western Ireland with a discussion of species composition of the group (Dipt.). – Entomol. Tidskr. 93: 148-155

R

Ferrarese, U., 1992. Chironomids of Italian rice fields. – Neth. J. Aquat. Ecol. 26: 341-346.
Ferrarese, U. & B. Rossaro, 1981. Chironomidi, 1 (Diptera, Chironomidae: Generalità, Diamesinae, Prodiamesinae). – Guide per il riconoscimento delle specie animali delle acque interne Italiane 12: 1-96.
Ferrington, L.C., 1987. Microhabitat preferences of larvae of three Orthocladiinae species (Diptera: Chironomidae) in Big Springs, a sandbottom spring in the high plains of western Kansas. – Ent. scand. Suppl. 29: 361-368.
Ferrington, L.C., 2000. Hibernal emergence of Chironomidae in lotic habitats of Kansas versus ambient air and water temperatures. - In: Hoffrichter, E.O. (ed.): Late 20th century research on Chironomidae. Shaker Verlag, Aachen: 375-382.
Ferrington, L.C. & O.A. Saether, 2011. A revision of the genera *Pseudosmittia* Edwards, 1932, *Allocladius* Kieffer, 1913, and *Hydrosmittia* gen. n. (Diptera: Chironomidae, Orthocladiinae). – Zootaxa 2849: 1-314.
Fesl, C., 2002. Biodiversity and resource use of larval chironomids in relation to environmental factors in a large river. – Freshw. Biol. 47: 1065-1087.
Fittkau, E.-J. & J. Lehmann, 1970. Revision der Gattung *Microcricotopus* Thien. et Harn. (Dipt. Chironomidae). – Int. Rev. ges. Hydrobiol. 55: 391-402.
Fittkau, E.-J. & F. Reiss, 1978. Chironomidae. – In: Illies, J. (ed.): Limnofauna europaea. 2. Aufl. Stuttgart: 404-440.
Fretwell, S.D., 1972. Populations in a seasonal environment. – Monogr. in Population Biol. 5: 1-217.
Friday, L.E., 1987. The diversity of macroinvertebrate and macrophyte communities in ponds. – Freshw. Biol. 18: 87-104.
Fu, Y., O.A. Saether & X. Wang, 2009. *Corynoneura* Winnertz from East Asia, with a systematic review of the genus (Diptera: Chironomidae: Orthocladiinae). – Zootaxa 2287: 1-44.
Fu, Y., O.A. Saether & X. Wang, 2010. *Thienemanniella* Kieffer from East Asia, with a systematic review of the genus (Diptera: Chironomidae: Orthocladiinae). – Zootaxa 2431: 1-42.
Gaevskaya, N.S., 1969. The role of higher aquatic plants in the nutrition of the animals of fresh-water basins. – Translation by D.G. Maitland-Muller. – Mann, K.A. (ed.): Natn. Lending Libr. Sci. Technol., Boston Spa, Yorkshire. 629 pp.
Garcia, X.-F. & H. Laville, 2001. Importance of floodplain waters for the conservation of chironomid (Diptera) biodiversity in a 6th order section of the Garonne river (France). – Ann. Limnol. 37: 35-47.
Geijskes, D.C., 1935. Faunistisch-ökologische Untersuchungen am Röserenbach bei Liestal im Basler Tafeljura. – Tijdschr. Entomol.78: 249-382.
Gendron, J.-M. & H. Laville, 1992. Diel emergence patterns of drifting chironomid (Diptera) pupal exuviae in the Aude River (Eastern Pyrenees, France). – Neth. J. Aquatic Ecol. 26: 273-279.
Gendron, J.-M. & H. Laville, 1997. Les Chironomidés (Diptera) de lÁude, rivière méditerranéenne des Pyrénées orientales: étude biocénotique et typologique. – Ann. Limnol. 33: 93-106.
Gendron, J.-M. & H. Laville, 2000. Impact of a catastrophic flood on the chironomid populations (Diptera: Chironomidae) of the hyporhthral (4th order) zone of the river Aude (Eastern Pyrenees). – In: Hoffrichter, E.O. (ed.): Late 20th century research on Chironomidae. Shaker Verlag, Aachen: 383-391.
Gibson, N.H.E., 1945. On the mating swarms of certain Chironomidae (Diptera). – Trans. R. ent. Soc. London 95: 263-294.
Gijsen, M.E.A. van & T.H.L. Claassen, 1978. Integraal structuurplan Noorden des lands, landsdelig mikieu-onderzoek. Deelrapp. 2: Biologisch wateronderzoek macrofyten en macrofauna. – Rijksinst. Natuurbeheer, Leersum: 1-121 + app.
Goddeeris, B.R., 1989. A methodology for the study of the life cycle of aquatic Chironomidae (Diptera). – Verh. Symposium Invertebr. België 1989: 379-385.
Goedkoop, W. & R.K. Johnson, 1992. Modelling the importance of sediment bacterial carbon for profundal macroinvertebrates along a lake nutrient gradient. – Neth. J. Aq. Ecol. 26: 477-483.
Goetghebuer, M. 1914. Contribution à l'ètude des Chironomides de Belgique. – Annls. Biol. lacustre 7: 165-229.
Goetghebuer, M., 1932. Diptères (Nématocères). Chironomidae IV. Orthocladiinae, Corynoneurinae, Clunioninae, Diamesinae. – Faune de France 23: 1-204.
Goetghebuer, M., 1939. Tendipedidae (Chironomidae). e) Subfamilie Corynoneurinae. A. Die Imagines. – In: Lindner, E. (ed.): Die Fliegen der palaearktischen Region 13f: 1-14.
Goetghebuer, M., 1940-50. Tendipedidae (Chironomidae). f) Subfamilie Orthocladiinae. A. Die Imagines. – In: Lindner, E. (ed.): Die Fliegen der palaearktischen Region 13g: 1-208.
Gołowin, S., 1968. Dead flow zones evaluation of pollution degree of rivers. – Polskie Archiwum Hydrobiologii 15: 263-267.
Golubeva, G.V., 1980. Novyj vid komarov-zvontsov roda *Metriocnemus* v.d. Wulp (Diptera: Chironomidae) iz Yaroslavskoj oblasti. – Ent. Obozr. 59: 654-659.
Golubeva, G.V., 1986. Povedenie lichinok

komarov *Psectrocladius obvius* Walker (Diptera) v protsesse metamorfoza. – In: Gagarin, V.G. (red.): Povedenie vodnykh bespozvonochnykh. Materialy IV Vsecoyuznogo simposiuma Borok. 1983: 136-138. Andropov.

Gouin, F., 1936. Métamorphoses de quelques chironomides d' Alsace et de Lorraine. – Revue fr. Ent. 3: 151-173.

Gresens, S.E., E. Stur & T. Ekrem, 2012. Phenotypic and genetic variation within the *Cricotopus sylvestris* species-group (Diptera, Chironomidae), across a Nearctic – Palaearctic gradient. – Fauna norv. 31: 137-149.

Gripekoven, H., 1913. Minierende Tendipediden. – Arch. Hydrobiol. Suppl. 2: 129-230.

Grodhaus, G., 1980. Aestivating chironomid larvae associated with vernal pools. - In: Murray, D.A. (ed.): Chironomidae. Ecology, Systematics, Cytology and Physiology. Oxford, Pergamon Press: 315-322.

Grzybkowska, M. & M. Dukowska, 2002. Communities of Chironomidae (Diptera) above and below a reservoir on a lowland river: long-term study. – Ann. Zool. (Warsawa) 52: 235-247.

Grzybkowska, M. & J.Witczak, 1990. Distribution and production of Chironomidae (Diptera) in the lower course of the Grabia river (Central Poland). – Freshw. Biol. 24: 519-531.

Hall, R.E., 1951. Comparative observations on the chironomid fauna of a chalk stream and a system of acid streams. – J. Soc. Brit. Entomol. 3: 253-262.

Hamerlik, L. & K.P. Brodersen , 2010. Non-biting midges (Diptera: Chironomidae) from fountains of two European cities: micro-scale island biogeography. – Aquatic Insects 32: 67-79.

Hannesdóttir, E.R., G. M. Gislason & J.S. Olafsson, 2012. Life cycles of *Eukiefferiella claripennis* (Lundbeck 1898) and *Eukiefferiella minor* (Edwards 1929) (Diptera: Chironomidae) in spring-fed streams of different temperatures with reference to climate change. – Fauna norv. 31: 35-46.

Hawczak, A., P.D. Armitage & J.H. Blackburn, 2009. The macroinvertebrate fauna and environmental quality of the Oakers Stream, a small tributary of the River Frome, Dorset. – Proc. Dorset Nat. Hist. Archaeol. Soc. 130: 17-24.

Hawtin, E., 1998. Chironomid communities in relation to physical habitat. – In: Bretschko, G. & J. Helešic (eds.): Advances in river bottom ecology: 175-184. Backhuys, Leiden.

Hawtin, E. , 2000. Chironomid communities characteristic of lowland rivers in the U.K. – In: Hoffrichter, E.O. (ed.): Late 20th century research on Chironomidae. Shaker Verlag, Aachen: 497-505.

Hayes, B.P. & D.A. Murray, 1989. Seasonal and yearly variation in chironomid emergence as determined by collections of pupal exuviae in a lowland river in Ireland. – Acta Biol. Debr. Oecol. Hung. 2: 229-235.

Hershey, A.E., 1987. Tubes and foraging behavior in larval Chironomidae: implications for predator avoidance. – Oecologia (Berlin) 73: 236-241.

Hershey, A.E. & S. I. Dodson, 1987. Predator avoidance by *Cricotopus*: cyclomorphosis and the importance of being big and hairy. – Ecology 68: 913-920.

Higler, L.W.G., 1975. Reactions of some caddis larvae (Trichoptera) to different types of substrate in an experimental stream. – Freshw. Biol. 5: 151-158.

Higler, L.W.G., 1977. Macrofauna-cenoses on *Stratiotes* plants in Dutch broads. – Verh. R.I.N. 11: 1-86.

Higler, B. & B. Statzner, 1988. A simplified classification of freshwater bodies in the world. – Verh. Internat. Verein. Limnol. 23: 1495-1499.

Hirabayashi, K. & R.S. Wotton, 1999. Organic matter processing by chironomid larvae (Diptera: Chironomidae). - Hydrobiologia 382 : 151-159.

Hirvenoja, M., 1973. Revision der Gattung *Cricotopus* van der Wulp und ihrer Verwandten (Diptera, Chironomidae). – Ann. Zool. Fennici 10: 1-363.

Hirvenoja, M. & E. Hirvenoja, 1988. *Corynoneura brundini* spec. nov. Ein Beitrag zur Systematik der Gattung *Corynoneura* (Diptera, Chironomidae). – Spixiana Suppl. 14: 213-238.

Hoffmann, A. & D. Hering, 2000. Wood-associated macro-invertebrate fauna in Central European streams. – Internat. Rev. Hydrobiol. 85: 25-48.

Holzer, M., 1980. Die Belebung der Gewässer von Sandkiesanschwemmungen unterhalb des aktiven Stromes des Flusses March in der Obermährischen Talsenkung. – Acta Univ. Palackianae Olomuc. Fac. Rerum Nat. 67: 107-129 (in Czech).

Huisman, J. & L.W.C.A. van Breemen, 1986. Distribution of macroinvertebrates in the shallow part of "De Gijster", a water storage lake in the Biesbosch (The Netherlands). – Hydrobiol. Bull. 19: 199-206.

Illies, J., 1952. Die Mölle. Faunistisch-ökologische Untersuchungen an einem Forellenbach im Lipper Bergland. – Arch. Hydrobiol. 46: 424-612.

Izvekova, E.I., 1996. Khironomidy nekotorykh Kubanskikh limanov Akhtarsko-Gribenskoj gruppy. - In: Shobanov, N.A. & T.D. Zinchenko: Ekologiya, Evolutsiya i Sistematika Khironomid. Togliatti: 138-139.

Izvekova, E.I., A.A. Kuz'minykh & S.G. Nikolaev, 1996. Khironomidy nekotorykh malykh rek bassejna reki Oki i vozmozhnost ispol'zovaniya ikh lichinok v kachestve indikatorov zagryaznjeniya. – In: Shobanov,

N.A. & T.D. Zinchenko: Ekologiya, Evolutsiya i Sistematika Khironomid. Togliatti: 132-137.
Janecek, B.F.U., 1995. *Tanytarsus niger* Andersen (Diptera: Chironomidae) and the chironomid community in Gebhartsteich, a carp pond in northern Austria. – In: Cranston, P. (ed.): Chironomids, from genes to ecosystems. CSIRO Publications, East Melbourne: 281-296.
Janecek, B.F.U., 2000. Changes of chironomid communities during iron chloride treatment of Alte Donau, the largest stagnant water of Vienna (Austria). – In: Hoffrichter, E.O. (ed.): Late 20th century research on Chironomidae. Shaker Verlag, Aachen: 449-461.
Janecek, B., 2007. Chironomidae (Zuckmücken). Larven Teil V. – In: Kursunterlagen zu Fauna Aquatica Austriaca. Univ. f. Bodenkultur, Abt. Hydrobiologie. 1-128.
Janeček, B., 2010. Bestimmung und Bearbeitung von Chironomidae – Imagines und Puppen – aus Lichtfängen während des Joint Danube Survey 2 (JDS 2). – Univ. f. Bodenkultur Wien, Inst. f. Hydrobiologie und Gewässermanagement. 1-68.
Jankovi , M. 1974. Feeding and food assimiliation in larvae of *Prodiamesa olivacea*. – Ent. Tidskr. 95 Suppl.: 116-119.
Janssens de Bisthoven, L., E. van Looy, R. Ceusters, F. Gullentops & F. Ollevier, 1992. Densities of Prodiamesa olivacea (Meigen) (Diptera: Chironomidae) in a second order stream, the Laan (Belgium): relation to river dynamics. – Neth. J. Aq. Ecol. 26: 485-490.
Janzen, L., 2003. Typisierung und Bewertung von Fliessgewässern mit Hilfe der Chironomidae (Zuckmücken)-Fauna anhand des AQEM Datensatzes. – Diplomarbeit Univ. Duisburg-Essen Abt. Hydrobiol. 1-117.
Kaiser, T.S., D. Neumann, D.G. Heckel & T.U. Berendonk, 2010. Strong genetic differentiation and postglacial origin of populations in the marine midge *Clunio marinus* (Chironomidae, Diptera). – Molecular ecology 19: 2845-2857.
Kajak, Z., 1987. Determinants of maximum biomass of benthic Chironomidae (Diptera). – Ent. scand. Suppl. 29: 303-308.
Karlström, U., 1978. Role of the organic layer on stones in detrital metabolism in streams. – Verh. Internat. Verein. Limnol. 20: 1463-1470.
Kawecka, B., 1977. The food of dominant species of bottom fauna larvae in the River Raba (Southern Poland). – Acta Hydrobiol., Kraków 19: 191-213.
Kawecka, B. & A. Kownacki, 1974. Food conditions of Chironomidae in the River Raba. – Ent. tidskr. 95: 120-128.
Kawecka, B., A. Kownacki & M. Kownacka, 1978. Food relations between algae and bottom fauna communities in glacial streams. - Verh. Internat. Verein. Limnol. 20: 1527-1530.
Kesler, D.H., 1981. Grazing rate determination of Corynoneura scutellata Winnertz (Chironomidae: Diptera). – Hydrobiologia 80: 63-66.
Ketelaars, H.A.M., A.M.J.P. Kuijpers & L.W.C.A. van Breemen, 1992. Temporal and spatial distribution of chironomid larvae and oligochaetes in two Dutch storage reservoirs. – Neth. J. Aquat. Ecol. 26: 361-370.
Kettisch, J., 1936-1938. Zur Kenntnis der Morphologie und Ökologie der Larve von *Cricotopus trifasciatus*. – Konowia, 1936: 15: 248-263; 1937: 16: 153-163; 1938: 16: 193-204.
Kieffer, J.J. & A. Thienemann, 1908. Neue und bekannte Chironomiden und ihre Metamorphose. II. Chironomidenmetamorphosen. - Z. wissenschaftl. Insektenbiol. 4: 184-190, 277-286.
Kitching, R.L., 1972. Population studies of the immature stages of the tree-hole midge *Metriocnemus martinii* Thienemann (Diptera: Chironomidae). – J. Anim. Ecol. 41: 53-62.
Kleef, H. van, 2010. Identifying and crossing thresholds in managing moorland pool macroinvertebrates. – Thesis Radboud Univ., Nijmegen. 1-147.
Kleef, H. van & H. Esselink, 2005. Monitoring van abiotiek, vegetatie, dansmuggen en kokerjuffers in gerestaureerde zwakgebufferde oppervlaktewateren. Kolonisatie van diersoorten. – St. Bargerveen, Kath.Univ. Nijmegen. 1-52.
Klink, A., 1984. De Maas voor en na België. – Rapp. Meded. Hydrobiol. Adviesbur. Klink 10. 1-8.
Klink, A., 1985. Hydrobiologie van de Grensmaas. Huidig funktioneren, potenties en bedreigingen. – Rapp. Meded. Hydrobiol. Adviesbur. Klink 15: 1-38, app. pp.1-111.
Klink, A., 1986. Geschiedenis van de verzuring in Nederland. – Rapp. Meded. Hydrobiol. Adviesbur. Klink 27: 1-43, app. 1-10.
Klink, A., 1989. The lower Rhine: Palaeoecological analysis. – In: Petts, G.E. (ed.): Historical change of large alluvial rivers: Western Europe: 183-201.
Klink, A., 1990. Drift van makro-evertebraten in de Maas. – Rijkswaterstaat Dienst Binnenwateren / RIZA, nota 90.071. 45 pp. + app.
Klink, A., 1991. Maas 1986 – 1990. Evaluatie van 5 jaar hydrobiologisch onderzoek van makro-evertebraten. - Rapp. Meded. Hydrobiol. Adviesbur. Klink 39: 1-38 + app.
Klink, A., 2010. Macroinvertebrates of the Seine basin. - Rapp. Meded. Hydrobiol. Adviesbur. Klink 108: 1-62 + app.
Klink, A., 2010a. Paleoecologie en KRW-referentie voor laaglandbeken. Een pilotstudie in de Lunterse Beek (thans Oude Lunterse beek). – Rapp. Meded. Hydrobiol. Adviesbur. Klink 104: 1-36.

Klink, A., 2011. Macrofauna op bakenbomen in de bedijkte Maas. Een tussenstand na 4-5 jaar. – Rapp. Meded. Hydrobiol. Adviesbur. Klink 116: 1-28.

Klink, A., 2012. Egelbeek Vaassen: Rapportage van het hydrobiologisch onderzoek, uitgevoerd in 2010 en 2012. - Rapp. Meded. Hydrobiol. Adviesbur. Klink 122.

Klink, A.G. & H. Moller Pillot, 1982. Onderzoek aan de makro-evertebraten in de grote Nederlandse rivieren. – Wageningen/ Tilburg (private publ.). 57 pp.

Klink, A. & H. Moller Pillot, 1996. Lijst van de Nederlandse Chironomidae. – Werkgr. Ecol. Waterbeheer Themanr. 08. 10 pp. + table.

Klink, A.G. & H.K.M. Moller Pillot, 2003. Chironomidae larvae. Key to the higher taxa and species of the lowlands of Northwestern Europe. - CD-ROM, Expert center for Taxonomic Information, Amsterdam. (out of trade)

Klink, A. & J. Mulder, 1992. Inventarisatie van de ecologische waarden in oppervlaktewateren van de provincie Flevoland. - Rapp. Meded. Hydrobiol. Adviesbur. Klink 41. 1-42. + app.

Klink, A. & B. bij de Vaate, 1994. De Grensmaas en haar problemen zoals blijkt uit hydrobiologisch onderzoek aan makro-evertebraten. - Rapp. Meded. Hydrobiol. Adviesbur. Klink 53: 1-63.

Kobayashi, T., 1995. *Eurycnemus* sp. (Diptera: Chironomidae) larvae ectoparasitic on pupae of *Goera japonica* (Trichoptera: Limnephilidae). - In: Cranston, P. (ed.): Chironomids: From genes to ecosystems. CSIRO Publ., Melbourne: 317-322.

Kondo, S., 1996. Life cycle of *Hydrobaenus kondoi* Saether (Chironomidae) at the middle reaches of the Kiso river, Japan. – Hydrobiologia 318: 79-84.

Konstantinov, A.S., 1958. Biologiya khironomid i ikh razvedenie. - Trudy sarat. otdel. vses. nauchno-issled. Inst. ozer rech. ryb. khoz. 5: 1-362.

Konstantinov, A.S., 1971. Ecological factors affecting respiration in chironomid larvae. – Limnologica (Berlin) 8: 127-134.

Koskenniemi, E. & L. Paasivirta, 1987. The chironomid (Diptera) fauna in a Finnish reservoir during its first four years. – Ent. scand. Suppl. 29: 239-246.

Kouwets, F.A.C. & C. Davids, 1984. The occurrence of chironomid imagines in an area near Utrecht (the Netherlands) and their relations to water mite larvae. – Arch. Hydrobiol. 99: 296-317.

Kownacka, M. & A. Kownacki, 1972. Vertical distribution of zoocenoses in the streams of the Tatra, Caucasus and Balkan Mts. – Verh. Internat. Verein. Limnol. 18: 742-750.

Kownacki, A., 1989. Taxocens of Chironomidae as an indicator for assessing the pollution of rivers and streams. – Acta Biol. Debrecina Oecol. Hung. 3: 219-230.

Kraak, M.H.S., S.C. Stuijfzand & W. Admiraal, 2000. Interaction between organic matter and toxicants in polluted river water: beneficial and inhibiting effects on the midge *Chironomus riparius*. - In: Hoffrichter, E.O. (ed.): Late 20th century research on Chironomidae. Shaker Verlag, Aachen: 507-511.

Krasheninnikov, A.B., 2011. New data on chironomids (Diptera, Chironomidae) of the Middle Urals. - Vladimir Ya. Levanidov's Biennial Mem. Meetings 5: 247-264.

Krebs, B.P.M., 1981. Aquatische macrofauna van binnendijkse wateren in het Deltagebied. I. Zuid-Beveland. – Delta Inst. Hydrobiol. Onderz., Rapp. & Versl. 1981-8: 1-158.

Krebs 1984. Aquatische macrofauna van binnendijkse wateren in het Deltagebied. II. Zeeuws-Vlaanderen, oostelijk deel. – Delta Inst. Hydrobiol. Onderz., Rapp. en Versl. 1984-2: 1-124.

Krebs, B.P.M., 1990. Aquatische macrofauna van binnendijkse wateren in het Delta-gebied. IV: Schouwen-Duiveland. – Delta Inst. Hydrobiol. Onderz., Rapp. en versl. 1990-07: 1-124.

Krebs, B.P.M. & H.K.M. Moller Pillot, in prep. Influence of some environmental factors on the abundance of Chironomidae in a predominantly brackish water area.

Kretzschmar, A.U. & K. Böttger, 1994. Die Chironomidae (Diptera, Nematocera) eines kleinen naturnahen Baches im Nord-deutschen Tiefland (Kremper Au, Schleswig-Holstein). – Lauterbornia 19: 161-169.

Kreuzer, R., 1940. Limnologisch-ökologische Untersuchungen an holsteinischen Kleingewässern. – Arch. Hydrobiol. Suppl. 10: 359-572.

Kroon, H. de , H. de Jong & J.T.A. Verhoeven, 1985. The macrofauna distribution in brackish inland waters in relation to chlorinity and other factors. – Hydrobiologia 127: 265-275.

Kuijpers, A.M.J.P., H.A.M. Ketelaars & L.W.C.A. van Breemen, 1992. Chironomid pupal exuviae and larvae of two storage reservoirs in the Netherlands. – Neth. J. Aquat. Ecol. 26: 379-383.

Kuper, J.T. & H.K.M. Moller Pillot, 2012. *Metriocnemus carmencitabertarum* Langton & Cobo 1997: Een nieuwe dansmug voor Nederland (Chironomidae: Diptera). – Ned. Faun. Meded. 38: 49-54.

Kurazhskovskaya, T.N., 1969. Stroennie slyunnykh zhelez lichinok khironomid. – Trudy Inst. Biol. vnutr. Vod AN SSSR 19: 185-195.

Ladle, M., D.A. Cooling, J.S. Welton & J.A.B. Bass, 1985. Studies in Chironomidae in experimental recirculating stream systems. II. The growth, development and production of a spring generation of *Orthocladius (Euorthocladius) calvus* Pinder. – Freshw. Biol. 15: 243-255.

R

Ladle, M., J.S.Welton & J.A.B. Bass, 1984. Larval growth and production of three species of Chironomidae from an experimental recirculating stream. – Arch. Hydrobiol. 102: 201-214.
Lamberti, G.A. & J.W. Moore, 1984. Aquatic insects as primary consumers. – In: Resh, V.H. & D.M. Rosenberg (eds.): The ecology of aquatic insects. New York, Praeger Publ.: 164-195.
Langdon, P.G., Z. Ruiz, K.P. Brodersen & I.D.L. Foster, 2006. Assessing lake eutrophication using chironomids: understanding the nature of community response in different lake types. – Freshw. Biol. 51: 562-577.
Langton, P.H., 1980. The genus *Psectrocladius* Kieffer (Diptera: Chironomidae) in Britain. – Entomologist's Gazette 31: 75-88.
Langton, P.H., 1984. A key to pupal exuviae of British Chironomidae. – P.H. Langton, March (private publ.). 1-324.
Langton, P.H., 1991. A key to pupal exuviae of West Palaearctic Chironomidae. – P.H. Langton, Huntingdon (private publ.). 386 pp.
Langton, P.H., 1997. *Corynoneurella paludosa* Brundin, 1949: *Thienemanniella majuscula* Langton 1991 nec Edwards (1924) (Diptera, Chironomidae), new to Britain. – Dipterists Digest 1997: 20-21.
Langton, P.H. & J.J. Casas, 1997. Changes in chironomid community composition in two Mediterranean mountain streams over a period of extreme hydrological conditions. – Unpublished paper presented at chironomid symposium Freiburg.
Langton, P.H. & P.S. Cranston, 1991. Pupae in nomenclature and identification: West Palaearctic *Orthocladius* s.str. (Diptera: Chironomidae) revised. – Syst. Entomol. 16: 239-252.
Langton, P.H. & F. Cobo, 1997. *Metriocnemus (Inermipupa) carmencitabertarum* subgen. n., sp. n. (Diptera: Chironomidae) from Spain and Portugal. – Entomologist's Gaz. 48: 263-271.
Langton, P.H. & L.A. McLarnon, 1998. *Monodiamesa ekmani* Brundin (Diptera: Chironomidae), confirmed new to Britain and Ireland. – Br. J. Ent. Nat. Hist. 10: 196-202.
Langton, P.H. & H. Moller Pillot, 1997. The pupa and biology of *Metriocnemus picipes* (Meigen) (Diptera: Chironomidae). – Entomologist's Gaz. 48: 178-180.
Langton, P.H. & L.C.V. Pinder, 2007. Keys to the adult male Chironomidae of Britain and Ireland. – Freshw. Biol. Ass. Sc. Publ. 64: 239 + 168 pp., 276 figs.
Langton, P.H. & H. Visser, 2003. Chironomidae exuviae – a key to pupal exuviae of the West Palaearctic Region. – CD-ROM, Expert center for Taxonomic Information, Amsterdam. (out of trade)
Laville, H., 1971. Recherches sur les Chironomides (Diptera) lacustres du massif de Néouvielle (Hautes-Pyrénées). – Ann. Limnol. 7: 173-332.
Laville, H. & N. Giani, 1974. Phénologie et cycles biologiques des Chironomides de la zone littorale (0-7 m) du lac de Port-Bielh (Pyrénées centrales). – Ent. Tidskr. 95, 1974 Suppl.: 139-155.
Laville, H. & P. Lavandier, 1977. Les chironomides (Diptera) d'un torrent pyreneen de haute montagne: l'Estaragne. – Annales de Limnologie 13: 57-81.
Laville, H. & & M. Viaud-Chauvet, 1985. Etude comparée de la structure des peuplemants de Chironomidés dans cinq rivières du Massif Central: relation entre cette structure et la qualité des eaux. – Verh. Internat. Verein. Limnol. 22: 2261-2269.
Laville, H. & G.Vinçon, 1991. A typological study of Pyrenean streams: Comparative analysis of the Chironomidae (Diptera) communities in the Ossau and Aure valleys. – Verh. Internat. Verein. Limnol. 24: 1775-1784.
Learner, M.A., P.R. Wiles & J.G. Pickering, 1989. The influence of aquatic macrophyte identity on the composition of the chironomid fauna in a former gravel pit in Berkshire, England. – Aquatic Insects 11: 183-191.
Learner, M.A., R. Williams, M. Harcup & B. D. Hughes, 1971. A survey of the macro-fauna of the River Cynon, a polluted tributary of the River Taff (South Wales). – Freshw. Biol. 1: 339-367.
Lechthaler, W., 1993. Gesellschaften epiphytischer Macroevertebraten in überschwemmten Wiesen an der March (Niederösterreich). Synökologische Studien an Zoozönosen unter variierenden Flutungsverhältnissen, mit besonderer Berücksichtigung der Chironomidae (O. Diptera, UO. Nematocera). – Thesis Wien. 1-217.
Leentvaar, P., 1959. Hydrobiologische waarnemingen in de Leersumse Plassen. – De Levende Natuur 62: 52-58.
Lehmann, J., 1971. Die Chironomiden der Fulda. – Arch. Hydrobiol. Suppl. 37: 466-555.
Lehmann, J., 1972. Revision der europäischen Arten (Puppen ♂♂ und Imagines ♂♂) der Gattung *Eukiefferiella* Thienemann. – Beitr. Ent. 22: 347-405.
Lenz, F., 1939. Tendipedidae (Chironomidae). d) Subfamilie Podonominae. B. Die Metamorphose der Podonominae. – In: Lindner, E. (ed.): Die Fliegen der palaearktischen Region 13e : 5-16.
Lenz, F., 1939a. Tendipedidae (Chironomidae). e) Subfamilie Corynoneurinae. B. Die Metamorphose der Corynoneurinae. – In: Lindner, E. (ed.): Die Fliegen der palaearktischen Region 13f: 14-19.
Lenz, F., 1950. Tendipedinae (Chironomidae). h) Subfamilie Clunioninae. B. Die

Metamorphose der Clunioninae. – In: Lindner, E. (ed.): Die Fliegen der palaearktischen Region 13h: 8-23.
LeSage, L. & A.D. Harrison, 1980. The biology of *Cricotopus* (Chironomidae: Orthocladiinae) in an algal-enriched stream: Part I. Normal biology. – Arch. Hydrobiol. Suppl. 57: 375-418.
LeSage, L. & A.D. Harrison, 1980a. The biology of *Cricotopus* (Chironomidae: Orthocladiinae) in an algal-enriched stream: Part II. Effects of parasitism. – Arch. Hydrobiol. Suppl. 58: 1-25.
Leuven, R.S.E.W., J.A. van der Velden, J.A.M. Vanhemelrijk & G. van der Velde, 1987. Impact of acidification on chironomid communities in poorly buffered waters in the Netherlands. – Ent. scand. Suppl. 29: 269-280.
Lindegaard, C., 1980. Bathymetric distribution of Chironomidae (Diptera) in the oligotrophic Lake Thingvallavatn, Iceland. - In: Murray, D.A. (ed.): Chironomidae. Ecology, Systematics, Cytology and Physiology. Oxford, Pergamon Press: 225-232.
Lindegaard, C., 1995. Chironomidae (Diptera) of European cold springs and factors influencing their distribution. – J. Kansas Ent. Soc. 68 (2) suppl.: 108-131.
Lindegaard, C. & K.P. Brodersen, 1995. Distribution of Chironomidae (Diptera) in the river continuum. – In: Cranston, P. (ed.): Chironomids: From genes to ecosystems. CSIRO Publ., Melbourne: 257-271.
Lindegaard, C. & K.P. Brodersen, 2000. The influence of temperature on emergence periods of Chironomidae (Diptera) from a shallow Danish lake. – In: Hoffrichter, E.O. (ed.): Late 20th century research on Chironomidae. Shaker Verlag, Aachen: 313-324.
Lindegaard, C. & & P.M. Jónasson, 1979. Abundance, population dynamics and production of zoobenthos in Lake Myvatn, Iceland. – Oikos 32: 202-227.
Lindegaard, C. & E. Mortensen, 1988. Abundance, life history and production of Chironomidae (Diptera) in a Danish lowland stream. – Arch. Hydrobiol. Suppl. 81: 563-587.
Lindegaard, C., J. Thorup & M. Bahn, 1975. The invertebrate fauna of the moss carpet in the Danish spring Ravnkilde and its seasonal , vertical and horizontal distribution. – Arch Hydrobiol. 75: 109-139.
Lindegaard-Petersen, C., 1972. An ecological investigation of the Chironomidae (Diptera) from a Danish lowland stream (Linding Å). Arch. Hydrobiol. 69: 465-507.
Lloyd, L., 1937. Observations on sewage flies: their seasonal incidence and abundance. – J. Proc. Inst. Sewage Purification 1: 1-16.
Lloyd, L., 1941. The seasonal rhythm of a fly (*Spaniotoma minima*) and some theoretical considerations. – Trans. R. Soc. trop. Med. Hyg. 35: 93-104.
Lloyd, L., 1943. Materials for a study in animal competition. II. The fauna of the sewage bacteria beds. – Ann. appl. Biol. 30: 47-60.
Lloyd, L., I.F. Graham & T.B. Reynoldson, 1940. Materials for a study in animal competition. The fauna of the sewage bacteria beds. – Ann. appl. Biol. 27: 122-150.
Loo, D. van der, 2005. Sleutelfactoren voor het voorkomen van *Lasiodiamesa gracilis* en *Somatochlora arctica* in het Wooldse Veen. – Stichting Bargerveen, Nijmegen. 15 pp.
Luferov, V.P., 1971. The role of light in the populating of water bodies by epibiotic chironomid larvae. – Limnologica (Berlin) 8 (1): 139-140.
Lundström, J.O., Y. Brodin, M.L. Schäfer, T.Z. Persson Vinnersten & Ö. Östman, 2010. High species richness of Chironomidae (Diptera) in temporary flooded wetlands associated with high species turn-over rates. – Bull. Entomol. Res. 100: 433-444.
Maasri, A., S. Fayolle & E. Franquet, 2008. Epilithic chironomid community structure: is it a matter of algae? – Bol. Mus. Mun. Funchal. Suppl. 13: 133-140.
Macan, T.T., 1949. Survey of a moorland fishpond. – J. Anim. Ecol. 18: 160-186.
Mackey, A.P., 1976. Quantitative studies on the Chironomidae (Diptera) of the rivers Thames and Kennet. I. The *Acorus* zone. – Arch. Hydrobiol. 78: 240-267.
Mackey, A.P., 1977. Growth and development of larval Chironomidae. – Oikos 28: 270-275.
Mackey, A.P., 1979. Trophic dependencies of some larval Chironomidae (Diptera) and fish species in the River Thames. – Hydrobiologia 62: 241-247.
Madder, C.A., D.M. Rosenberg & A.P. Wiens, 1977. Larval cocoons in *Eukiefferiella claripennis* (Diptera: Chironomidae). – The Canadian Entomologist 109: 891-892.
Makarchenko, E.A. & M.A. Makarchenko, 2010. New data on the fauna and taxonomy of *Corynoneura* Winnertz (Diptera, Chironomidae, Orthocladiinae) for the Russian Far East and bordering territories. – Euroasian Ent. J. 9: 353-370 + II. (in Russian)
Markošová, R., 1979. Development of the periphytic community on artificial substrates in fish ponds. – Int. Revue ges. Hydrobiol. 64: 811-825.
Marziali, L., V. Lencioni, P. Parenti & B. Rossaro, 2008. Benthic macroinvertebrates as water quality indicators in Italian lakes. – Bol. Mus. Mun. Funchal Suppl. 13: 51-59.
Mason, C.F. & R.J. Bryant, 1975. Periphyton production and grazing by chironomids in Alderfen Broad, Norfolk. – Freshw. Biol. 5: 271-277.
Mat na, J., 1982. Benthos development in planktonic and nursery ponds. – Buletin VÚRH Vod any 4: 11-15.
Mat na, J., 1989. Seasonal dynamics of

a *Chironomus plumosus* (L.) (Diptera, Chironomidae) population from a fish pond in Soutern Bohemia. – Int. Revue ges. Hydrobiol. 74: 599-610.

Menzie, C.A., 1980. The chironomid (Insecta: Diptera) and other fauna of a *Myriophyllum spicatum* L. plant bed in the Lower Hudson River. – Estuaries 3: 38-54.

Meuche, A., 1938. Die Fauna im Algenbewuchs. – Thesis Stuttgart. Arch. Hydrobiol. 34: 349-520.

Michailova, P., 1982. External morphological and karyological characteristics of *Orthocladius bipunctellus* Zett., 1850 (Diptera, Chironomidae). – Zool. Anz. 208: 82-91.

Michailova, P., 1985. Cytotaxonomic review of some species of the genus *Orthocladius* van der Wulp (Diptera, Chironomidae). – Entomol. Abh. Staatl. Mus. Tierkunde Dresden 48: 149-165.

Michiels, S., 1999. Die Chironomidae (Diptera) der unteren Salzach. – Lauterbornia 36: 45-53.

Michiels, S., 2004. Die Zuckmücken (Diptera: Chironomidae) der Elz – ein Beitrag zur Limnofauna des Schwarzwaldes. – Mitt. bad. Landesver. Naturkunde u. Naturschutz N.F. 18: 111-128.

Millet, X., I. Munoz & N. Prat, 1987. The use of a transect sampling in estimating chironomid abundance and distribution. – Ent. scand. Suppl. 29: 323-330.

Milošević, D., V. Simić, M. Stojković, D. Čerba, D. Mančev, A. Petrović & M. Paunović, 2012. Spatio-temporal pattern of the Chironomidae community: toward the use of non-biting midges in bioassessment programs. – Aquat. Ecol. (publ. on line).

Minshall, G.W., 1984. Aquatic insect-substratum relationships. – In: Resh, V.H. & D.M. Rosenberg: The ecology of aquatic insects. New York, Praeger Publ.: 358-400.

Mol, A.W.M., M. Schreijer & P. Vertegaal, 1982. De makrofauna van de Maarsseveense plassen. – Rapp. Rijksinst. Natuurbeheer: 1-134, 1-187.

Moller Pillot, H.K.M., 1971. Faunistische beoordeling van de verontreiniging in laaglandbeken. – Tilburg. 285 pp.

Moller Pillot, H.K.M., 1984. De larven der Nederlandse Chironomidae (Diptera). Orthocladiinae sensu lato. – Nederl. Faun. Meded. 1B. 1-175.

Moller Pillot, H., 2003. Hoe waterdieren zich handhaven in een dynamische wereld. – St. Noordbr. Landsch., Haaren. 182 pp.

Moller Pillot, H., 2005. Influence of flooding on terrestrial chironomids in grassland (Diptera: Chironomidae). – Ned. Faun. Meded. 23: 113-123 (in Dutch).

Moller Pillot, H.K.M. 2008. Identification and ecology of the genus *Smittia* Holmgren in the Netherlands (Diptera: Chironomidae). – Tijdschrift voor Entomologie 151: 245-270.

Moller Pillot, H.K.M., 2009. Chironomidae Larvae of the Netherlands and adjacent lowlands. Biology and Ecology of the Chironomini. – KNNV Publ. Zeist. 1-270.

Moller Pillot, H.K.M. & R.F.M. Buskens, 1990. De larven der Nederlandse Chironomidae (Diptera). Autoekologie en verspreiding. – Nederl. faun. Meded. 1C. 87 pp.

Moller Pillot, H. & M.D. Moroz, 2002. Issledovanie dvukrylykh (Insecta: Diptera) iz rodnikovykh sistem Belarusi. – Vestsi Natsyyan. Akad. Navuk Belarusi, Ser. Biyal. Navuk 4: 93-97.

Moog, O. (ed.), 1995. Fauna aquatica Austriaca. Katalog zur autökologischen Einstufung aquatischer Organismen Österreichs. – Wien, Bundesministerium Land- und Forstwirtschaft, loose-leaf.

Moore, J.W. , 1979. Factors influencing algal consumption and feeding rate in *Heterotrissocladius changi* Saether and *Polypedilum nubeculosum* (Meigen) (Chironomidae: Diptera). – Oecologia 40: 219-227.

Moore, J.W., 1979a. Some factors influencing the distribution, seasonal abundance and feeding of subarctic Chironomidae (Diptera). – Arch. Hydrobiol. 85: 302-325.

Móra, A., 2008. Faunistic peculiarities of the chironomid assemblages (Diptera: Chironomidae) of the Upper-Tisza, NE Hungary. – Bol. Mus. Mun. Funchal Suppl. 13: 5-12.

Morris, D.L. & M.P. Brooker, 1980. An assessment of the importance of the Chironomidae (Diptera) in biological surveillance. - In: Murray, D.A. (ed.): Chironomidae. Ecology, Systematics, Cytology and Physiology. Oxford, Pergamon Press: 195-202.

Mossberg, P. & P. Nyberg, 1980. Bottom fauna of small acid forest lakes. – Report Inst. Freshw. Res. 58: 77-87.

Moubayed-Breil, J., 2008. Non-biting midges from Continental France: new records, faunal and biogeographical outline [Diptera, Chironomidae]. – Ephemera 9: 17-32.

Mozley, S.C., 1970. Morphology and ecology of the larva of *Trissocladius grandis* (Kieffer) (Diptera, Chironomidae) , a common species in the lakes and rivers of northern Europe. – Arch. Hydrobiol. 67: 433-451.

Mundie, J.H., 1957. The ecology of Chironomidae in storage reservoirs. – Trans. R. ent. Soc. Lond. 109: 149-232.

Munsterhjelm, G., 1920. Om Chironomidernas Ägglägging och Äggrupper. – Acta Soc. Fauna Flora Fenn. 47: 1-174.

Murray, D.A. & P. Ashe, 1985. A description of the adult female of *Buchonomyia thienemanni* Fittkau and a re-assessment of the phylogenetic position of the subfamily Buchonomyinae (Diptera, Chironomidae). – Spixiana Suppl. 11: 149-160.

Neumann, D., 1976. Adaptations of

chironomids to intertidal environments. – Ann. Rev. Entomol. 21: 387-414.
Nijboer, R. & P. Verdonschot (red.), 2001. Zeldzaamheid van de macrofauna van de Nederlandse binnenwateren. – Werkgr. Ecol. Waterbeheer, themanr. 19: 1-77. (Basal data unpublished).
Nolte, U., 1993. Egg masses of Chironomidae (Diptera). A review, including new observations and a preliminary key. – Ent. scand. Suppl. 43: 1-75.
Nolte, U., 1989. Chironomid communities of lotic mosses. – Acta Biol. Debrecina Oecol. hung. 3: 249-255.
Nolte, U., 1991. Seasonal dynamics of moss-dwelling chironomid communities. – Hydrobiologia 222: 197-211.
Nolte, U. & T. Hoffmann, 1992. Life cycle of *Pseudodiamesa branickii* (Chironomidae) in a small upland stream. - Neth. J. Aq. Ecol. 26: 309-314.
Odum, H.T., 1956. Primary production in flowing waters. – Limnol. Oceanogr. 1: 102-117.
Oliver, D.R., 1983. The larvae of Diamesinae (Diptera: Chironomidae) of the Holarctic region – Keys and diagnoses. – Ent. scand. Suppl. 19: 115-138.
Oliver, D.R. & M.E. Dillon, 1988. Review of *Cricotopus* (Diptera: Chironomidae) of the nearctic arctic zone with description of two new species. – Can. Ent. 120: 463-496.
Oliver, D.R. & B.J. Sinclair, 1989. Madicolous Chironomidae (Diptera), with a review of *Metriocnemus hygropetricus* Kieffer. – Acta Biol. Debrecina Suppl. Oecol. Hung. 2: 285-293.
Orendt, C., 1993. Vergleichende Untersuchungen zur Ökologie litoraler, benthischer Chironomidae und anderer Diptera (Ceratopogonidae, Chaoboridae) in Seen des nördlichen Alpenvorlandes. – Thesis München Univ. 315 pp.
Orendt, C., 1999. Chironomids as bioindicators in acidified streams: a contribution to the acidity tolerance of chironomid species, with a classification in sensitivity classes. – Internat. Rev. Hydrobiol. 84: 439-449.
Orendt, C., 2002. Die Chironomidenfauna des Inns bei Mühldorf (Oberbayern). – Lauterbornia 44: 109-120.
Orendt, C., 2002a. Biozönotische Klassifizierung naturnaher Flussabschnitte des nördlichen Alpenvorlandes auf der Grundlage der Zuckmücken-Lebensgemeinschaften (Diptera: Chironomidae). – Lauterbornia 44: 121-146.
Ortal, R. & F.D. Por, 1978. Effect of hydrological changes on aquatic communities in the Lower Jordan River. – Verh. Internat. Verein. Limnol. 20: 1543-1551.
Otto, C.-J., 1991. Benthonuntersuchungen am Belauer See (Schleswig-Holstein): eine ökologische, phaenologische und produktionsbiologische Studie unter besonderer Berücksichtigung der merolimnischen Insekten. – Thesis Kiel. 139 pp.
Paasivirta, L., 1974. Abundance and production of the larval Chironomidae in the profundal of a deep, oligotrophic lake in southern Finland. – Ent. Tidskr. 95: 188-194.
Paasivirta, L.. 2000. Chironomidae (Diptera) of the northern Baltic Sea. - In: Hoffrichter, E.O. (ed.): Late 20th century research on Chironomidae. Shaker Verlag, Aachen: 589-598.
Paasivirta, L., 2012. Chironomidae (Diptera: Nematocera) in the biogeographical provinces of Finland. – www.ymparisto.fi/vesihyonteisrhyma.
Pagast, F., 1931. Chironomiden aus der Bodenfauna des Usma-Sees in Kurland. – Folia Zool. Hydrobiol. 3: 199-248.
Pagast, F., 1936. Chironomidenstudien II. – Stett. Ent. Ztg. 97: 270—278.
Pagast, F., A. Thienemann & F. Krüger, 1941. Terrestrische Chironomiden VIII. *Metriocnemus fuscipes* Mg. und *Metriocnemus terrester* n.sp. Pagast. – Zool. Anz. 133: 202-213.
Palmén, E., 1959. Microcricotopus balticus n.sp. (Dipt., Chironomidae) aus dem Brackwasser des Finnischen Meerbusens. – Suom. Hyönt. Aikak. 25: 61-65.
Palomäki, R., 1989. The chironomid larvae in the different depth zones of the littoral in some Finnish lakes. – Acta Biol. Debr. Oecol. Hung. 3: 257-266.
Pankratova, V.Ya., 1964. Lichinki Tendipedid reki Oki. – Trudy Zool. Inst. AN SSSR 32: 189-207.
Pankratova, V.Ya., 1968. Lichinki khironomid (tendipedid) reki Nevy. – In: Ivanova, M.B. (ed.): Zagryaznenie i samoochishchenie reki Nevy. Trudy Zool. Inst. AN SSSR 45: 247-257.
Pankratova, V.Ya., 1970. Lichinki i kukolki komarov podsemejstva Orthocladiinae fauny SSSR (Diptera, Chironomidae = Tendipedidae). – Opred. po faune SSSR 102: 1-343.
Pedersen, B.V., 1971. Diptera Nematocera. – In: Tuxen, S.L. (ed.): Zoology of the Faroes at the expense of the Carlsberg-Fund. XLIIb. Copenhagen. 1-71.
Peeters, E.T.H.M., 1988. Hydrobiologisch onderzoek in de Nederlandse Maas. Macrofauna in relatie tot biotopen. – Rapp. LH Vakgr. Natuurbeheer. 150 pp.
Peters, A.J.G.P., R. Gijlstra & J.J.P. Gardeniers, 1988. Waterkwaliteitsbeoordeling van genormaliseerde beken met behulp van macrofauna. – STORA-rapport 88-06: 1-56 + app.
Petran, M. & P. Kothé, 1978. Influence of bedload transport on the macrobenthos of running waters. – Verh. Internat. Verein. Limnol. 20: 1867-1872.
Pinder, L.C.V., 1974. The Chironomidae of a

R

small chalk-stream in Southern England. – Ent. Tidskr. 95 Suppl.: 195-202.
Pinder, L.C.V., 1977. The Chironomidae and their ecology in chalk streams. – Freshw. Biol. Ass. Ann. Rep. 45: 62-69.
Pinder, L.C.V., 1978. A key to the adult males of the British Chironomidae (Diptera) the non-biting midges. – Freshw. Biol. Ass. Sci. Publ. 37: 1-169 + 189 figs.
Pinder, L.C.V., 1980. Spatial distribution of Chironomidae in an English chalk stream. - In: Murray, D.A. (ed.): Chironomidae. Ecology, Systematics, Cytology and Physiology. Oxford, Pergamon Press: 153-161.
Pinder, L.C.V., 1983. Observations on the life cycles of some Chironomidae in Southern England. - Mem. Amer. Ent. Soc. 34: 249-265.
Pinder, L.C.V., 1985. Studies in Chironomidae in experimental recirculating stream systems. I. *Orthocladius (Euorthocladius) calvus* sp. nov. – Freshw. Biol. 15: 235-241.
Pinder, L.C.V., 1986. Biology of freshwater Chironomidae. – Ann. Rev. Entomol. 31: 1-23.
Pinder, L.C.V. & P.D. Armitage, 1985. A description of the larva and pupa of *Chaetocladius melaleucus* (Meigen) (Diptera: Chironomidae). – Entomologist's Gazette 36: 119-124.
Pinder, L.C.V. & P.S. Cranston, 1976. Morphology of the male imagines of *Orthocladius (Pogonocladius) consobrinus* and *O. glabripennis* with observations on the taxonomic status of *O. glabripennis* (Diptera: Chironomidae). – Ent. scand. 7: 19-23.
Pinder, L.C.V. & I.S. Farr, 1987. Biological surveillance of water quality – 3. The influence of organic enrichment on the macroinvertebrate fauna of small chalk streams. – Arch. Hydrobiol. 109: 619-637.
Pinder, L.C.V., M. Ladle, T. Gledhill, J.A.B. Bass & A.M. Matthews, 1987. Biological surveillance of water quality – 1. A comparison of macroinvertebrate surveillance methods in relation to assessment of water quality, in a chalk stream. – Arch. Hydrobiol. 109: 207-226.
Potthast, A., 1914. Über die Metamorphose der *Orthocladius*-Gruppe. Ein Beitrag zur Kenntnis der Chironomiden. – Arch. Hydrobiol. Suppl. 2: 243-376.
Prat, N., M.-A. Puig, G. Gonzalez & X. Millet, 1983. Chironomid longitudinal distribution and macroinvertebrate diversity along the Llobregat River (NE Spain). – Mem. Amer. Entomol. Soc. 34: 267-275.
Przhiboro, A.A. & O.A. Saether, 2010. *Tvetenia vitracies* (Saether, 1969), a synonym of *Tvetenia tshernovskii* (Pankratova, 1968), comb. n. (Diptera: Chironomidae). – Zootaxa 2675: 57-64.
Punti, T., M. Rieradevall & N. Prat, 2009. Environmental factors, spatial variation, and specific requirements of Chironomidae in Mediterranean reference streams. – J. N.Am. Benthol. Soc. 28: 247-265.
Raddum, G.G. & O.A. Saether, 1981. Chironomid communities in Norwegian lakes with different degrees of acidification. – Verh. Internat. Verein. Limnol. 21: 399-405.
Raddum, G.G., G. Hagenlund & G.A. Halvorsen, 1984. Effects of lime treatment on the benthos of Lake Søndre Boksjø. – Report Inst. Freshw. Res., Drottningholm 61: 167-176.
Ramcharan, V. & C.G. Paterson, 1978. A partial analysis of ecological segregation in the chironomid community of a bog lake. – Hydrobiologia 58: 129-135.
Reiff, N., 1994. Chironomiden (Diptera: Nematocera) oberbayerischer Seen und ihre Eignung zur Trophieindikation. – Thesis München. 297 pp.
Reiss, F., 1968. Ökologische und systematische Untersuchungen an Chironomiden (Diptera) des Bodensees. Ein Beitrag zur lakustrischen Chironomidenfauna des nördlichen Alpenvorlandes. – Arch. Hydrobiol. 64: 176-323.
Reiss, F., 1984. Die Chironomidenfauna (Diptera, Insecta) des Osterseengebietes in Oberbayern. – Berichte Akad. Naturschutz u. Landschaftspflege 8: 186-194.
Reiss, F., 1984a. Chironomiden (Diptera, Insecta) aus dem Ampertal bei Schöngeising, Oberbayern. – Mitt. Zool. Ges. Braunau 4: 211-220.
Remmert, H. 1953. Zwei neue Chironomiden (Diptera) von der schleswig-holsteinischen Küste. – Kieler Meeresforschungen 9: 235-237.
Remmert, H., 1955. Ökologische Untersuchungen über die Dipteren der Nord- und Ostsee. - Arch. Hydrobiol. 51: 1-53.
Ring, R.A., 1989. Intertidal Chironomidae of British Columbia, Canada. – Acta Biol. Debrecina Oecol. Hung. 3: 275-288.
Ringe, F., 1974. Chironomiden-Emergenz 1970 in Breitenbach und Rohrwiesenbach. Schlitzer Produktionsbiologischen Studien (10). –-Arch. Hydrobiol. Suppl. 45: 212-304.
Ringe, F., 1976. *Heleniella serratosioi* n. sp., eine neue Orthocladiine (Dipt., Chir.) aus der Emergenz von Rohrwiesenbach und Kalkbach. – Arch. Hydrobiol. 77: 254-266.
Rodova, R.A., 1966. Razvitie *Cricotopus sylvestris* (Diptera, Chironomidae). – Trudy Inst. Biol. vnutr. Vod 12:199-213.
Rodova, R. A. & Yu. I. Sorokin, 1965. Kolichestvennye dannye o pitanii *Cricotopus sylvestris* F. – Trudy Instituty Biologii vnutrennykh vod 8: 110-112.
Röser, B. & A. Neumann , 1985. The chironomid fauna in a self-purifying stretch of brook. – Arch. Hydrobiol. 102: 367-378.
Rossaro, B., 1979. Description of the larva

of Paratrichocladius rufiventris (Diptera, Chironomidae. – Notulae Entomologicae 59: 75-78.
Rossaro, B., 1982. Chironomidi, 2. (Diptera Chironomidae: Orthocladiinae). – Ist. Zool. Univ. Milano, guide 16: 1-80.
Rossaro, B., 1991. Chironomids and water temperature. – Aquatic Insects 13: 87-98.
Rossaro, B., V. Lencioni & C. Casalegno, 2003. Revision of West Palaearctic species of *Orthocladius* s.str. van der Wulp, 1874 (Diptera: Chironomidae: Orthocladiinae), with a new key to species. – Studi Trentini Sc. Nat. – Acta Biol. 79: 213-241.
Rossaro, B. & S. Prato, 1991. Description of six new species of the genus *Orthocladius* (Diptera, Chironomidae). – Fragm. Entomol., Roma: 23: 59-68.
Ruse, L., 2002. Chironomid pupal exuviae as indicators of lake status. – Arch. Hydrobiol. 153: 367-390.
Ruse, L., 2002a. Colonisation of gravel lakes by Chironomidae. – Arch. Hydrobiol. 153: 391-407.
Ruse, L., 2012. Trait-based surveillance of flood channel effects on the River Thames. – Fauna norv. 31: 109-116.
Russev, B., 1972. Über die Migration der Rheobionten in Flieszgewässern. – Verh. Internat. Verein. Limnol. 18: 730-734.
Saether, O.A., 1967. Descriptions of *Lasiodiamesa bipectinata* n.sp. and *Parochlus kiefferi* (Garret) Brundin. – Beitr. Entomol. 17: 235-249.
Saether, O.A., 1969. Some Nearctic Podonominae, Diamesinae and Orthocladiinae (Diptera: Chironomidae). – Bull. Fish. Res. Board Canada 170: 1-154.
Saether, O.A., 1975. Nearctic and Palaearctic Heterotrissocladius (Diptera: Chironomidae). – Bull. Fish. Res. Board Canada 193: 1-67.
Saether, O.A., 1975a. Two new species of *Protanypus* Kieffer, with keys to Nearctic and Palaearctic species of the genus (Diptera: Chironomidae). – J. Fish. Res. Board Canada 32: 367-388.
Saether, O.A., 1976. Revision of *Hydrobaenus, Trissocladius, Zalutschia, Paratrissocladius,* and some related genera (Diptera: Chironomidae). – Bull. Fish. Res. Board Canada 195: 1-287.
Saether, O.A., 1977. Taxonomic studies on Chironomidae: *Nanocladius, Pseudochironomus* and the *Harnischia* complex. – Bull. Fish. Res. Bd. Canada 196: 1-143.
Saether, O.A., 1977a. Female genitalia in Chironomidae and other Nematocera. – Bull. Fish. Res. Board Canada 197: 1-209.
Saether, O.A., 1979. Chironomid communities as water quality indicators. – Holarctic ecology 2: 65-74.
Saether, O.A., 1979a. Hierarchy of the Chironomidae with special emphasis on the female genitalia. – Ent. scand. Suppl. 10: 17-26.
Saether, O.A., 1980. The females and immatures of *Paracricotopus* Thienemann & Harnisch, 1932, with the description of a new species (Diptera: Chironomidae). – Aquatic Insects 2: 129-145.
Saether, O.A., 1980a. The influence of eutrophication on deep lake benthic invertebrate communities. – Proc. Wat. Tech. 12: 161-180.
Saether, O.A., 1980b. The female and immatures of *Trissocladius brevipalpis* Kieffer, 1908 (Diptera: Chironomidae). – Ent. scand. 11: 467-472.
Saether, O.A., 1985. A review of the genus *Rheocricotopus* Thienemann & Harnisch, 1932, with the description of three new species. – Spixiana Suppl. 11: 59-108.
Saether, O.A., 1985a. Redefinition and review of *Thienemannia* Kieffer, 1909 (Diptera: Chironomidae), with the description of *T. pilinucha* sp.n. – Aquatic Insects 7: 111-131.
Saether, 1985b. The females of *Compteromesa oconeensis* Saether, 1981, and *Prodiamesa olivacea* (Meigen, 1818) (syn. *Trichodiamesa autumnalis* Goetghebuer, 1926, n.syn.). – Spixiana Suppl. 11: 7-13.
Saether, 1985c. A review of *Odontomesa*. – Spixiana Suppl. 11: 15-29.
Saether, O.A., 1989. *Metriocnemus* van der Wulp: a new species and a revision of species described by Meigen, Zetterstedt, Staeger, Holmgren, Lundström and Strenzke (Diptera: Chironomidae). – Ent. scand. 19: 393-430.
Saether, O.A., 1990. A review of the genus *Limnophyes* Eaton from the Holarctic and Afrotropical regions (Diptera: Chironomidae, Orthocladiinae). – Ent. scand. Suppl. 35: 1-139.
Saether, O.A., 1995. *Metriocnemus* van der Wulp: Seven new species, revision of species, and new records (Diptera: Chironomidae). – Annls. Limnol. 31: 35-64.
Saether, O.A. , 1997. First description of the imagines and pupa of *Propsilocerus jacuticus* (Zvereva) (Diptera: Chironomidae). – Acta Zool. Acad. Sci. Hung. 43: 241-249.
Saether, O.A., 2000. Phylogeny of the subfamilies of Chironomidae (Diptera). – Systematic Entomology 25: 393-403.
Saether, O. A., 2003. A review of *Orthocladius* subgen. *Symposiocladius* Cranston (Diptera: Chironomidae). – Aquatic Insects 25: 281-317.
Saether, O.A., 2005. A new subgenus and new species of *Orthocladius* van der Wulp, with a phylogenetic evaluation of the validity of the subgenera of the genus (Diptera: Chironomidae).
Saether, O.A. & T. Ekrem, 1999. *Molleriella*, a new terrestrial orthoclad genus from the Netherlands (Diptera: Chironomidae). – Acta Zool. Acad. Sci. Hung. 45: 161-168.
Saether, O.A. & L.C. Ferrington, 2003.

Nomenclature notes on some orthoclads (Diptera: Chironomidae). – Zootaxa 322: 1-7.
Saether, O.A. & G.A. Halvorsen, 1981. Diagnoses of *Tvetenia* Kieff. emend., *Dratnalia* n. gen., and *Eukiefferiella* Thien. emend., with a phylogeny of the *Cardiocladius* group (Diptera: Chironomidae). – Ent. scand. Suppl. 15: 269-285.
Saether, O.A. & L. Kristoffersen, 1996. Chironomids with "M-fork". A reevaluation of the wing venation of the *Corynoneura*-group. - Spixiana 19: 229-232.
Saether, O.A. & Ø. A. Schnell, 1988. Heterotrissocladius brundini spec. nov. from Norway (Diptera, Chironomidae). – Spixiana Suppl. 14: 57-64.
Saether, O.A. & Ø. A. Schnell, 1988a. Two new species of the *Rheocricotopus (R.) effusus* group (Diptera, Chironomidae). Spixiana Suppl. 14: 65-74.
Saether, O.A. & M. Spies, 2010. Fauna Europaea: Chironomidae. In: Beuk, P. & T. Pape (eds.) (2010) Fauna Europaea: Diptera, Nematocera. Fauna Europaea version 2.4, http://www.faunaeur.org.
Saether, O.A. & J. E. Sublette, 1983. A review of the genera *Doithrix* n. gen., *Georthocladius* Strenzke, *Parachaetocladius* Wülker and *Pseudorthocladius* Goetghebuer (Diptera: Chironomidae, Orthocladiinae). – Ent. scand. Suppl. 20: 1-100.
Saether, O.A. & X. Wang, 1995. Revision of the genus *Paraphaenocladius* Thienemann, 1924 of the world (Diptera: Chironomidae, Orthocladiinae). – Ent. scand. Suppl. 48: 1-69.
Saether, O.A. & X. Wang, 1996. Revision of the orthoclad genus *Propsilocerus* Kieffer (= *Tokunagayusurika* Sasa) (Diptera: Chironomidae). – Ent. scand. 27: 441-479.
Särkkä, J., 1983. A quantitative ecological investigation of the littoral zoobenthos of an oligotrophic Finnish lake. – Ann. Zool. Fennici 20: 157-178.
Saunders, L.G., 1924. On the early stages of *Cardiocladius* (Diptera: Chironomidae). – Entomologist's mon. Mag. 60: 227-231.
Scheibe, M.A., 2002. Beitrag zur Artenliste der aquatischen Zuckmücken (Diptera: Chironomidae) des Taunus. – Lauterbornia 44: 99-107.
Schiemer, F., 1968. Die Generationsentwicklung einiger Chironomiden im Litoral des Lunzer Untersees. – Ann. Zool. Fenn. 5: 126.
Schiffels, S., 2009. Commensal and parasitic Chironomidae. – Lauterbornia 68: 9-33.
Schlee, D., 1968. Vergleichende Merkmalsanalyse zur Morphologie und Phylogenie der *Corynoneura*-Gruppe (Diptera: Chironomidae). – Stuttg. Beitr. Naturkunde 180: 1-150.
Schleuter, A. 1985. Untersuchung der Makroinvertebratenfauna stehender Kleingewässer des Naturparkes Kottenforst-Ville unter besonderer Berücksichtigung der Chironomidae. – Thesis Bonn Univ. 217 pp.
Schmale, J.C., 1999. Hydrobiologisch onderzoek Berkheide 1994 – 1995 – 1996 – 1997. N.V. Duinwaterbedrijf Zuid-Holland. 73 pp. + 80 app.
Schmale, J.C. & H.G.J.M. van der Hagen, 1999. Biorestauratie Pan 13, Meijendel. Evaluatie ecologische aspecten. – N.V. Duinwaterbedrijf Zuid-Holland. 1-42. + 11 app.
Schmid, P.E., 1992. Habitat preferences as patch selection of larval and emerging chironomids (Diptera) in a gravel brook.. – Neth. J. Aquat. Ecol.26: 419-429.
Schmid, P.E., 1992a. Population dynamics and resource utilization by larval Chironomidae (Diptera) in a backwater area of the River Danube. – Freshw. Biol. 28: 111-127.
Schmid, P.E., 1993. A key to the larval Chironomidae and their instars from Austrian Danube region streams and rivers. Part I. Diamesinae, Prodiamesinae and Orthocladiinae. – Wasser und Abwasser, Suppl. 3/93. 514 pp. (out of trade)
Schmid, P.E., 1993a. Random patch dynamics of larval Chironomidae (Diptera) in the bed sediments of a gravel stream. – Freshw. Biol. 30: 239-255.
Schnabel, S. & P.-M. A. Dettinger-Klemm, 2001. Chironomidenfauna temporärer Tümpel in der Lahnaue – ökologische Betrachtungen. – Deutsche Ges. Limnologie (DGL) – Tagunsbericht 2000 (Magdeburg), Tutzing 2001: 429-433.
Schnell, Ø. A., 1991. New records of Chironomidae (Diptera) from Norway (II), with two new species synonyms. – Fauna norv. Ser. B 38: 5-10.
Serra-Tosio, B., 1989. Ecologie et biogéographie des *Boreoheptagyia* (Diptera, Chironomidae, Diamesinae). – Acta Biol. Debr. Oecol. Hung. 3: 289-294.
Serra-Tosio, B., 1989a. Révision des espèces ouest-paléarctiques et néarctiques de *Boreoheptagyia* Brundin avec des clés pour les larves, les nymphes et les imagos. – Spixiana 11: 133-173.
Serra-Tosio, B. & H.Laville, 1991. Liste annotée des Diptères Chironomidés de France continentale et de Corse. – Annls Limnol. 27: 37-74.
Shcherbina, G.H., 1989. Ecology and production of monocyclic species of Chironomidae (Diptera) from Lake Vishtynetskoe of the Kaliningrad region (USSR). – Acta Biol. Debr. Oecol. Hung. 3: 295-303.
Shilova, A.I., 1976. Chironomidy Rybinskogo vodohranilishcha. – Izd. Nauka, Leningrad. 252 pp.
Shilova, A.I. & T.N. Kurazhskovskaya, 1980. Taksonomicheskie i adaptivnie ocobennosti stroeniya kishechnika lichinok khironomid.

- Trudy Inst. Biol. vnutr. vod 44 (47): 101-109.
Shilova, A.I. & N.I. Zelentsov, 1972. Vliyanie fotoperiodizma na diapauzu u khironomid. – Inf. Byull. Inst. Biol. vnutr. Vod 13: 37-42.
Silver, P., M.A. Palmer, C.M.Swan & D. Wooster, 2002. The small scale ecology of freshwater meiofauna. – In: Rundle et al. (eds.): Freshwater Meiofauna: Biology and Ecology: 217-239. Backhuys Publ., Leiden, The Netherlands.
Simpson, K.W., 1983. Communities of Chironomidae (Diptera) from an acid-stressed headwater stream in the Adirondack Mountains, New York. – Mem. Amer. Entomol. Soc. 34: 315-327.
Singh, M.P. & A.D. Harrison, 1984. The chironomid community (Diptera: Chironomidae) in a Southern Ontario stream and the annual emergence patterns of common species. – Arch. Hydrobiol. 99: 221-253.
Sládeček, V., 1973. System of water quality from the biological point of view. – Arch. Hydrobiol. Beiheft 7: 1-218.
Smissaert, H.R., 1959. Limburgse beken. – Natuurhist. Maandblad 48: 7-78.
Smit, H., 1982. De Maas. Op weg naar biologische waterbeoordeling van grote rivieren. – Rapp. LH Wageningen, Vakgr. Natuurbeheer 667: 1-100.
Smit, H., G. van der Velde & S. Dirksen, 1996. Chironomid larval assemblages in the enclosed Rhine-Meuse Delta: spatio-temporal patterns in an exposure gradient on a tidal sandy flat. – Arch. Hydrobiol. 137: 487-510.
Soponis, A.R., 1977. A revision of the Nearctic species of *Orthocladius (Orthocladius)* van der Wulp (Diptera: Chironomidae). – Mem. ent. Soc. Can. 102: 1-187.
Soponis, A.R., 1986. The transfer of *Orthocladius rusticus* Goetghebuer to *Chaetocladius* with a redesription of the type (Diptera: Chironomidae). – Ent. scand. 17: 299-300.
Soponis, A.R., 1987. Notes on *Orthocladius (Orthocladius) frigidus* (Zetterstedt) with a redescription of the species (Diptera: Chironomidae). – Ent. scand. Suppl. 29: 123-131.
Soponis, A.R., 1990.A revision of the Holarctic species of Orthocladius (Euorthocladius) (Diptera: Chironomidae). – Spixiana Suppl. 13: 1-68.
Sorokin, Ju.I., 1968. The use of ^{14}C in the study of nutrition of aquatic animals. – Mitt. int. Ver. Limnol. 16: 1-41.
Spänhoff, B., C. Alecke & E.I. Meyer, 2000. Colonization of submerged twigs and branches of different wood genera by aquatic macroinvertebrates. - Int. Revue Hydrobiol. 85: 49-66.
Speaker, R., K. Moore & S. Gregory, 1984. Analysis of the process of retention of organic matter in stream ecosystems. - Verh. Internat. Verein. Limnol. 22: 1835-1841.
Spies, M. & O.A. Saether, 2004. Notes and recommendations on taxonomy and nomenclature of Chironomidae (Diptera). – Zootaxa 752: 1-90.
Srokosz, K., 1980. Chironomidae communities of the river Nida and its tributaries. – Acta Hydrobiol. 22: 191-215.
Steenbergen, H.A., 1993. Macrofauna-atlas van Noord-Holland: verspreidingskaarten en responsies op milieufactoren van ongewervelde waterdieren. – Prov. Noord-Holland, Dienst Ruimte en Groen. Haarlem. 650 pp.
Steinhart, M., 1998. Einflüsse der saisonalen Überflutung auf die Chironomidenbesiedlung (Diptera) aquatischer und amphibischer Biotope des Unteren Odertals. – Thesis Berlin. Shaker Verlag, Aachen: 1-117.
Steinhart, M., 2000. How do Chironomidae (Diptera) cope with changing water levels in a floodplain? - In: Hoffrichter, E.O. (ed.): Late 20th century research on Chironomidae. Shaker Verlag, Aachen: 415-423.
Storey, A.W., 1987. Influence of temperature and food quality on the life history of an epiphytic chironomid. – Ent. scand. Suppl. 29: 339-347.
Strenzke, K., 1950. Systematik, Morphologie und Ökologie der terrestrischen Chironomiden. – Arch. Hydrobiol. Suppl. 18: 207-414.
Strenzke, K., 1950a. Die Pflanzengewässer von *Scirpus sylvaticus* und ihre Tierwelt. – Arch. Hydrobiol. 44: 123-170.
Strenzke, K., 1960. Terrestrische Chironomiden. XIX-XXIII. – Deutsche Ent. Z. N.F. 7: 414-441.
Stur, E., P. Martin & T. Ekrem, 2005. Non-biting midges as hosts for water mite larvae in spring hanitats in Luxembourg. – Ann. Limnol. – Int. J. Lim. 41: 225-236.
Stur, E. & O.A. Saether, 2004. A new hairy-winged *Pseudorthocladius* (Diptera: Chironomidae) from Luxemburg. – Aq. Insects 26: 79-83.
Stur, E. & M. Spies, 2011. Description of *Chaetocladius macrovirgatus* sp. n., with a review of *C. suecicus* (Kieffer) (Diptera: Chironomidae). – Zootaxa 2762: 37-48.
Stur, E. & S. Wiedenbrug, 2005. Two new Orthoclad species (Diptera: Chironomidae) from cold water springs of the Nationalpark Berchtesgaden, Germany. – Aq. Insects 27: 125-131.
Svensson, B.S., 1976. The association between *Epoicocladius ephemerae* Kieffer (Diptera: Chironomidae) and *Ephemera danica* Müller (Ephemeroptera). – Arch. Hydrobiol. 77: 22-36.
Svensson, B.S., 1979. Pupation, emergence and fecundity of phoretic *Epoicocladius*

ephemerae (Chironomidae). – Holarctic Ecol. 2: 41-50.
Svensson, B.S., 1986. *Eukiefferiella ancyla* sp. n. (Diptera: Chironomidae) a commensalistic midge on *Ancylus fluviatilis* Müller (Gastropoda: Ancylidae). – Ent. scand. 17: 291-298.
Swanson, S.M. & U.T. Hammer, 1983. Production of *Cricotopus ornatus* (Meigen) (Diptera: Chironomidae) in Waldsea Lake, Saskatchewan. – Hydrobiologia 105: 155-164.
Syrovátka, V., J. Bojková & V. Rádková, 2012. The response of chironomid assemblages to mineral richness gradient in the Western Carpathian helocrenes. – Fauna norv. 31: 117-124.
Syrovátka, V. & K. Brabec 2010. The response of chironomid assemblages (Diptera: Chironomidae) to hydraulic conditions: a case study in a gravel-bed river. – Fundam. Appl. Limnol. 178:43-57.
Syrovátka, V., J. Schenková & K. Brabec, 2009. The distribution of chironomid larvae and oligochaetes within a stony-bottomed river-stretch: the role of substrate and hydraulic characteristics. – Arch. Hydrobiol. 174: 43-62.
Tang, H., Song, M.-Y., Cho, W.-S., Park, Y.-S. & Chon, T.-S., 2010. Species abundance distribution of benthic chironomids and other macroinvertebrates across different levels of pollution in streams. – Ann. Limnol. – Int. J. Lim. 46: 53-66.
Tarkowska-Kukuryk, M. & Mieczan, T., 2008. Diet composition of epiphytic chironomids of the *Cricotopus sylvestris* group (Diptera: Chironomidae) in a shallow hypertrophic lake. – Aquatic Insects 30: 285-294.
Thienemann, A., 1926. Hydrobiologische Untersuchungen an Quellen. VII. Insekten aus norddeutschen Quellen mit besonderer Berücksichtigung der Dipteren. Deutsche ent. Z 1926: 1-50.
Thienemann, A., 1936. Chironomiden-Metamorphosen. XI. Die Gattung *Eukiefferiella*. – Stettiner Entomol. Z. 97: 43-65.
Thienemann, A., 1936a. Chironomiden - Metamorphosen. XIII. (Dipt.). Die Gattung *Dyscamptocladius* Thien. – Mitt. Deutsch. Ent. Ges. 7: 49-54.
Thienemann, A., 1937. Chironomiden-Metamorphosen (Diptera). XV. – Mitt. entomol. Ges. Halle 15: 22-36.
Thienemann, A., 1937a. Chironomiden aus Lappland. III. Beschreibung neuer Metamorphosen, mit einer Bestimmungstabelle der bisher bekannten *Metriocnemus*larven und –puppen. – Stetinn. ent. Ztg. 98: 165-185.
Thienemann, A., 1938. Chironomiden-Metamorphosen. XVI. – Encycl. ent. Diptera 9: 87-96.
Thienemann, A., 1944. Bestimmungstabellen für die bis jetzt bekannten Larven und Puppen der Orthocladiinen (Diptera Chironomidae). – Arch. Hydrobiol. 39: 551-664.
Thienemann, A., 1952. Bestimmungstabelle für die Larven der mit *Diamesa* verwandten Chironomiden. – Beitr. Entomol. 2: 244-256.
Thienemann, A., 1954. *Chironomus*. Leben, Verbreitung und wissenschaftliche Bedeutung der Chironomiden. – Binnengewässer 20: 1-834.
Titmus, G., 1979. The emergence of midges (Diptera: Chironomidae) from a wet gravel-pit. – Freshw. Biol. 9: 165-179.
Tokeshi, M., 1986. Population dynamics, life histories and species richness in an epiphytic chironomid community. – Freshw. Biol. 16: 431-441.
Tokeshi, M., 1986a. Resource utilisation, overlap and temporal community dynamics: a null model analysis of an epiphytic chironomid community. J. Anim. Ecol. 55: 491-506.
Tokeshi, M., 1986b. Population ecology of the commensal chironomid *Epoicocladius flavens* on its mayfly host *Ephemera danica*. – Freshw. Biol. 16: 235-243.
Tokeshi, M. & L.C.V. Pinder, 1985. Microhabitats of stream invertebrates on two submersed macrophytes with contrasting leaf morphology. – Holarctic Ecology 8: 313-319.
Tokeshi, M. & C.R. Townsend, 1987. Random patch formation and weak competition: coexistence in an epiphytic chironomid community. – J. Animal Ecol. 56: 833-845.
Tokunaga, M., 1935. Chironomidae from Japan (Diptera). IV. The early stages of a marine midge, *Telmatogeton japonicus* Tok. – Philipp. J. Sci. 57: 491-511.
Tolkamp, H. H., 1980. Organism-substrate relationships in lowland streams. – Thesis Wageningen. 211 pp.
Tõlp, Õ, 1971. Chironomid larvae in the brackish waters of Estonia. – Limnologica, Berlin 8: 95-97.
Tomlinson, T.G., 1946. Animal life in percolating filters. – Techn. Paper Dept. Sc. Industr. Res. Water Pollution Research. London 9: 1-19.
Toscano, R.J. & A.J. McLachlan, 1980. Chironomids and particles: microorganisms and chironomid distribution in a peaty upland river. – In: Murray, D.A. (ed.): Chironomidae. Ecology, Systematics, Cytology and Physiology. Oxford, Pergamon Press: 171-177.
Tourenq, J.-N., 1975. Recherches écologiques sur les chironomides (Diptera) de Camargue. – Thesis. Toulouse. 424 pp.
Townsend, C.R. & A.G. Hildrew, 1976. Field experiments on the drifting, colonization and ans continuous redistribution of stream benthos. – J. Anim. Ecol. 45: 759-772.
Townsend, C.R. & A.G. Hildrew, 1979. Resource partitioning by two freshwater invertebrate predators with contrasting

foraging strategies. – J. Anim. Ecol. 48: 909-920.
Townsend, C.R., A.G. Hildrew & K. Schofield, 1987. Persistence of stream invertebrate communities in relation to environmental varaibility. – J. Anim. Ecol. 56: 597-613.
Tshernovskij see Chernovskij.
Tuiskunen, J., 1985. *Tavastia australis*, a new genus and species (Diptera, Chironomidae, Orthocladiinae) from Finland. – Ann. Ent. Fenn. 51: 30-32.
Tuiskunen, J., 1986. The Fennoscandian species of *Parakiefferiella* Thienemann (Diptera, Chironomidae, Orthocladiinae). – Ann. Zool. Fenn. 23: 175-196.
Vala, J.-C., J. Moubayed & P. Langton, 2000. Chironomidae des rizières de Camargue, données faunistiques et écologiques (Diptera). – Bull. Soc. entomol. France 105: 293-300.
Valkanov, A., 1949. *Thalassomyia frauenfeldi* Schiner vom Schwarzen Meer. – Arb. Biol. Meer. Stat. Varna 14 (1948): 103-111. (in Bulgarian with summary in German)
Vallenduuk, H.J. & H.K.M. Moller Pillot, 2007. Chironomidae larvae of the Netherlands and adjacent lowlands. General ecology and Tanypodinae. – KNNV Publ. Zeist. 144 pp.
Van der Loo see: Loo, van der
Van Kleef see: Kleef, van
Van Orsouw, C. & I. Kolenbrander, 2004. *Lasiodiamesa gracilis*; indicatorsoort voor bijzondere gradiëntsituaties in hoogveengebieden. - Stichting Bargerveen, Nijmegen. Unpublished report, 16 pp.
Verberk, W.C.E.P., D.T. Bilton, P. Calosi & J.I. Spicer, 2011. Oxygen supply in aquatic ectotherms: Partial pressure and solubility together explain biodiversity and size patterns. – Ecology 92: 1565-1572.
Verberk, W.C.E.P., G.-J. A. van Duinen, H.K.M. Moller Pillot & H. Esselink, 2003. *Lasiodiamesa gracilis* (Chiornomidae: Podonominae) new for the Dutch fauna. – Entomol. Ber. 63: 40-42.
Verberk, W.C.E.P., H.H. van Kleef, M. Dijkman, P. van Hoek, P. Spierenburg & H. Esselink, 2005. Seasonal changes on two different spatial scales: response of aquatic invertebrates to water body and microhabitat. – Insect Sc. 12: 263-280.
Verberk W.C.E.P., R.S.E.W. Leuven, G.-J.A. van Duinen & H. Esselink, 2010. Loss of environmental heterogeneity and aquatic macroinvertebrate diversity following large-scale restoration management. - Basic and Applied Ecology 11: 440-449.
Verberk, W.C.E.P., P.F.M. Verdonschot, T. van Haaren, B. van Maanen, 2012. Milieu- en habitat-preferenties van Nederlandse zoetwatermacrofauna. - WEW Themanummer 23, Van de Garde-Jémé, Eindhoven. 32 pp. + table.
Verdonschot, P.F.M., L.W.G. Higler, W.F. van der Hoek & J.G.M. Cuppen, 1992. A list of macroinvertebrates in Dutch water types: a first step towards an ecological classification of surface waters based on key factors. – Hydrobiol. Bull. 25: 241-259.
Verdonschot, P. & W. Lengkeek, 2006. Habitat preferences of selected indicators. – Deliverable No. 92. Euro-limpacs. 40 pp.
Verdonschot, P.F.M. & J.A. Schot, 1987. Macrofaunal community types in helocrene springs. – Ann. Rep. Res. Inst. Nature Management 1986: 85-103. Arnhem, Leersum and Texel.
Verneaux, V. & L. Aleya, 1998. Spatial and temporal distribution of chironomid larvae (Diptera: Nematocera) at the sediment-water interface in Lake Abbaye (Jura, France). – Hydrobiologia 373/374: 169-180.
Vogels, J, A. van den Burg, E. Remke & H. Siepel, 2011. Effectgerichte maatregelen voor het herstel en beheer van faunagemeenschappen van heideterreinen. Evaluatie en ontwerp van bestaande en nieuwe herstelmaatregelen (2006-2010). – Bosschap, Driebergen. Report 2011/OBN152-DZ.
Waajen, G.W.A.M., 1982. Hydrobiologie van veenputten in de Mariapeel en de Liesselse Peel. – LH Wageningen, sektie Hydrobiol., verslag 82-1. 1-67.
Walentowicz, A.T. & A.J. McLachlan, 1980. Chironomids and particles: a field experiment with peat in an upland stream. - In: Murray, D.A. (ed.): Chironomidae. Ecology, Systematics, Cytology and Physiology. Oxford, Pergamon Press: 179-185.
Walker, I.R., D.R. Oliver & M.E. Dillon, 1992. The larva and habitat of *Parakiefferiella nigra* Brundin (Diptera: Chironomidae). – Neth. J. Aq. Ecol. 26: 527-531.
Walshe, B.M., 1948. The oxygen requirements and thermal resistance of chironomid larvae from flowing and from still waters. – J. Exp. Biol. 25: 35-44.
Wang, X., 2000. Nuisance chironomid midges recorded from China. - In: Hoffrichter, E.O. (ed.): Late 20th century research on Chironomidae. Shaker Verlag, Aachen: 653-658.
Wang, X. & O.A. Saether, 2001. The larvae of *Propsilocerus sinicus* Saether & Wang and *P. paradoxus* (Lundström) (Diptera: Chironomidae). – Aquatic Insects 23: 141-145.
Ward, G.M. & K.W. Cummins, 1979. Effects of food quality on growth of a stream detritivore, *Paratendipes albimanus* (Meigen) (Diptera: Chironomidae). – Ecology 60: 57-64.
Warmke, S. & D. Hering, 2000. Composition, microdistribution and food of the macroinvertebrate fauna inhabiting wood in low order mountain streams in Central Europe. – Internat. rev. Hydrobiol. 85: 67-78.
Werkgroep Hydrobiologie MEC Eindhoven, 1993. De Groote Peel als leefmilieu voor

aquatische macrofauna. – M.E.C. Eindhoven, 99 pp. + bijl.
Wesenberg-Lund, C., 1943. Biologie der Süsswasserinsekten. – Berlin, Wien: Springer. 1-682.
Whitman, R.L. & W.J. Clark, 1984. Ecological studies of the sand-dwelling community of an East Texas stream. – Freshw. Invertebr. Biol. 3: 59-79.
Wiley, M.J., 1981. Interacting influences of density and preference on the emigration rates of some lotic chironomid larvae (Diptera: Chironomidae). – Ecology 62: 426-438.
Wiley, M.J. & G. L. Warren, 1992. Territory abandonment, theft, and recycling by a lotic grazer: a foraging strategy for hard times. – Oikos 63: 495-505.
Williams, D.D. & H.B.N. Hynes 1974. The occurrence of benthos deep in the substratum of a stream. - Freshw. Biol. 4: 233-256.
Williams, K.A., 1981. Population dynamics of epiphytic chironomid larvae in a chalk stream. – Ph. D. thesis, Univ. of Reading. 1-317.
Wilson, R.S., 1977. Chironomid pupal exuviae in the River Chew. – Freshw. Biol. 7: 9-17.
Wilson, R.S., 1987. Chironomid communities in the River Trent in relation to water chemistry. – Ent. scand. Suppl. 29: 387-393.
Wilson, R.S., 1988. A survey of the zinc-polluted river Nent (Cumbria) and the East and West Allen (Northumberland), England, using chironomid pupal exuviae. – Spixiana Suppl. 14: 167-174.
Wilson, R.S., 1989. The modification of chironomid pupal exuvial assemblages by sewage effluent in rivers within the Bristol Avon catchment, England. – Acta Biol. Debrecina Oecol. Hung. 3: 367-376.
Wilson, R.S. & L.P. Ruse, 2005. A guide to the identification of genera of chironomid pupal exuviae occurring in Britain and Ireland and their use in monitoring lotic and lentic fresh waters. – Freshw. Biol. Ass., Special Publ. 13: 1-176.
Wilson, R.S. & S.E. Wilson, 1984. A survey of the distribution of Chironomidae (Diptera, Insecta) of the River Rhine by sampling pupal exuviae. – Hydrobiol. Bull. 18: 119-132.
Winner, R.W., 1984. The toxicity and bioaccumulation of cadmium and copper as affected by humic acid. – Aquatic Toxicology 5: 267-274.
Wolfram, G., 1996. A faunistic review of the chironomids of Neusiedler See (Austria) with a description of a new pupal exuviae. – Ann. Naturhist. Mus. Wien 98B: 513-523.
Wotton, R.S., P.D. Armitage, K. Aston, J.H. Blackburn, M. Hamburger & C.A. Woodward, 1992. Colonization and emergence of midges (Chironomidae: Diptera) in slow sand filter beds. – Neth. J. Aquat. Ecol. 26: 331-339.
Wülker, W., 1956. Zur Kenntnis der Gattung *Psectrocladius* Kieff. (Dipt., Chironom.). – Arch. Hydrobiol. Suppl. 24: 1-66.
Wülker, W., 1957. Über die Chironomiden der *Parakiefferiella*-Gruppe (Diptera, Tendipedidae, Orthocladiinae). – Beiträge z. Entomologie 7: 411-429.
Zavřel, J., 1939. Chironomidarum Larvae et Nymphae. II. (Genus Eukiefferiella Th.). – Acta Soc. Scient. nat. morav. 11 (10): 1-29.
Zelentsov, N.I., 1980. K sistematike roda *Psectrocladius* Kieff. – podrod *Psectrocladius* s.str. Wülk. (Diptera, Chironomidae). – Trudy Inst. Biol. vnutr. vod 41 (44): 192-231.
Zelentsov, N.I., 1980a. Rannie stadii pazvitiya i biologiya *Stackelbergina praeclara* Shilova et Zelentsov (Diptera, Chironomidae). - Trudy Inst. Biol. vnutr. vod 41 (44): 232-238.
Zelentsov, N.I., 1980b. Reviziya Pamirskikh ortokladiin roda *Psectrocladius* Kieff. (Diptera, Chironomidae). - Trudy Inst. Biol. vnutr. vod 44 (47): 110-135.
Zelentsov, N.I., 1983. Metamorphosis and biology of *Psectrocladius ventricosus* (Diptera, Chironomidae). – Zool. Zh. 62: 725-731. (In Russian)
Zelinka, M. & P. Marvan, 1961. Zur Präzisierung der biologischen Klassifikation der Reinheit fliessender Gewässer. – Arch. Hydrobiol. 10: 453-469.
Zinchenko, T. D. , 2002. Chironomids of surface waters in the Mid and Lower Volga basin (Samara district). Ecological and faunal review. – Togliatti, 174 pp. (in Russian)
Zinchenko, T.D., E.I. Izvekova & Yu.B. Semenov, 1986. Pishchevoe povedenie lichinok *Cricotopus bicinctus* Meig. i *Orthocladius oblidens* Walk. – khironomid-obrastateljej vodoprovodnogo kanala. – In: Gagarin, V.G. (red.): Povedenie vodnykh bespozvonochnykh. Materialy IV Vsesoyuznogo simposiuma Borok. 1983: 130-135. Andropov.

ACKNOWLEDGEMENTS

This book was made possible by the material or information sent by many colleagues, especially Christine Becker (Germany), Hub Cuppen (Eerbeek), Albert Dees, Hein van Kleef and Jan Kuper (Stichting Bargerveen, Nijmegen), Niels Evers (Limnodata), Ton van Haaren, Amy Storm and David Tempelman (Grontmij | Aqua Sense, Amsterdam), Berthold Janecek (Austria), Alexander Klink (Wageningen), Peter Langton (Northern Ireland), Evgeniy Makarchenko (Russia), Susanne Michiels (Germany), Lauri Paasivirta (Finland), Andrey Przhiboro (Russia), Nicola Reiff (Germany), Mark Scheepens (GWL, Boxtel), Ole Sæther (Norway), Martin Spies (Germany), Elisabeth Stur (Norway), G erard van der Velde (Nijmegen) and Piet Verdonschot (Alterra). Maria Sanabria made the mentum drawings and helped with some of the field work. During the field work Sabine Schiffels, Mieke Moeleker, David Tempelman, Monique Korsten and Barend van Maanen also rendered assistance. Jarmila Lešková translated some texts from Slovak language. The many discussions with Ronald Buskens and Henk Vallenduuk were very valuable and I received helpful comments on the text from Djuradj Milošević (Serbia), Boudewijn Goddeeris (Belgium), Mieke Moeleker, Gert-Jan van Duinen and Wilco Verberk. Godard Tweehuysen and Danny Boomsma helped with the literature search. Derek Middleton (Zevenaar) corrected my English and copy-edited the text.

INDEX

Only the first page of the descriptions in Chapter 3-8 and the pages of the tables in Chapter 9 are given. Synonyms are given in italics.

OTHER AVAILABLE TITLES BY KNNV PUBLISHING

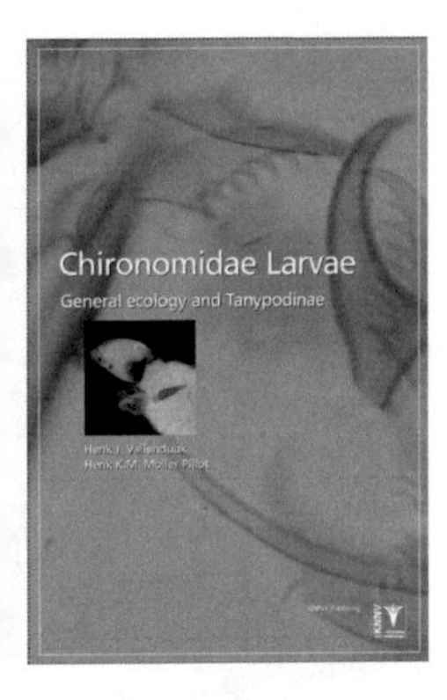

CHIRONOMIDAE LARVAE

General ecology and Tanypodinae
Henk J. Vallenduuk
Henk K.M. Moller Pillot

This book shows identification keys and additional information on the biology and ecology of Tanypodinae Larvae and contains information on the general ecology of the whole family Chironomidae Larvae.

Hardcover, 2007, 16,5 x 24 cm,
144 pp, € 69,50, English,
978-90-5011-295-8

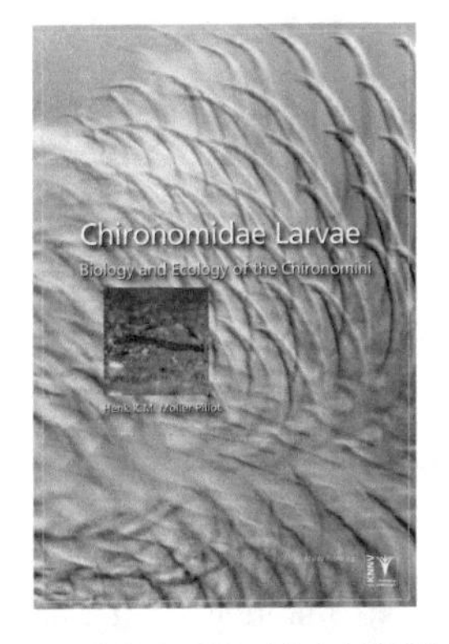

CHIRONOMIDAE LARVAE II

Biology and Ecology of the Chironomini
Henk K.M. Moller Pillot

This 2nd part of a series on Chironomidae larvae covers the most important tribes, Chironomini and Pseudochironomini, the well-known red bloodworms.

Hardcover, 2009, 16,5 x 24 cm
174 pp, € 79,95, English
ISBN: 978-90-5011-303-8

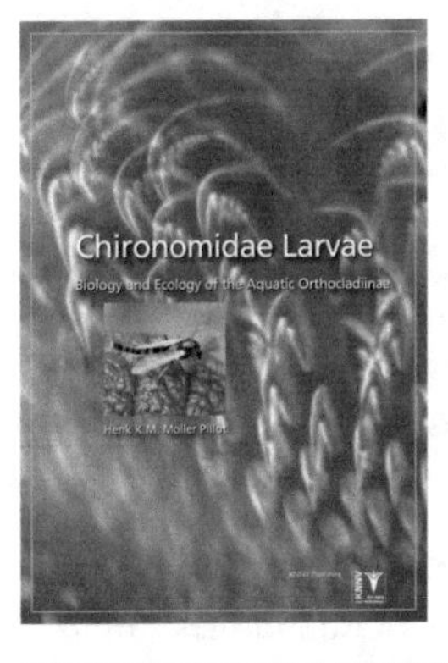

CHIRONOMIDAE LARVAE III

Biology and Ecology of the aqauatic Orthocladiinae
Henk K.M. Moller Pillot

3rd part of a series on Chironomidae larvae covers the subfamily Orthocladiinae which is especially well represented in flowing water.

Hardcover, 2013, 16,5 x 24 cm
314 pp, € 89,95, English
ISBN: 978-90-5011-459-2

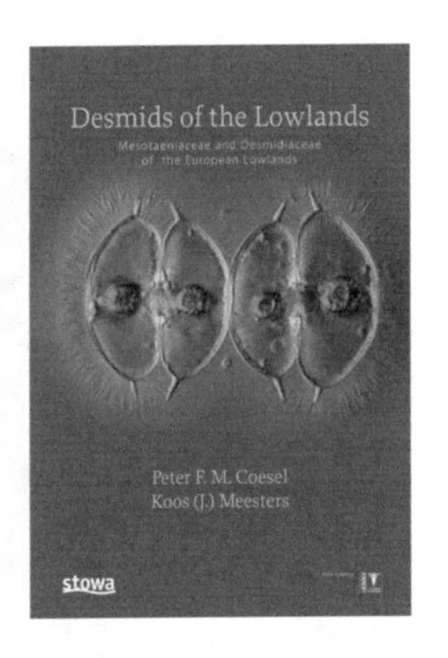

DESMIDS OF THE LOWLANDS I

Mesotaeniaceae and Desmidiaceae of the European Lowlands
Peter F.M. Coesel,
Koos (J.) Meesters

This flora represents all desmid taxa known from the Netherlands and adjacent lowland areas: morphology, ecology and taxonomy. Covers more than 500 species and more than 150 additional varieties.

Hardcover + CD Rom, 2007, 16,5 x 24 cm
352 pp, € 89,50, English
ISBN: 978-90-5011-265-9

DESMIDS OF THE LOWLANDS II

European flora of the desmid genera Staurastrum and Staurodesmus
Peter F.M. Coesel,
Koos (J.) Meesters

This flora represents the European species of het desmid genera Staurastrum and Staurodesmus. Contains reliable identification keys and general information on the morphology, taxonomy, ecology and geographical distribution.

Hardcover, 2013, 16,5 x 24 cm
400 pp, € 99,50, English

AQUATIC OLIGOCHAETA

Aquatic Oligochaeta of the Netherlands and Belgium
Ton van Haaren,
Jan Soors

This book presents a comprehensive overview, focusing on morphology, collecting and preservation, identification, and ecology and offers a new, practical key for identification to species level.

Hardcover, 2013, 19,2 x 26 cm
302 pp, € 195,-, English
ISBN: 978-90-5011-378-6

COLOPHON

Author
Henk K.M. Moller Pillot

Mentum drawings
Maria Sanabria

Graphic layout and design
Erik de Bruin, Varwig Design, Hengelo

Cover illustrations
Background illustration: Orthocladus glabripennis-front claws, David Tempelman
Inset: Criotopus spec., © Hedy Jansen, Zeist

Printed by
Koninklijke Wöhrmann BV

ISBN 978 90 5011 4592
NUR 432
www.knnvpublishing.nl